CISM COURSES AND LECTURES

The series presents lecture notes, monographs, edited works and proceedings in the field of Mechanics, Engineering, Computer Science and Applied Mathematics.
Purpose of the series is to make known in the international scientific and technical community results obtained in some of the activities organized by CISM, the International Centre for Mechanical Sciences.

INTERNATIONAL CENTRE FOR MECHANICAL SCIENCES

COURSES AND LECTURES - No. 513

PATTERN FORMATION AT INTERFACES

EDITED BY

PIERRE COLINET
UNIVERSITE LIBRE DE BRUXELLES, BRUSSELS, BELGIUM

ALEXANDER NEPOMNYASHCHY
TECHNION-ISR. INST. OF TECHNOLOGY, HAIFA, ISRAEL

SpringerWienNewYork

This volume contains 113 illustrations

Printed in Italy
SPIN 12821907

All contributions have been typeset by the authors.

ISBN 978-3-7091-0124-7 SpringerWienNewYork

PREFACE

Interfacial pattern formation occurs in several natural, technological and medical contexts such as, e.g., water films and droplets flowing along inclined surfaces, drying of paint films or coatings, electro-deposition, semiconductor processing, micro-fluidics, mucus flow in the lungs, dry eye syndrome... Modeling these processes requires an understanding of their physics and the knowledge of length and time scales that characterize them. Interfaces play a dominant role at small scales, and their correct modeling is therefore also crucial in the rapidly expanding field of nano-technology (e.g. self-organized nano-particle deposition patterns, quantum dots, ...).

The need to compare various descriptions of the physics of multi-scale interfacial phenomena and the desire to help newcomers in the field in mastering their modeling and identifying modern scientific and technological challenges, inspired the organization of the CISM Advanced School "Pattern Formation at Interfaces with Applications to Materials Science, Biomedical and Physico-Chemical Processes" which took place in Udine, Italy, October 16-20, 2006. Six lecturers from Belgium, Germany, Israel and the USA gave series of lectures to an audience of graduate students, postdocs and young researchers. The lectures have covered most modern methods allowing to treat interfacial instabilities, such as multi-scale asymptotic expansions, linear stability, weakly nonlinear methods and bifurcation theory. Analytical and fully numerical techniques have been discussed, and experimental results have been presented either to confirm theory, or to illustrate directions for further research.

The present book consists of six chapters inspired from the lectures given during the CISM Advanced School. All contributions are presented within the framework of continuous theories and transport phenomena, i.e. mass, momentum and heat transfer. Yet, the investigation of phenomena taking place at interfaces and fronts of phase transition needs the combination of two different kinds of models. In the first approach, the interfaces or the fronts are considered as objects of zero thickness, while the second approach recognizes the internal "diffuse" structure of the transition zone.

The dynamics of interfaces and fronts are characterized by numerous kinds of instabilities leading to nonlinear patterns and waves.

Several challenging physical problems such as phase transition instabilities, front velocity selection and transitions to chaotic spatio-temporal regimes, are discussed with a view of identifying the relevant physico-chemical processes and taking into account multi-scale couplings at theoretical and numerical level, in a rigorous manner. Generic aspects of nonlinear phenomena are also emphasized and to complete the picture, experimental evidence of pattern formation at interfaces is provided to illustrate and validate modeling approaches.

The book is intended to graduate students, researchers and lecturers in physics and engineering, interested in mastering the modern methods of nonlinear stability theory applied to the problems of continuous media mechanics in the presence of interfaces, with applications to materials science, chemical engineering, heat transfer technologies, as well as in combustion and other reaction-diffusion systems. The readers are expected to have followed a first level course in fluid mechanics or transport phenomena and some background in linear algebra, ordinary and partial differential equations. Other goals of this volume are to allow the reader to identify key problems of scientific value, the methods to resolve modeling issues, and to see the similarity between a variety of seemingly different physical problems. Reading the book and some of the references cited therein should allow the reader to quickly move into an area of physics and engineering that is rich in phenomena and replete in applications.

The idea of carrying out this CISM Advanced School on pattern formation at interfaces belongs to Professor Manuel G. Velarde, while he was actually Rector of CISM. Its realization would not have been possible without his encouragement and help. This book is dedicated to him, and to his essential contributions to the field.

Tragically, one of the authors of the present book, Professor Alexander (Sasha) Golovin, passed away in September 2008, while this book was in preparation. This is a great loss for his friends and for the whole "nonlinear science community". Many of us benefited much from scientific discussions and collaboration with this exceptional person who never stopped smiling. According to the students and colleagues who followed the School, his lectures were an example of combined rigor, great enthusiasm for Science, and pedagogy. We will never forget him.

Alexander Nepomnyashchy and Pierre Colinet

CONTENTS

Interfacial patterns and waves in liquid layers and thin films

Pierre Colinet*

Laboratory TIPs (Transfers, Interfaces and Processes) – Fluid Physics Unit, Université Libre de Bruxelles, C.P. 165/67, 50 av. F.D. Roosevelt, B-1050 Bruxelles, Belgium

Abstract This chapter describes the phenomenology and modeling of Bénard-like patterns and waves in liquid layers, including the case of thin films for which surface tension effects are dominant. Attention is also paid to the (generalized) one-sided description of instabilities in the presence of evaporation, with or without an inert component in the gas phase. Then, the focus is on simplified models of patterned structures, and on the role of symmetry properties of the physical system considered.

Contents

*Research Associate (Fonds de la Recherche Scientique - FNRS)

1 Introduction

Natural convective flows in fluid layers submitted to temperature and/or concentration gradients can be induced either by density gradients and associated buoyancy, or by surface tension gradients. As most of this text is devoted to the latter mechanism, and for readers unfamiliar with the topic, it is worth describing first the origin of the so-called Marangoni effect in greater details. Interfacial (or surface) tension at the boundary between two different fluid phases originates from the long-range attractive forces between molecules, i.e. the forces which are responsible for the cohesion of a condensed phase (Israelachvili, 1992). In the case of a liquid-gas interface for instance, molecules in the liquid bulk are attracted by neighboring molecules in an isotropic way, while those located in the interfacial region are globally attracted towards the liquid phase, as they practically do not feel the presence of much more distant gas molecules. Consequently, surface molecules are in an energetically unfavorable situation, which the liquid will tend to avoid as far as possible, by adopting a spherical shape (in the absence of other forces such as gravity or adhesion forces with a solid). Surface tension can be defined as the energy necessary to create a unit surface area, indeed positive since it requires bringing molecules from the bulk to the surface.

Surface tension is generally decreasing with temperature, for most liquids like water or oils. Therefore, when a free surface[1] is non-isothermal, the resulting inbalance of intermolecular forces leads to a macroscopic tangential stress, which occurs from the warmer portions of the surface to the colder parts, hence generating a flow. This thermocapillary (or Marangoni) flow may be seen as a way for the system to reduce regions of high surface tension by enlarging regions of low surface tension. Note that in case gradients of surface tension result from concentration differences along the surface of a liquid mixture, one rather speaks about a "solutal" Marangoni (or solutocapillary) effect.

Marangoni flows are commonly observed in everyday life, e.g. via the motion of dusts in the wax of a candle (see Fig. 1, left), the proximity of the flame being at a much higher temperature than the periphery of the candle. Another famous example is the phenomenon of "tears of wine" (see Fig. 1, right) whose mechanism is however far more complicated, involving preferential evaporation of the alcohol, wetting properties with glass, thermal and solutal Marangoni effects, and gravitational instability of fluid rising along glass walls. A tentative explanation of this phenomenon was

[1] We will generally use the term "free" surface for an interface between a liquid and a gas, or a liquid and its vapor.

given as early as in 1855 by James Thomson (Thomson, 1855), even though the Marangoni effect was given its name following the later work of Carlo Marangoni in the 1870's.

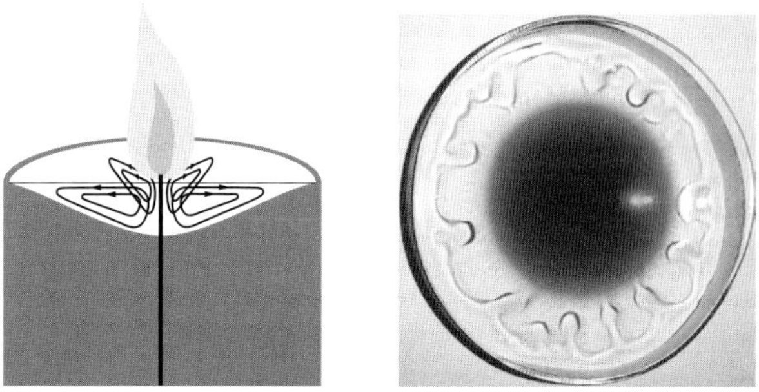

Figure 1. Left : thermocapillary flow in candle wax. Right : tears of wine in a glass viewed from above (Courtesy of John Bush and Anette Hosoi, MIT).

In these lectures, we will mostly be concerned with the phenomenology and modeling of thermal Marangoni flows generated in a liquid layer heated from below (see Fig. 2) or from above. In the former case, hexagonal structures were first observed by Henri Bénard (Bénard, 1901) at the beginning of the century (hence the name of Marangoni-Bénard flows), even though their correct interpretation in terms of the Marangoni effect was given much later (Block, 1956).

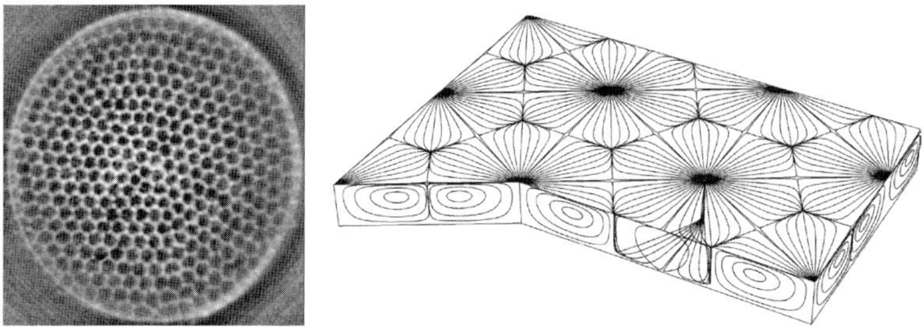

Figure 2. Left : hexagonal pattern observed by shadowgraphy, perpendicularly to the free surface of a liquid layer heated from below (Eckert et al., 1998), in a circular dish (Courtesy of Kerstin Eckert, Dresden University). Right : Theoretically calculated three-dimensional particle paths within steady hexagonal convection cells.

Generally speaking, several levels of modeling are possible : starting from the smallest time and length scales, molecular dynamics simulations can be undertaken, which certainly provide the most accurate description of interface structure and microscopic kinetics, though the computational cost of these techniques remains extremely high, especially for polyatomic molecules and large system sizes. At an intermediate (mesoscopic) level, methods based on the so-called diffuse-interface or phase-field theory have proven quite flexible and adequate to numerically simulate problems involving complex interface shapes and dynamical events such as rupture and coalescence. At the same level, lattice Boltzmann methods, proceeding on the basis of discrete models of the celebrated Boltzmann equation and its variants, also describe interfaces as non-zero thickness objects across which fluid quantities (density, concentration, ...) undergo sharp variations. Finally, at a macroscopic level, interfaces are modelled as zero-thickness surfaces of discontinuity along which adequate interfacial (jump) boundary conditions must be expressed. These lectures will exclusively be concerned with the latter approach, not only because significant progresses have been accomplished along these lines in the last decades (e.g., in connection with Nonlinear Physics), but also because it is now commonly admitted that macroscopic models may indeed be used down to very small scales, even in the submicrometer range.

Importantly, there exists a formal analogy between the problems of heat transfer involving thermocapillarity, and mass transfer associated with concentration gradients generating solutocapillary flows. Actually, the mathematical descriptions of such phenomena are quite similar, provided some complications specific with mass transfer through an interface are ignored, such as solute accumulation at the interface, adsorption kinetics, ... Consequently, at least in the simplest descriptions, it will be formally equivalent to say that a pure liquid layer is heated from below, or that a two-component isothermal liquid layer undergoes desorption through its upper free surface of the component which lowers the surface tension (hereafter called a surfactant). In the second situation considered in these lectures, a liquid layer is heated from the top (or undergoes absorption of a soluble surfactant). In these cases, one typically observes a self-organization of the system in the form of propagating wave trains, e.g. azimuthally in an annular container as in Fig. 3(left). Occasionally, simultaneous propagation of waves in both azimuthal directions may also be observed, such as in Fig. 3(right). Such wave trains have strong analogies with solitary waves, though their weak dissipation on a large time scale can here be balanced by energy input from an underlying instability. These interesting aspects have been studied in details by Velarde and co-workers (see e.g. Christov and Velarde, 1995;

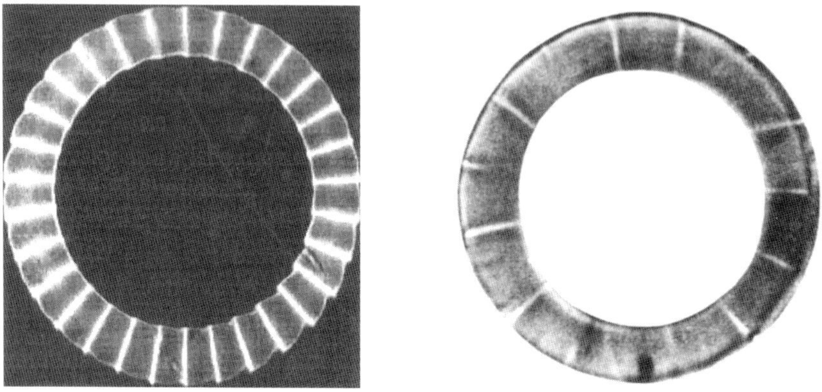

Figure 3. Left : azimuthally-propagating wave trains visualized by a shadowgraph technique during absorption of pentane vapor into toluene (Courtesy of Andreas Wierschem). Right : counter-propagating wave trains observed in a heat transfer experiment with heating from the top (after Weh and Linde, 1997).

Rednikov and Velarde, 2000).

Both cases of heating from above or below (absorbing or desorbing a surfactant) therefore lead to a spontaneous breaking of the natural symmetries of the Bénard layer, either in the form or waves, or of steady convection patterns. Indeed, in case a flux (either of heat or of matter) occurs perpendicularly to an interface, surface tension is expected to be uniform, i.e. no flow is expected *a priori*. In mathematical terms, the system is invariant with respect to translations along its interface (in the idealized case of a laterally infinite layer), and enjoys isotropy as well, i.e. any rotation with respect to an axis orthogonal to the layer leaves the latter unaffected. Symmetry-breaking associated with the emergence of ordered structures in non-equilibrium systems is typical of instability phenomena (see other examples in the same volume), among which surface-tension-driven (and buoyancy-driven) instabilities discussed in these lectures.

2 Phenomenology of Bénard instabilities

2.1 Physical mechanisms of patterns and waves

Let us first consider the so-called Rayleigh–Bénard instability (Fig. 4, left), associated with buoyancy-driven (rather than surface-tension-driven) convection. A layer of pure liquid (which will be considered as incompressible and Newtonian) is confined between two horizontal plates maintained at controllable temperatures.

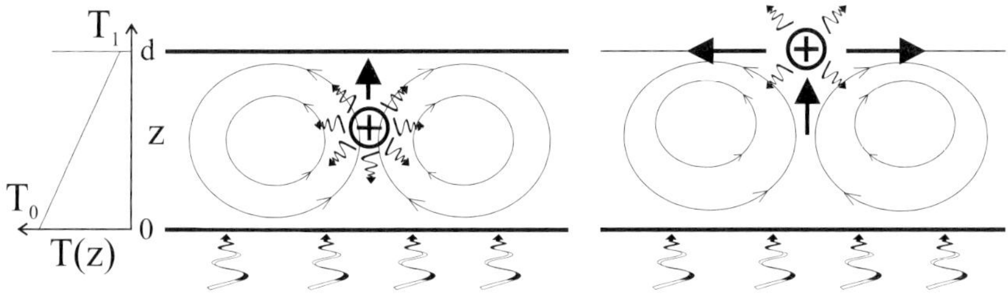

Figure 4. Sketches illustrating the mechanisms at the origin of the Rayleigh–Bénard (left) and the Marangoni–Bénard (right) instabilities. In both cases, the temperature of the bottom plate is T_0 while that of the upper surface (either solid or open to air) is $T_1 < T_0$ (the layer is heated from below).

Assume for the moment that the dissipative effects of molecular transport of heat and momentum are negligible (i.e. associated with very long time scales, see §2.3). Then, when a localized fluctuation of temperature (say, a hot spot) arises in a certain volume element, an upward force on this fluid particle is induced by buoyancy, as its density is decreased compared with its environment (differential buoyancy force). As molecular diffusion is assumed to be negligible, the fluid particle will set into motion in the upward direction, without friction, and will remain at the same temperature during this motion. The buoyancy force it will experience will then increase, because the particle moves towards colder regions, such that the velocity is increased, ... This is the origin of the amplification (instability) phenomenon.

The Marangoni-Bénard mechanism may be explained similarly (Fig. 4, right) : if a fluctuation of temperature (say, a hot spot) occurs at the free surface (or interface), a local surface tension gradient is created, directed radially away from the fluctuation. The associated tangential stresses then induce radially divergent surface fluid motion (the interface may be seen as an elastic membrane which relaxes at the point where the disturbance is created). Now, continuity of the fluid requires a vertical ascending flow to take place below the disturbance. If the layer of fluid is heated from below, this uprising fluid, being hotter, will make the free surface temperature increase at the initial location of the disturbance. Consequently, the surface tension at that point decreases, thus increasing the tangential stresses.

Clearly, thermal diffusivity will act against these instability processes, by favoring thermalization of fluid particles with their environment, i.e. damping of fluctuations. Viscosity will also generate stresses, preventing macroscopic motions to develop within the fluid. As a result, the temperature difference $\Delta T = T_0 - T_1$ has to exceed a threshold value for Bénard instabilities to develop, in order for the destabilizing mechanisms to overcome the dissipative effects. This will be quantified in terms of time scales and dimensionless numbers in §2.3.

Note that contrary to the Marangoni-Bénard instability leading to hexagons (see Fig. 2) when the instability threshold is exceeded, the Rayleigh-Bénard instability typically yields roll patterns, the study of which has motivated important progresses in the theory of pattern formation (see for instance Cross and Hohenberg, 1993). Note that both rolls and hexagons are very seldom perfectly regular, and various kinds of defects have indeed been observed, and studied extensively for their interesting universal properties. For instance, hexagonal patterns very often display isolated pentagon-heptagon defects, which can act as mediators in the transition from hexagonal to square patterns observed at increasing temperature difference ΔT, via a process of nucleation of lines of pentagons (Eckert and Thess, 1999).

Considering now the opposite case of a liquid layer heated from above, first in the buoyancy-driven case, it is quite natural to expect an oscillatory behavior of fluctuations, again in the case where the effect of dissipation is not too strong. Indeed, during its upward motion, a fluid particle initially warmer than its environment will now move towards *warmer* regions. This means that the upward buoyancy force it experiences will first decrease, then eventually change sign when the particle overshoots its neutrally buoyant position, due to inertia. The particle is then accelerated downwards, towards colder regions, and may oscillate around its neutrally buoyant position several times before dissipation eventually damps out any fluid motion. This oscillatory mechanism is known in the literature as (Brunt–Väisälä) internal waves, observed in situations of stable density stratification (e.g. in the atmosphere or the ocean, see Turner, 1973).

Similarly, surface waves due to thermocapillarity in a layer heated from above are also expected, and may be explained considering once again a hot spot at the free surface. As usual the Marangoni effect drives a surface flow away from the spot, which now brings *colder* liquid from the bulk. Because of fluid inertia, the initially hot spot eventually becomes colder than its neighborhood (overshoot), thus reversing the flow and leading to oscillations. These waves involve motions essentially parallel to the free surface, and may be called longitudinal waves for this reason.

Note that both internal and surface waves are usually damped, though it has been shown recently (Rednikov et al., 2000) that their interaction near resonance may lead to amplification and mode-mixing (see section 3.4). Actually, internal waves excited by the Marangoni effect are indeed thought (Wierschem et al., 2000) to be responsible for the propagating wave trains shown in Fig. 3(left), i.e. in a situation where pentane is absorbed in toluene. Given that pentane is lighter than toluene, absorption indeed generates a stable density stratification prone to triggering of internal waves by external disturbances, just as in the thermal case discussed above. As pentane also lowers the surface tension of the mixture, longitudinal surface waves may also occur and resonantly interact with (transverse) internal waves.

Finally, the well-known capillary-gravity waves are yet another oscillatory mode, which occurs whenever a free surface is perturbed by an external force. Ripples propagating along a free surface involve an essential deformation orthogonal to the latter, and are transverse waves in this respect. Interestingly, such capillary-gravity waves may also be excited in layers heated from the top, via resonant interaction and mixing with the surface longitudinal waves just discussed (Rednikov et al., 2001). A review of experimental findings on waves generated by heat or mass transfer through an interface, their interpretation in terms of internal, surface and capillary-gravity waves, as well as their links with solitary waves, are given in the book of Nepomnyashchy et al., 2002.

2.2 Application-oriented aspects

When a layer of fluid is heated from below, the state of pure conduction prevailing at small ΔT is first replaced by a cellular convective structure at a certain ΔT_c, as explained in the previous section. Convective heat transport now adding to the effect of heat conduction, the global heat transfer through the layer is enhanced, which may be seen as an increase of the apparent thermal conductance (or heat transfer coefficient) of the liquid layer (defined, e.g., by $\Phi/\Delta T$ where Φ is the mean heat flux crossing the layer). Subsequent transitions, e.g. to time-dependent regimes, generally result in further increases of the heat transfer coefficient. This has been extensively studied for the case of Rayleigh-Bénard convection, even in the regimes of so-called weak and strong thermal turbulence setting in at very high $\Delta T/\Delta T_c$ (see for instance Guyon et al., 1991).

In surface-tension-driven situations, much less is known, even though interfacial turbulence has been observed as early as in the 50's, in the context of Chemical Engineering. In many situations involving interfacial heat and mass transfer (for instance in techniques such as liquid/liquid extraction, liquid/gas absorption or desorption, distillation, ...), order of magnitude changes of transfer rates from one phase to another have been measured and correlated with empirical relationships. An extensive review of such phenomena and a rather complete bibliography are provided in the thesis of T. Molenkamp (Molenkamp, 1998). Note also that in chemical reactors, mass transfer through interfaces is often accompanied by chemical reactions, and interesting couplings with surface-tension-driven convection can occur (Eckert and Grahn, 1999; Eckert et al., 2004).

Marangoni–Bénard convection is also important in processes involving evaporation, such as in the drying of paint films, in the coating industry, and for heat transfer devices. In Fig. 5, a thin layer of ethyl alcohol is evaporated under an inert gas (nitrogen) flow. Even though no external heating is imposed in this experiment, cellular convection develops as soon as the flow is imposed, due to the cooling of the free surface induced by evaporation (consumption of latent heat). The fact that these convective patterns are sustained by evaporation is further confirmed by their sudden disappearance when vapor removal by the gas flow is interrupted (images 8 and 9). In turn, convective transport of heat reduces the overall thermal resistance of the liquid layer, therefore contributing to an increase of the evaporation rate.

In addition to the increase of the interfacial exchange coefficients (i.e. per unit surface), Marangoni flows often induce significant interfacial deformation, leading to important variations of the total surface of exchange itself. In the case of thin liquid films, specific instabilities (see §2.4) may occur, leading in some cases to dry regions. While such dry spots are generally unfavorable (for instance, in thin-film evaporators, they lead to possible overheating and destruction of the substrate), they also have a positive effect since the heat transfer resistance scales proportionally to the local film thickness, as a first approximation. As a consequence, the local heat flux can become locally very high in the transition region between the film and the dry spot (i.e. at the contact line), as was quantified by P. Stephan in the context of heat-transfer devices such as heat pipes and boilers (Stephan and Busse, 1992).

Generally speaking, Marangoni effects and associated instabilities are expected to be dominant at small length scales, where gravity and buoyancy are ineffective (as will be seen from dimensionless groups in the next section). There is therefore a renewed interest in recent years, especially in connection with the rapidly developing fields of microfluidics and even nanotechnologies. At such scales, self-organization phenomena such as the surface-tension-driven Bénard instability could for instance be used as a way to design special types of solid surfaces, by deposition of nanocrystals in the forms of polygonal arrays and rings (see e.g. Maillard et al., 2000).

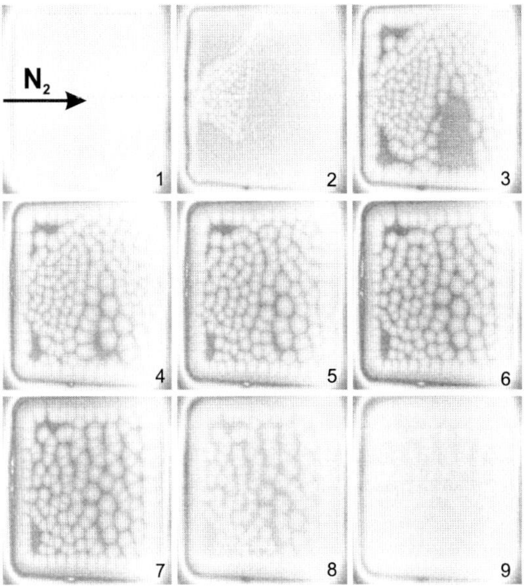

Figure 5. Sequence of infrared camera images of the free surface of a thin liquid layer of ethyl alcohol, forced to evaporate under a flux of nitrogen (indicated by the arrow) at room temperature and atmospheric pressure.

2.3 Dimensionless numbers and time scales

Several time scales can be defined from physical properties of the fluid and geometrical characteristics such as its depth d. We first have the thermal time scale

$$\tau_{\text{th}} = \frac{d^2}{\kappa} \tag{1}$$

i.e. the typical time scale it takes for temperature fluctuations to be damped over a distance d (κ is the thermal diffusivity[2]). Similarly, velocity (or

[2] In this section, we will generally refer to a pure fluid submitted to a temperature difference ΔT. However, for an isothermal binary fluid, similar time scales and dimensionless numbers may be defined, replacing ΔT by ΔN (a typical mass fraction difference between bottom and top), κ by D (the isothermal diffusion coefficient), and defining $\gamma = -(\partial\sigma/\partial N)$ and $\alpha = -(\partial\rho/\partial N)/\rho$, where σ and ρ are the surface tension and the volumic mass, respectively.

vorticity) fluctuations decay on a viscous time scale

$$\tau_{\text{visc}} = \frac{d^2}{\nu} \tag{2}$$

where ν is the kinematic viscosity.

Now, it is possible to construct two other time scales which do not depend on any molecular diffusivity mechanism. The first one is the "buoyancy time scale"

$$\tau_{\text{buoy}} = \sqrt{\frac{d}{\alpha g \Delta T}} \tag{3}$$

which is roughly the time it takes for a fluid particle at the bottom of the layer to be accelerated by buoyancy up to its upper surface (see Fig. 4), i.e. the typical time scale of the instability leading to convective patterns. In Eq. (3), α is the thermal expansion coefficient, g is the gravity acceleration, and $\Delta T = T_0 - T_1$ is the temperature drop across the layer.

Similarly, a typical time scale of the thermocapillary (or Marangoni) instability can be defined as

$$\tau_{\text{ma}} = \sqrt{\frac{\rho d^3}{\gamma \Delta T}} \tag{4}$$

where ρ is the liquid volumic mass, and $\gamma = -\partial\sigma/\partial T$ is the coefficient of variation of surface tension with temperature (positive for usual liquids).

Note that when the layer is heated from the top, $|\tau_{\text{buoy}}|^{-1}$ and $|\tau_{\text{ma}}|^{-1}$ provide order of magnitude estimates of the frequencies of internal and surface waves (see §2.1), respectively.

Comparing the different time scales given by Eqs (1–4), it may be expected that instability will develop if the time for a fluid particle to travel over a distance of order d is shorter than the times necessary for the particle either to be slowed down by viscosity, or thermally equilibrated with its environment. Thus, buoyancy-driven instability would occur typically if

$$Ra = \frac{g\alpha \Delta T d^3}{\nu\kappa} = \frac{\tau_{\text{visc}}\tau_{\text{th}}}{\tau_{\text{buoy}}^2} \gg 1 \tag{5}$$

while surface-tension-driven instability will occur when

$$Ma = \frac{\gamma \Delta T d}{\rho\nu\kappa} = \frac{\tau_{\text{visc}}\tau_{\text{th}}}{\tau_{\text{ma}}^2} \gg 1 \tag{6}$$

These equations define both the Rayleigh number Ra and the Marangoni number Ma, which are the usual measures for the relative importance of destabilizing and stabilizing effects. Typically, the critical value of the Rayleigh number above which instability sets in is of order 10^3, while the critical Marangoni number is of order 10^2. Their actual values depend on the nature of upper and lower plates (e.g. rigid and heat-conducting, or free and poorly-conducting, ...), and must be determined by linear stability analysis (see §3.4).

Note that other usual dimensionless parameters are the Prandtl number $Pr = \nu/\kappa = \tau_{\text{th}}/\tau_{\text{visc}}$ and the Galileo number $Ga = gd^3/\nu\kappa = \tau_{\text{th}}\tau_{\text{visc}}/\tau_{\text{grav}}^2$ where $\tau_{\text{grav}} = (d/g)^{1/2}$, i.e. the typical time scale for a body to be accelerated over a distance d under the action of gravity.

Finally, in case the liquid layer presents an upper free surface, it is common to describe heat transfer through the latter using a constant heat transfer coefficient h (or a constant Biot number $Bi = hd/\lambda$ in dimensionless form, where λ is the liquid thermal conductivity). Even though this is a crude approximation since the heat transfer coefficient generally depends on convective motions in the gas phase, and therefore on the temperature of the interface itself, it does provide the correct tendency for the influence of heat transfer at the upper surface (namely an increase of the critical Marangoni number with increasing Biot number). A more rigorous definition of a "generalized Biot number" will be presented in §3.

2.4 Other instability mechanisms in very thin liquid films

Yet another instability mode can occur, specifically for thin liquid films of viscous liquids. This is illustrated in Fig. 6 : consider that an initially flat heated liquid film (top figure) is perturbed, resulting in a non-uniform liquid depth (middle figure). At the thinner parts of the film, the free surface is warmer (since it is closer to the heated plate) than at thicker regions. A surface tension gradient therefore arises, and generates a flow dragging fluid away from the depression, hence depleting the thinner parts even more. This may result in the formation of dry spots (bottom figure), even though the correct description of the film evolution near rupture should involve intermolecular forces between the fluid and the substrate, typically when the local thickness decreases below 100 nm (Israelachvili, 1992; Oron et al., 1997).

Note that the horizontal length scale of convective motions (and of surface deformations) generated by this mechanism is generally much larger than the fluid depth, hence the name long-wave deformational mode for this instability mechanism. In contrast, polygonal convection cells typical of Marangoni-Bénard convection turn out to scale proportionally to the depth of the liquid layer (see Fig. 2) in case the rigid bottom plate is a good thermal conductor (such as a metal). In case the bottom plate has a low heat transfer coefficient, just as the upper free surface, the Marangoni-Bénard convection cells become large-scale as well, which offers a way to describe them in terms of long-wave asymptotic theories.

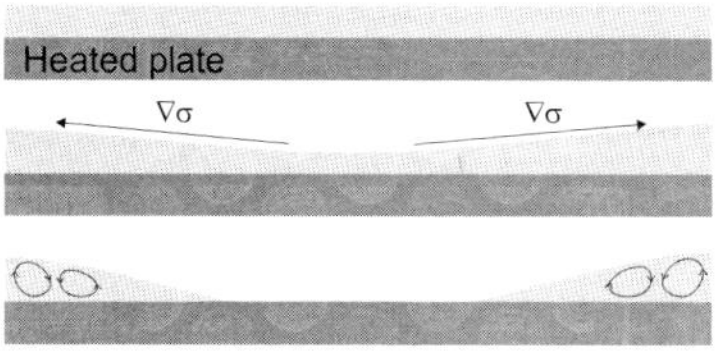

Figure 6. Long-wave deformational instability of heated thin liquid film.

The long-wave deformational mode occurs typically for $Ma > Ma_0 \sim Ga$, for a purely insulating upper free surface (a good approximation for liquids in contact with a gas phase). As introduced in the previous section, the Galileo number $Ga = gd^3/\nu\kappa$ measures the stabilizing influence of gravity on surface deformations, and is usually extremely large ($\sim 10^8$!) for usual fluids with depths in the centimeter range. In these cases, cellular convection mechanisms will always occur first, since their critical Marangoni number is typically 10^2 (actually, for depths larger than about 1 cm, buoyancy-driven convection dominates, while surface-tension-driven convection is most effective at depths below some millimeters).

For very thin layers of highly viscous liquids however, long-wave deformational modes may become active, and a fortiori the primary instability mode when the long-wave threshold $Ma_0 \sim Ga$ becomes lower than the usual threshold $Ma_c \sim 10^2$ for cellular Marangoni-Bénard convection. While for water this corresponds to unusually thin liquid films ($d \sim 100$ μm or less), this becomes experimentally accessible with highly viscous liquids such as silicone oils [e.g. for a 200 cSt oil, $Ma_0 \sim Ga \sim 10^2$ for $d \simeq 0.6$ mm]. Recent experimental results of VanHook et al. (VanHook et al., 1997) have indicated that the cross-over between the two modes may indeed be observed, possibly resulting in coexistence of large-scale dry spots with depth-scaled convection cells such as sketched in Fig. 6 (bottom). VanHook et al. have further observed and explained that the long-wave deformational mode indeed generally leads to film rupture (see also Oron et al., 1997), but more complex situations may occur (surface elevations or "high spots", cascades of structures, ...). Note that situations involving essential surface deformation will not be considered further in these notes, as they will be described in details in the lectures of M. Bestehorn and L. Pismen, including the description of rupture of ultrathin films due to molecular interactions with the substrate, droplets, ...

3 Basic equations and boundary conditions

Even though the description of Bénard instabilities in layers with a free surface generally involves fluid motions and temperature fluctuations in both liquid and gas phases (two-layer system), it is often possible to make some simplifying assumptions (given hereafter) in order to describe them in terms of liquid quantities alone (one-sided model), with suitable boundary conditions. Using the thermal time scale $\tau_{\rm th} = d^2/\kappa$, the length scale d, the pressure scale $\mu\kappa/d^2$, and a yet undefined temperature range θ, the basic dimensionless equations generally used to describe thermally-driven Bénard instabilities of a liquid layer are

$$\vec{\nabla} \cdot \vec{V} = 0 \tag{7}$$

$$\Delta\vec{V} - \vec{\nabla}p - Ga\vec{1}_z[1 - \alpha\theta(T - \tilde{T}_{\rm r})] = Pr^{-1}\left(\frac{\partial\vec{V}}{\partial t} + (\vec{V}\cdot\vec{\nabla})\vec{V}\right) \tag{8}$$

$$\Delta T = \frac{\partial T}{\partial t} + (\vec{V}\cdot\vec{\nabla})T \tag{9}$$

where $\vec{V} = (U, V, W)$, T, and p are respectively the dimensionless fields of velocity, temperature and pressure, $\vec{1}_z$ is the unit vector directed vertically upwards, $\tilde{T}_{\rm r} = (T_0 + T_1)/2\theta$ is the dimensionless reference (mean) temperature at which the physical properties are evaluated (note that their possible variation is neglected, in the temperature range θ considered), and other dimensionless quantities have been defined earlier. Equations (7-9) respectively express the conservation of mass (for incompressible liquids), of momentum (in the presence of a buoyancy force), and of energy (neglecting viscous heating).

Importantly, this form of the equations makes use of the Boussinesq approximation, i.e. $|\alpha\theta| \ll 1$. This means that the variations of density may be neglected everywhere, unless when multiplied by large quantities such as the Galileo number Ga, as it indeed occurs for the body force term of the Navier-Stokes equations (8). Actually, the Rayleigh number (defined using the temperature range θ) is the product $Ra = \alpha\theta Ga$, i.e. the product of a small and a large number, and can therefore be sufficient to trigger instability of the Bénard layer.

Boundary conditions on the bottom plate $z = 0$ are usually taken as the no-slip and constant temperature conditions

$$\vec{V} = T - T_{bot} = 0 \quad \text{at } z = 0 \tag{10}$$

where $T_{bot} = \theta^{-1}T_0$, while those at the upper free surface should in principle be written at a unknown position, which should be found as a solution of the full problem. However, we will not deal with such deformable free surfaces here (the reader is referred to the book of Colinet et al., 2001, for general interfacial conditions) and assume the free surface to be located at $z = 1$, i.e. disregarding the case of very thin liquid films (the effect of gravity, quantified by the Galileo number, is sufficiently strong to prevent free surface deformations).

In addition, as mentioned above, a coupling of liquid and gas phases is expected at $z = 1$. We now underline the main hypotheses necessary to reach a one-sided description, i.e. to write the correct number of free surface boundary conditions, though in terms of liquid quantities alone.

First, the liquid velocity orthogonal to the interface is zero, since it is assumed motionless. Hence,

$$W = 0 \quad \text{at } z = 1 \tag{11}$$

Second, in view of the fact that the dynamic viscosity of the gas is generally much lower than that of the liquid, we may neglect gas viscous stresses in the tangential momentum balance at $z = 1$. Taking into account surface tension gradients and in dimensionless form, this reads

$$\partial_z U + Ma\,\partial_x T = \partial_z V + Ma\,\partial_y T = 0 \quad \text{at } z = 1 \tag{12}$$

where $Ma = \gamma\theta d/\rho\nu\kappa$ is the Marangoni number. Note that an equivalent form is obtained by differentiating the first of these boundary conditions with respect to x, the second with respect to y, adding the results and using the incompressibility condition (7). This yields

$$\partial_z^2 W = Ma\,\Delta_{\mathrm{h}} T \quad \text{at } z = 1 \tag{13}$$

where $\Delta_{\mathrm{h}} = \partial^2/\partial x^2 + \partial^2/\partial y^2$ is the horizontal Laplacian operator.

A straightforward way to close the problem would be to neglect the thermal conductivity of the gas, and therefore to assume $\partial_z T = 0$ at the free surface (zero heat transfer coefficient). Under some assumptions mentionned in section 3.2, evaporation can also be included in the energy conservation at the interface, which yields

$$\partial_z T = -EJ \quad \text{at } z = 1 \tag{14}$$

where $E = L/c_p\theta$ is an evaporation number, comparing the latent heat of evaporation L to $c_p\theta$, i.e. the heat needed to warm up the liquid over the selected temperature range θ. Note that Eq. (14) introduces the evaporation rate J (here scaled by $\rho\kappa/d$), which is often written as

$$J = \beta\sqrt{\frac{M}{2\pi R T_{sat}}} p'_{eq}(T_{sat}) \frac{\theta d}{\rho\kappa}(T - \theta^{-1}T_{sat}) \tag{15}$$

i.e. one of the possible forms of the Hertz-Knudsen kinetic law (see also Burelbach et al., 1988), in which β is the accomodation coefficient, M is the liquid molecular mass, R is the universal gas constant, $p_{eq}(T)$ is the equilibrium pressure at temperature T (e.g. the Clausius-Clapeyron coexistence curve), and T_{sat} is the (dimensional) saturation pressure at the local vapor pressure p_v, i.e.

$$p_{eq}(T_{sat}) = p_v \tag{16}$$

Hence, for a pure vapor phase where the vapor pressure p_v may be assumed constant, T_{sat} is a constant and Eqs (14) and (15) combine to give a mixed thermal condition

$$\partial_z T + Bi_{ev,p}(T - \theta^{-1}T_{sat}) = 0 \tag{17}$$

where the effective heat transfer coefficient (i.e. the Biot number, in dimensionless form) reads

$$Bi_{ev,p} = \beta\sqrt{\frac{M}{2\pi RT_{sat}}}\, p'_{eq}(T_{sat})\frac{Ld}{\lambda} \tag{18}$$

and the index p stands for "pure vapor phase". Note that the values calculated for $Bi_{ev,p}$ are generally very high apart for very thin liquid films or very small accomodation coefficient[3], such that the interfacial temperature remains everywhere very close to T_{sat} (local thermodynamic equilibrium between the vapor and the liquid), and the Marangoni effect is very weak.

Hence, the problem is now closed, either for the case where the liquid is non-volatile and the vapor phase has very small thermal conductivity ($Bi_{ev,p}$ has to be set to zero in Eq. (17)), or when the liquid can evaporate into its own vapor (in which case $Bi_{ev,p}$ is a very large quantity). However, there are cases where one or both of these assuptions fail, and a more general approach is needed.

[3] As also discussed by Colinet et al., 2001, the cross-over between the reaction-limited regime $Bi_{ev,p} \ll 1$ and the heat transfer-limited regime $Bi_{ev,p} \gg 1$ occurs for a depth d =O(100 nm), for water near the normal boiling point and an accomodation coefficient $\beta = 1$. This depth increases when β decreases, however.

3.1 Non-negligible gas thermal conductivity

As the assumption of vanishing gas thermal conductivity may not be justified in all cases (e.g. helium is equally conducting as a silicone oil), it seems preferable to generalize the rough approach of the previous section, though first in the non-volatile case.

In general, the thermal diffusivity κ_g of a gas is about one hundred times larger than that of a liquid. Therefore, the time scale of thermal fluctuations in the gas, i.e. d_g^2/κ_g (where d_g is the gas depth), is much smaller than the liquid thermal time scale $\tau_{th} = d^2/\kappa$. In addition, we may also expect the energy transport due to gas flows to be negligible compared with thermal diffusion (small Péclet number assumption[4]), and the energy equation in the gas may be written in the quasi-steady approximation

$$\Delta T_g = 0 \tag{19}$$

which may in principle be solved using boundary conditions $T_g = T(x, y, z = 1, t)$ at $z = 1$ and $T_g = T_{up}$ at $z = 1 + H$ (where $H = d_g/d$). Given such a formal solution $T_g[T(z = 1)]$, we may then express the continuity of heat flux as

$$\partial_z T = \bar{\lambda}\, \partial_z T_g[T(z = 1)] \quad \text{at } z = 1 \tag{20}$$

where $\bar{\lambda} = \lambda_g/\lambda$ is the ratio of gas and liquid thermal conductivities.

We have now reached a generalized one-sided description of the problem, i.e. the system of equations (7-9) with boundary conditions (10, 11, 12, 20) is closed. However, such a problem is non-local, since the heat flux (20) depends on the instantaneous temperature *distribution* $T(z = 1)$ at the free surface (and not only on its local value).

Fortunately, this problem may easily be handled in Fourier space. Denoting the horizontal average by $\langle\cdot\rangle$, the horizontal Fourier transforms by tilded quantities, and realizing that from Eq. (19), $\langle T_g\rangle$ must vary linearly from $\langle T(z = 1)\rangle$ at $z = 1$ to T_{up} at $z = 1 + H$, the horizontal average of Eq. (20) yields

$$\partial_z \langle T\rangle = H^{-1}\bar{\lambda}\,(T_{up} - \langle T\rangle) \quad \text{at } z = 1 \tag{21}$$

which shows that the increase of heat transfer due to possible convection in the liquid is directly linked with the increase of the mean surface temperature.

Now, for Fourier components with wavenumber $k \neq 0$, Eq. (19) yields $(\partial_z^2 - k^2)\tilde{T}_g = 0$, which has to be solved with boundary conditions $\tilde{T}_g = \tilde{T}$

[4] Note that for this development to hold, it is necessary to assume that the gas phase is *confined*, i.e. its depth d_g must not be taken much larger than the liquid depth d. Moreover, the case of externally imposed fast gas flows is also excluded.

at $z = 1$ and $\tilde{T}_g = 0$ at $z = 1 + H$. Then, using this result in the Fourier transform of Eq. (20), we get

$$\partial_z \tilde{T} = -Bi_c(k)T \quad \text{at } z = 1 \tag{22}$$

where $Bi_c(k) = \bar{\lambda} k \coth(kH)$ is a generalized, *wavenumber-dependent* Biot number (i.e. dimensionless heat transfer coefficient, see end of §2.3). Non-locality therefore manifests itself here through the fact that the different Fourier components of the temperature field evolve with different heat transfer coefficients at the upper free surface.

3.2 Generalized one-sided modeling of evaporation

When the liquid is volatile and evaporates into an inert gas such as air, the boundary condition (17) cannot be applied anymore, because the partial pressure of vapor cannot be assumed constant in general. Assuming the gas mixture to be perfect, we have $p_v = p_g N_g/(r + (1-r)N_g)$ where p_g is the total gas pressure (assumed constant here, as can be justified *a posteriori*), $r = M/M_i$ is the ratio of molecular masses of the vapor and of the inert gas, and N_g is the vapor mass fraction for which a mass diffusion equation should be solved. The latter will here be written

$$\Delta N_g = 0 \tag{23}$$

i.e. in the quasi-steady approximation, just as for the temperature field in the previous section (these assumptions are coherent since the Lewis number $Le = D/\kappa$, where D is the isothermal diffusion coefficient, is of order unity for a gaseous mixture – hence the relaxation times in the gas are generally much smaller than those in the liquid, provided the depth of the gas is not too large).

In order to understand the basic role of an inert gas without entering into detailed descriptions of the gas phase dynamics, we will here consider that the mass fraction, just as the temperature, is imposed at the upper boundary, i.e.

$$N_g = N_{up} \quad \text{at } z = 1 + H \tag{24}$$

while at the interface, it will be assumed that the inert gas is not absorbed by the liquid, yielding

$$(1 - N_g)J + \bar{\rho}\bar{\kappa} Le \partial_z N_g = 0 \quad \text{at } z = 1 \tag{25}$$

where $\bar{\rho} = \rho_g/\rho$ and $\bar{\kappa} = \kappa_g/\kappa$ are the ratios of densities and of thermal diffusivities, respectively. Again, a Hertz-Knudsen kinetic equation is needed for the evaporation rate J, which will here be written under the form

$$QJ = p_s(T_i) - \frac{N_i}{r + (1-r)N_i} \tag{26}$$

where $T_i = T(z=1)$ is the dimensionless interfacial temperature, $N_i = N_g(z=1)$ is the mass fraction of vapor at the interface, $p_s(T_i) = p_{eq}(T_i\theta)/p_g$ is the scaled saturation pressure, and the kinetic resistance to evaporation is quantified by

$$Q = \frac{\rho\kappa}{\beta d p_g}\sqrt{\frac{2\pi RT}{M}} \tag{27}$$

i.e. inversely proportional to the usual accomodation coefficient β.

Note that in the present treatment of evaporation, either for a pure vapor or for a gas mixture, possible motions of the interface are neglected, i.e. $z = h(t) \simeq 1$. This quasi-steady assumption is valid provided its relative motion during a typical relaxation time d^2/κ is small, i.e. $|\partial_t h| \ll 1$ in our choice of length and time scales. Actually, the total (jump) mass balance at the interface reads

$$J = W - \partial_t h = \bar{\rho}(W_g - \partial_t h) \qquad \text{at } z = h(t) \tag{28}$$

where W and W_g are vertical velocities of the liquid and of the gas mixture, respectively. Averaging in the horizontal direction and taking into account that $\langle W \rangle = 0$ because of Eqs (7) and (10), we get $\partial_t h = -\langle J \rangle$. Hence, the quasi-static approach will be valid provided the mean evaporation rate $\langle J \rangle$ is small enough compared to unity.

Another assumption is that the vertical velocity of the liquid at the interface is neglected, i.e. Eq. (11) is assumed, even though we should actually have $W(z=1) = J - \langle J \rangle$. This assumption, which may be checked a posteriori, should be valid provided typical velocities induced by the Marangoni (or Rayleigh) effect are much higher than the liquid velocity induced by evaporation[5].

In order to avoid nonlinearities in the mass transfer boundary conditions, and to remain coherent with the assumption of a small evaporation rate J, it

[5] In fact, in our choice of scales, $|J|$ is a typical Péclet number based on the liquid velocity induced by evaporation, and we assume it to be small in the present analysis. In turn, the Marangoni number $Ma = U_\gamma d/\kappa$ is another Péclet number based on the typical thermocapillary velocity $U_\gamma = \gamma\theta/\rho\nu$. As Ma is of order unity (or more), the ratio of thermocapillary and evaporation velocities is of order $Ma/|J| \gg 1$, which justifies Eq. (11), since the bulk value of W will always be much larger than $W(z=1)$. Note that the reasoning also holds for buoyancy-driven convection.

is convenient to assume the system to be close to a global equilibrium state defined by the overall gas pressure p_g (e.g., atmospheric) and the externally imposed vapor mass fraction N_{up}. At equilibrium, $J = 0$ and the mass fraction is equal to N_{up} everywhere in the gas, while both the liquid and the gas are at a temperature T^* determined by Eq. (26), i.e.

$$p_s(T^*) = \frac{N_{up}}{r + (1-r)N_{up}} \tag{29}$$

Then, assuming small deviations around this equilibrium state, boundary conditions (25) and (26) can be linearized, leading to

$$(1 - N_{up})J + \bar{\rho}\bar{\kappa}Le\partial_z N_g = 0 \quad \text{at } z = 1 \tag{30}$$

and

$$QJ = p_s'(T^*)(T_i - T^*) - \frac{r}{(r + (1-r)N_{up})^2}(N_i - N_{up}) \tag{31}$$

where Q is evaluated at the saturation temperature $T_{sat} = \theta T^*$.

Now, separating the horizontal average from the Fourier components with wavenumber k, as in section 3.1, we have to solve $\partial_z^2 \langle N_g \rangle = 0$ and $(\partial_z^2 - k^2)\tilde{N}_g = 0$ subject to corresponding boundary conditions. After some algebra, and taking into account that the complete interfacial energy balance reads $\partial_z T = \bar{\lambda}\partial_z T_g - EJ$ and that the thermal problem in the gas can be solved as before, we again get a mixed boundary condition

$$\partial_z \tilde{T} + Bi(k)\tilde{T} = 0 \quad \text{at } z = 1 \tag{32}$$

for the Fourier components with wavenumber $k \neq 0$, where the effective Biot number is now given by

$$Bi(k) = Bi_c(k) + Bi_{ev}(k) = \bar{\lambda}k\coth(kH) + \frac{Ep_s'(T^*)}{Q + \frac{H(1-N_{up})}{\bar{\rho}\bar{\kappa}Le}\frac{r(kH)^{-1}\tanh(kH)}{(r+(1-r)N_{up})^2}} \tag{33}$$

and the Fourier components of the evaporation rate are

$$\tilde{J} = E^{-1}Bi_{ev}(k)\tilde{T}_i = kH\coth(kH)\frac{\bar{\rho}\bar{\kappa}Le}{H(1-N_{up})}\tilde{N}_i \tag{34}$$

In addition, the horizontal average of the interfacial energy balance gives the boundary condition for $\langle T \rangle$, i.e.

$$\partial_z \langle T \rangle + Bi_{c,0}(\langle T \rangle - T_{up}) + Bi_{ev,0}(\langle T \rangle - T^*) = 0 \quad \text{at } z = 1 \tag{35}$$

where $Bi_{c,0} = Bi_c(k \to 0)$ and $Bi_{ev,0} = Bi_{ev}(k \to 0)$, while the averaged evaporation rate reads

$$\langle J \rangle = E^{-1} Bi_{ev,0}(\langle T_i \rangle - T^*) = \frac{\bar{\rho}\bar{\kappa} Le}{H(1 - N_{up})}(\langle N_i \rangle - N_{up}) \tag{36}$$

To summarize, this section has demonstrated that, in some vicinity of a global equilibrium state defined by the total gas pressure p_g and the vapor mass fraction N_{up} at some distance from the interface, convective evaporation of a pure liquid into an inert gas may be described by the system of equations (7-9), together with boundary conditions (10-12), (32) and (35). Note that additional reasonable assumptions are i) negligible gas viscosity; ii) sufficiently large gas thermal diffusivity; iii) not too large gas to liquid depth ratio H. This problem is not only nonlinear, but also non-local, as could be realized by applying an inverse Fourier transform to Eqs (32), leading to a convolution product. However, it is most convenient to write (and solve) it in Fourier space.

3.3 Reference states

Due to its symmetries, the full system of equations and boundary conditions always admits a horizontally homogeneous solution. On a long time scale, this solution evolves because of the decrease of the layer thickness in time. However, in the quasi-steady approximation described in the previous section, the free surface may be fixed at $z = 1$, and a steady solution exists, namely

$$\vec{V}_{ref} = 0 \ , \qquad T_{ref} = T_{bot} - z \ , \qquad p_{ref} = p_{bot} - Gaz + Ra\frac{z(1-z)}{2} \tag{37}$$

where the dimensionless temperature drop across the layer has been set to unity, by a suitable choice of the (yet undetermined) temperature scale

$$\theta = \frac{Bi_{c,0}(T_0 - T_H) + Bi_{ev,0}(T_0 - T_{sat})}{1 + Bi_{c,0} + Bi_{ev,0}} \tag{38}$$

as can be seen by solving Eq. (35) with $T_{bot} = T_0/\theta$, $T_{up} = T_H/\theta$ and $T^* = T_{sat}/\theta$. Note that T_{sat} depends on p_g and N_{up} according to our definition of the global equilibrium state. Now, using Eq. (36) and after some algebra, the evaporation rate in the non-equilibrium reference state (37) may be written as

$$J_{ref} = E^{-1} \frac{T_{i,c} - T^*}{Bi_{ev,0}^{-1} + (1 + Bi_{c,0})^{-1}} \tag{39}$$

where

$$T_{i,c} = \frac{T_{bot} + Bi_{c,0}T_{up}}{1 + Bi_{c,0}} = \frac{HT_{bot} + \bar{\lambda}T_{up}}{H + \bar{\lambda}} \tag{40}$$

is the interfacial temperature in a (fictive) purely conducting state, where evaporation is prevented (i.e. by assuming a vanishing accomodation coefficient leading to $Bi_{ev,0} = 0$).

It will be useful in the following to define

$$R_i = Q\ , \quad R_d = \frac{H(1 - N_{up})}{\bar{\rho}\bar{\kappa}Le} \frac{r}{(r + (1 - r)N_{up})^2}\ , \quad R_c = \frac{Ep'_s(T^*)H}{H + \bar{\lambda}} \tag{41}$$

and to recast Eq. (39) in the form

$$(R_i + R_d + R_c)J_{ref} = p'_s(T^*)(T_{i,c} - T^*) \tag{42}$$

which clearly shows (see also Haut and Colinet, 2005) that the evaporation flux is limited by three resistances "in series", namely the interfacial (kinetic) resistance R_i, the resistance R_d to diffusion in the gas mixture, and the resistance R_c to heat conduction towards the interface from both liquid and gas phases.

Other useful expressions for the reference evaporation rate can be found from Eqs (31), (36) and (42) :

$$(R_i + R_d)J_{ref} = p'_s(T^*)(T_{i,ref} - T^*) \tag{43}$$

$$(R_d + R_c)J_{ref} = p'_s(T^*)(T_{i,c} - T_{i,ref}) + \frac{r}{(r + (1 - r)N_{up})^2}(N_{i,ref} - N_{up}) \tag{44}$$

$$(R_i + R_c)J_{ref} = p'_s(T^*)(T_{i,c} - T^*) - \frac{r}{(r + (1 - r)N_{up})^2}(N_{i,ref} - N_{up}) \tag{45}$$

from which various limiting cases can be identified : i) when $R_c \ll R_d$ or $R_c \ll R_i$, comparing Eqs (42) and (43) readily shows that $T_{i,ref} = T_{i,c}$, i.e. the interfacial temperature is not affected by evaporation and reaches its value in the purely conducting state; ii) when $R_i \ll R_d$ or $R_i \ll R_c$, Eqs (42) and (44) lead to $p'_s(T^*)(T_{i,ref} - T^*) = r(N_{i,ref} - N_{up})/(r + (1 - r)N_{up})^2$, which means that local thermodynamic equilibrium prevails at the interface, according to Eq. (31); iii) when $R_d \ll R_c$ or $R_d \ll R_i$, Eqs (42) and (45) readily yield $N_{i,ref} = N_{up}$, and the gas composition is uniform. Clearly, two of these three situations may be realized simultaneously. For instance, in the case $R_i \gg R_d$ and $R_i \gg R_c$, the evaporation rate $J_{ref} = Q^{-1}p'_s(T^*)(T_{i,c} - T^*)$ is *reaction-limited*[6], even though this case should only

[6]Even though a phase transition is not a reaction, this term is used here to mean that the limitating step is linked to kinetic effects, i.e. the interface is not at equilibrium.

apply to extremely thin liquid and gas phases (note that $Q^{-1}p'_s(T^*)$ is equal to $E^{-1}Bi_{ev,p}$ – see Eq. (18) for the pure vapor case – though $Bi_{ev,p}$ is here very small). When R_c dominates, evaporation is *heat-diffusion-limited*, i.e. entirely determined by conduction towards the interface, which remains at the saturation temperature T^*. Finally, when R_d dominates, one has the *mass-diffusion-limited* regime, where evaporation is driven by the mass fraction difference $N_i^* - N_{up}$, in which N_i^* is the equilibrium mass fraction corresponding to the interface temperature $T_{i,c}$ in the purely conducting state. Note that these considerations are valid for the reference state, but do not necessarily apply to fluctuations around the latter.

3.4 Linear stability analysis

The stability of the steady reference state (37) is studied by adding infinitesimal perturbation fields to all variables, i.e.

$$\vec{V} = \vec{V}_{ref} + \vec{V}' \ , \quad T = T_{ref} + T' \ , \quad p = p_{ref} + p' \tag{46}$$

and linearizing the basic equations with respect to the primed quantities. Then, omitting the primes, we can obtain a simpler system by first forming the divergence of Eq. (8) and using Eq. (7), leading to $\Delta p = Ra\,\partial_z T$. This allows eliminating p (as well as U and V) from the full system, by applying the Laplacian operator to Eq. (8), and using the relation just found. We then obtain

$$\Delta^2 W + Ra\Delta_{\rm h} T = Pr^{-1}\frac{\partial}{\partial t}\Delta W \tag{47}$$

$$\Delta T + W = \frac{\partial T}{\partial t} \tag{48}$$

where D stands for the dimensionless z-derivative throughout in the following, while $\Delta_{\rm h} = \partial_x^2 + \partial_y^2$ is the horizontal Laplacian operator. This system (and associated boundary conditions discussed in previous sections) can be Fourier-transformed in the horizontal plane, and admits solutions of the form

$$\begin{pmatrix} \tilde{W} \\ \tilde{T} \end{pmatrix} = \exp[\sigma t] \begin{pmatrix} w(z) \\ \vartheta(z) \end{pmatrix} \tag{49}$$

The corresponding eigenvalue problem for σ then reads

$$(D^2 - k^2)^2 w - k^2 Ra\vartheta = \sigma Pr^{-1}(D^2 - k^2)w \tag{50}$$

$$(D^2 - k^2)\vartheta + w = \sigma\vartheta \tag{51}$$

with boundary conditions equivalent to Eqs (10-12) and (32), using Eq. (7) to eliminate horizontal velocity components. This yields

$$w = Dw = \vartheta = 0 \qquad \text{at } z = 0 \tag{52}$$

$$w = D^2 w + k^2 Ma\vartheta = D\vartheta + Bi(k)\vartheta = 0 \quad \text{at } z = 1 \tag{53}$$

Due to symetries of the original problem (invariance with respect to rotations and translations in the horizontal plane), the eigenvalue problem only depends on the modulus $k = (k_x^2 + k_y^2)^{1/2}$ of the wavevector $\vec{k} = (k_x, k_y)$ of the Fourier mode considered.

The above spectral problem may be solved exactly by a superposition of solutions of the form $\exp[r_i z]$, with $r_{i=1,\dots,6}$ given by the roots of the characteristic equation

$$(r^2 - k^2 - \sigma)(r^2 - k^2 - \sigma Pr^{-1})(r^2 - k^2) + k^2 Ra = 0 \tag{54}$$

Thus, the z-dependencies of temperature and velocity perturbations read (excluding degenerate cases) as

$$\vartheta = \sum_{i=1}^{6} c_i \exp[r_i z] \ , \quad w = \sum_{i=1}^{6} c_i (k^2 + \sigma - r_i^2) \exp[r_i z] \tag{55}$$

and the six independent integration constants have to be determined from the boundary conditions. This leads to a linear homogeneous system of the form $A.C = 0$, where C is the vector of unknowns c_i and A is a 6×6 matrix. Then, the compatibility relation of the system (which guarantees the existence of non-trivial solutions) is $\det(A) = 0$, which leads to the dispersion relation

$$f[\sigma, k, Pr, Bi(k), Ma, Ra] = 0 \tag{56}$$

Given Pr and $Bi(k)$, this relation may in principle be solved for the dispersion relation $\sigma = \sigma(Ra, Ma, k)$. However, this is technically impossible (the reason for this being the complicated dependency of the roots r_i upon σ), and the determinant (56) in general has to be calculated numerically, and combined with an iterative method for root finding (e.g. Newton's method). There are however particular cases, for which analytical results can be obtained.

For given values of the control parameters Ra and Ma and fixed values of Pr and $Bi(k)$, the reference state will be unstable if there exists at least a value of k for which $\Re(\sigma) > 0$, while it will be stable to infinitesimal perturbations if $\Re(\sigma) < 0$ for all k. The threshold for instability is then given by $\Re(\sigma) = 0$. If $\omega = \Im(\sigma) = 0$, one speaks of neutral stability (monotonic modes), while if $\omega \neq 0$ at criticality, one has overstability (oscillatory modes, or waves).

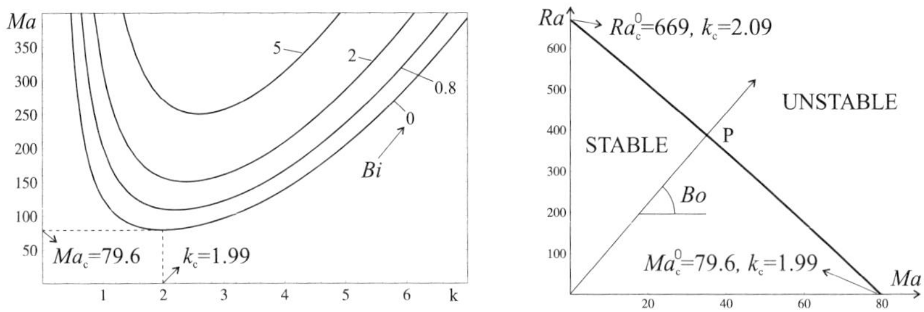

Figure 7. Left : neutral stability curves for the Marangoni-Bénard problem ($Ra = 0$) for various *constant* Biot numbers Bi. Right : locus of minima of neutral stability curves for the Rayleigh-Marangoni-Bénard problem, for $Bi = 0$.

Monotonic modes In case $\sigma = 0$ (neutral stability), the problem (50-53) can be solved analytically for $Ra = 0$, as done by Pearson (Pearson, 1958). Clearly, for $\sigma = 0$, the problem is independent of the Prandtl number Pr, and the compatibility relation can be solved for Ma, leading to

$$Ma = \frac{16k(k\cosh[k] + Bi(k)\sinh[k])(2k - \sinh[2k])}{4k^3\cosh[k] + 3\sinh[k] - \sinh[3k]} \tag{57}$$

This relationship is represented in Fig. 7(left) for various Biot numbers, here taken as constant. It is seen that the neutral stability curves present a minimum Ma_c at finite $k = k_c$, providing the critical conditions (when $Ma \sim \theta$ increases, the threshold is first reached at Ma_c, where $\sigma(k_c)$ changes from negative to positive). Above Ma_c, a finite band of wavenumbers have positive growth rate. An increase of the Biot number leads to an increase of Ma_c, therefore stabilizing the reference state.

For $Ra \neq 0$, the problem (50-53) is the one solved by Nield in 1964. Its solution is represented graphically in Fig. 7(right), for $Bi = 0$ (no evaporation, negligible gas thermal conductivity). The locus of threshold conditions appears close to a straight line given by

$$\frac{Ra_c}{Ra_c^0} + \frac{Ma_c}{Ma_c^0} = 1 \tag{58}$$

and allowing to determine the critical conditions (point P) for a given value of the dynamic Bond number $Bo = Ra/Ma = \alpha\rho g d^2/\gamma$.

Even though these linear stability results were originally obtained in the non-volatile case, the discussion of the previous sections showed that for sufficiently small evaporation fluxes, the stability of a volatile liquid layer can be studied on the basis of the same spectral problem, using a generalized Biot number $Bi(k)$ given by Eq. (33). Note that in this frame, such a non-volatile limit formally corresponds to a vanishing accomodation coefficient β, or $Q \to \infty$ in the interfacial kinetic law (26). In such a *reaction-limited* regime, Eq. (33) leads to $Bi(k) = Bi_c(k) = \bar{\lambda} k \coth(kH)$, i.e. heat transfer at the free surface is limited to weak heat conduction in the gas phase (typically, $\bar{\lambda} \sim 0.1$). Moreover, according to Eq. (38), a non-zero temperature difference θ can only be created when externally controlled temperatures at the bottom T_0 and at the top T_H are different.

When the kinetic resistance to evaporation $R_i = Q$ is finite, evaporation can generate a finite θ, hence possible instability, even in the absence of external driving force $T_0 - T_H$. Moreover, the effective Biot number increases due to the additional cooling effect generated by evaporation, which quickly becomes larger than pure heat conduction in the gas. Actually, in the case of a liquid evaporating into its *pure vapor* ($N_{up} = 1$), the effective Biot number may turn out to be extremely large (depending on the accomodation coefficient, it can be of the order 10^2–10^4). As the corresponding critical Marangoni number is then also very high, the reference state is then generally stable to the Marangoni effect. This is easily understandable since in such a *heat-diffusion-limited* regime (where $R_c \gg R_i$ and $R_d = 0$), the interface should remain close to the saturation temperature corresponding to the vapor pressure which, apart for small hydrodynamic fluctuations, should be roughly constant. This means that surface tension gradients will generally remain quite small, preventing thermocapillary flows. However, even though not considered here, it can be conjectured that the system could then be highly unstable to buoyancy-driven convection.

Now, when adding an inert gas such as nitrogen in the vapor phase, the effective Biot number drops down quite rapidly (the second term in the denominator of Eq. (33), proportional to R_d, quickly exceeds $R_i = Q$). The interface remains near local thermodynamic equilibrium, but fluctuations of the *partial* vapor pressure, linked to concentration fluctuations, are now allowed. Clearly, the increase of R_d is due to the additionnal diffusion barrier (stagnant film of inert gas though which the vapor must diffuse), but this effect is somehow counter-balanced by the fact that the partial pressure imposed at some distance from the interface also decreases, therefore increasing the driving force for evaporation (the right-hand-side of Eq. (42) can increase, since according to Eq. (29), T^* decreases when N_{up} decreases). Accordingly, the net effect is in general a destabilization of the evaporative reference state, as illustrated in Fig. 8.

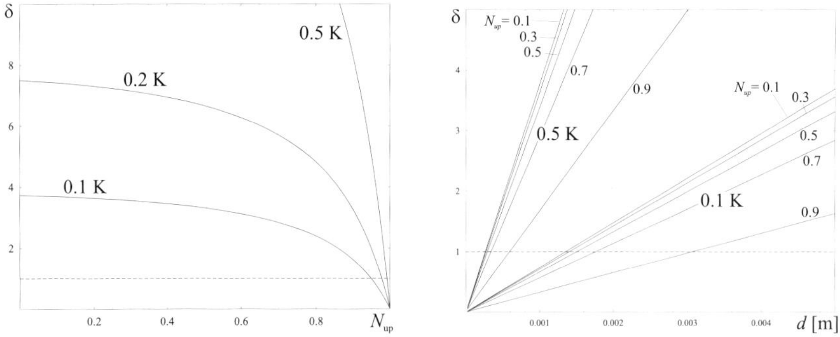

Figure 8. Linear stability results for ethanol (ethyl alcohol) evaporating into nitrogen, for $T_0 = T_H = 298$ K, an accomodation coefficient $\beta = 0.1$, and negligible buoyancy effect. Left: $\delta = Ma/Ma_c$ as a function of the vapor mass fraction N_{up}, for depths $d = d_g = 5$ mm of liquid and gas phases, and $\theta = 0.1, 0.2, 0.5$ K. Right: $\delta = Ma/Ma_c$ as a function of d (with $d_g = d$) for two temperature drops $\theta = 0.1, 0.5$ K and several vapor mass fractions N_{up}. The instability threshold ($\delta = 1$) is indicated by a dashed line.

In this figure, the "supercriticality" $\delta = Ma/Ma_c$ is defined as the ratio between the actual value of the Marangoni number (proportional to the temperature drop θ between the bottom plate and the interface), and its critical value obtained by minimizing (57) with respect to k. In order to remain close to a global thermodynamic state for all conditions of Fig. 8, the total pressure has been adjusted in each case, such as to maintain a constant temperature drop θ, i.e. a constant driving force for the instability. For the curves corresponding to $\theta = 0.1$ K on the right plot, the corresponding values are $p_g = 0.086, 0.12, 0.19, 0.35, 1.14$ atm, respectively for $N_{up} = 0.9, 0.7, 0.5, 0.3, 0.1$, and almost independently of d. Note that for $N_{up} = 1$, $\delta \ll 1$ for all d (stable reference state). Actually, the values of δ in Fig. 8(left) for $N_{up} = 1$ and $\theta = 0.1, 0.2, 0.5$ K are $\delta = 0.008, 0.017, 0.042$, respectively. Note finally that, in these calculations, the gas mixture properties (density and thermal diffusivity) are estimated using perfect gas laws and mixing rules described in (Bird et al., 1960).

Oscillatory modes As mentioned in section 2.1, oscillatory modes are not expected when both Ra and Ma are positive (heating from the bottom), but rather in the opposite case Ra, $Ma < 0$ (heating from the top). The case of negative dynamic Bond numbers $Bo = Ra/Ma$ is rather exotic and not considered hereafter. In order to demonstrate that the coupling between internal and surface (longitudinal) waves can lead to instability, as shown first in (Rednikov et al., 2000), one has to seek for solutions of the compatibility relation (56) with $\sigma = i\omega$, where ω is the (real) frequency.

In general, such calculation of marginal stability conditions of oscillatory modes should be performed numerically, because of the complicated dependency of Eq. (56) upon σ. This can be done in several ways, e.g. by first solving Eq. (56) with respect to Ma, leading to a function $Ma(\sigma, Ra, k)$ which is generally complex if $\Im(\sigma) \neq 0$. Then, the values of $\omega(k, Ra)$ satisfying $\Im[Ma(\sigma = i\,\omega, Ra, k)] = 0$ are numerically tracked[7] (note that $\omega(k, Ra)$ can be multi-valued). Having cancelled the imaginary part, the real part directly provides the Marangoni number for which a wave mode is marginally stable at these values of k and Ra. Moreover, taking into account that $Ra/Ma = Bo$, the equation $Ra = Bo\,Ma(i\omega(k, Ra), Ra, k)$ must be solved with respect to Ra. The marginal stability curves are represented in Fig. 9(left) as a function of Ma, for $Bi = 0$ (no evaporation, negligible gas thermal conductivity). In Fig. 9, the locus of critical conditions is represented in the plane (Ra, Ma). Note that critical values of the Rayleigh and Marangoni numbers appear to be much higher than for monotonic modes, though realizable in principle, for low viscosity liquids such as water and depths in the centimeter range.

As shown in (Rednikov et al., 2000), analytical results can also be obtained, in the limit $Ra \to -\infty$ and $Ma \to -\infty$ (still with $Bo = Ra/Ma$ finite) using matched asymtotic expansions. Only part of the theory is presented here. First, it will be useful to change the time scale from the thermal one $\tau_{\text{th}} = d^2/\kappa$ to the buoyant one $\tau_{\text{buoy}} = (-d/\alpha g\theta)^{1/2}$ (note that $\theta < 0$ here). This amounts to multiplying both σ and w by $\tau_{\text{th}}/\tau_{\text{buoy}} = (-PrRa)^{1/2}$ in Eqs (50-53). Defining $\varepsilon = (-Pr/Ra)^{1/4}$, this system can be rewritten

$$\varepsilon^2(D^2 - k^2)^2 w + k^2\vartheta = \sigma(D^2 - k^2)w \tag{59}$$

$$\varepsilon^2 Pr^{-1}(D^2 - k^2)\vartheta + w = \sigma\vartheta \tag{60}$$

$$w = Dw = \vartheta = 0 \quad \text{at } z = 0 \tag{61}$$

$$w = \varepsilon^2 D^2 w - k^2 Bo^{-1}\vartheta = D\vartheta + Bi\vartheta = 0 \quad \text{at } z = 1 \tag{62}$$

As dissipative time scales are much longer than the buoyancy (and thermocapillarity) time scales for $|Ra| \gg 1, Bo = O(1)$, it is not surprising that

[7] Note that the dependency of the dispersion relation upon other parameters, such as Pr and $Bi(k)$, is omitted here for clarity of notation.

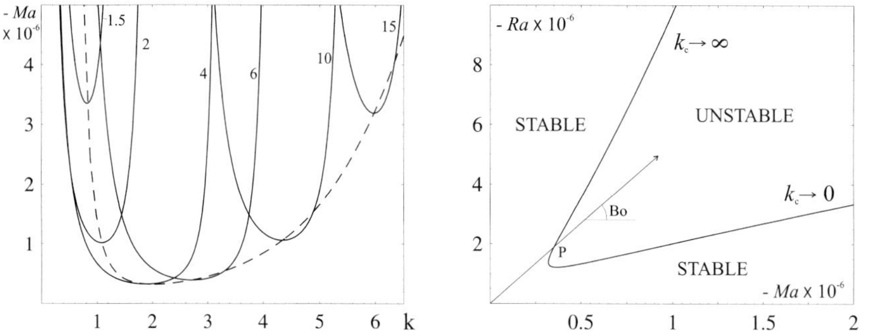

Figure 9. Onset of waves in a liquid layer heated from above, for $Pr = 6$ and $Bi = 0$. Left: marginal stability curves for several values of the dynamic Bond number $Bo = Ra/Ma$. The dashed line is the locus of minima (critical conditions). Right: stability diagram in the "Nield's plane". The critical wavenumber varies from 0 to ∞ when Bo increases.

highest order (dissipative) terms in the bulk equations (59) and (60) are multiplied by $\varepsilon^2 \ll 1$. However, not all boundary conditions at $z = 0$ and $z = 1$ can be satisfied if these terms are simply dropped out in the limit $\varepsilon \to 0$, and the correct resolution of the problem requires the use of stretched vertical coordinates in thin boundary layers located near the boundaries. In the bottom boundary layer, we define $\tilde{z} = z/\varepsilon$, such that $D = \varepsilon^{-1}\partial_{\tilde{z}}$. At the leading order, the equations to be solved are therefore

$$\partial_{\tilde{z}}^4 \tilde{w} = \sigma \partial_{\tilde{z}}^2 \tilde{w} \tag{63}$$

$$Pr^{-1}\partial_{\tilde{z}}^2 \tilde{\vartheta} + \tilde{w} = \sigma \tilde{\vartheta} \tag{64}$$

where tildes are used to distinguish variables in this bottom boundary layer. Note that identical equations have to be solved for $\bar{w}$ and $\bar{\vartheta}$ in the upper boundary layer, with the stretched coordinate here defined by $\bar{z} = (1-z)/\varepsilon$. With these definitions, the boundary conditions are

$$\tilde{w} = \partial_{\tilde{z}}\tilde{w} = \tilde{\vartheta} = 0 \quad \text{at } \tilde{z} = 0 \tag{65}$$

$$\bar{w} = \partial_{\bar{z}}^2 \bar{w} - k^2 Bo^{-1}\bar{\vartheta} = \partial_{\bar{z}}\bar{\vartheta} - \varepsilon Bi \bar{\vartheta} = 0 \quad \text{at } \bar{z} = 0 \tag{66}$$

and asymptotic matching conditions between the boundary layers and the bulk read

$$\tilde{w}|_{\tilde{z}\to\infty} \equiv w|_{z\to 0}\ , \qquad \tilde{\vartheta}|_{\tilde{z}\to\infty} \equiv \vartheta|_{z\to 0} \tag{67}$$

$$\bar{w}|_{\bar{z}\to\infty} \equiv w|_{z\to 1}\ , \qquad \bar{\vartheta}|_{\bar{z}\to\infty} \equiv \vartheta|_{z\to 1} \tag{68}$$

where w and ϑ are the (potential flow) solutions obtained by solving Eqs (59) and (60) for $\varepsilon = 0$. This directly leads to $\vartheta = a\cosh[rz] + b\sinh[rz]$ and $w = \sigma\vartheta$ with $r = k(1+\sigma^{-2})^{1/2}$. In turn, the resolution of Eqs (63) satisfying boundary conditions $\tilde{w} = \partial_{\tilde{z}}\tilde{w} = 0$ yields $\tilde{w} = f(\tilde{z} + \sigma^{-1/2}(\exp[-\sigma^{1/2}\tilde{z}] - 1))$, where it is assumed that $\Re[\sigma^{1/2}] > 0$. Note that the term proportional to $\exp[\sigma^{1/2}\tilde{z}]$ is omitted, because diverging exponentially for $\tilde{z} \to \infty$, hence preventing to satisfy matching conditions (67). Actually, the same matching conditions also require $f = 0$ at the lowest order, since $w|_{z\to 0} = \sigma a$, while $\tilde{w}|_{\tilde{z}\to\infty} \to f\tilde{z}$. Then, $a = 0$ as well and the solution in the bulk satisfying all matching conditions (67) is

$$\vartheta = b\sinh[rz] \ , \quad w = b\sigma\sinh[rz] \tag{69}$$

while $\tilde{w} = \tilde{\vartheta} = 0$ everywhere in the bottom boundary layer. In the top boundary layer, a similar resolution yields

$$\bar{w} = b\,\sigma\sinh[r]\left(1 - \exp[-\sigma^{1/2}\bar{z}]\right) \tag{70}$$

already satisfying $\bar{w}|_{\bar{z}=0} = 0$ and the first of Eqs (68). Then, solving for $\bar{\vartheta}$ and taking into account that the heat transfer boundary condition reduces to $\partial_{\bar{z}}\bar{\vartheta}|_{\bar{z}=0} = 0$ at the lowest order, we obtain

$$\bar{\vartheta} = b\sinh[r]\left(1 + \frac{\exp[-(Pr\sigma)^{1/2}\bar{z}]}{Pr^{1/2}(1 - Pr^{-1})} + \frac{\exp[-\sigma^{1/2}\bar{z}]}{Pr^{-1} - 1}\right) \tag{71}$$

Finally, it remains to express the last (Marangoni) boundary condition, which can easily be rearranged as

$$b\sinh[r]\left[\sigma^2 + \frac{k^2}{Bo(1 + Pr^{1/2})}\right] = 0 \tag{72}$$

Taking into account that $r = k(1 + \sigma^{-2})^{1/2}$, non-trivial solutions $b \neq 0$ are possible only for dissipationless wave modes $\sigma = i\omega$, where either

$$\omega = \omega_{\text{int}} = \sqrt{\frac{k^2}{k^2 + n^2\pi^2}} \ , \qquad n = 1, 2, \ldots \tag{73}$$

or when

$$\omega = \omega_{\text{sf}} = \frac{k}{\sqrt{Bo(1 + Pr^{1/2})}} \tag{74}$$

The first of these wave modes is the (Brunt–Väisälä) internal wave, which, in addition to the wavenumber k, is parametrized by the number or vortices

n in the vertical direction ($w, \vartheta \sim \sin[n\pi z]$). The second mode is genuinely related to the Marangoni effect, and corresponds to the surface (longitudinal) wave, the mechanism of which was described in section 2.1. Note that contrary to the internal waves for which $\bar{w} = 0$ at the lowest order in the upper boundary layer, surface waves are characterized by $\bar{w} \neq 0$, hence by an intense horizontal velocity field $\bar{u} = ik^{-1}\varepsilon^{-1}\partial_{\bar{z}}\bar{w}$, i.e. ε^{-1} times larger than other quantities in the limit $\varepsilon \to 0$. Both these wave modes are neither damped nor amplified on the (fast) time scale considered, and only next order corrections to the complex growth rate σ can lead to information on their long-term behaviour. Omitting details for conciseness, we have to expand $\sigma = i\omega_0 + \varepsilon\sigma_1$, as well as to consider $O(\varepsilon)$ corrections to velocity and temperature fields in each asymptotic region. Selecting either ω_{int} or ω_{sf} for the lowest-order frequency ω_0, we finally obtain the corresponding σ_1 for each wave, proceeding similarly (but more tediously) as above. Interestingly, both values of σ_1 indeed possess a non-zero real part, and are found to diverge when

$$Bo = Bo_{\text{res}} = \frac{k^2 + n^2\pi^2}{1 + Pr^{1/2}} \tag{75}$$

indicating that the asymtotic analysis fails when the frequencies of both waves are equal (linear resonance). An improved asymtotics is actually needed in the vicinity of this point (Rednikov et al., 2000). More reliable results obtained from σ_1 are the marginal stability thresholds $\Re[\sigma_1] = 0$, which yield

$$Bo = Bo_{\text{int}} = \frac{1 + 2Pr^{1/2}}{Pr + Pr^{1/2}}(k^2 + n^2\pi^2) \tag{76}$$

for the internal wave, and

$$Bo = Bo_{\text{sf}} = \frac{k^2 + x_n^2}{1 + Pr^{1/2}} \tag{77}$$

for the surface wave, where the $x_{n=1,2,\dots}$ are the (real) roots of the transcendental equation $x \cot[x] + Bi/(1 + Pr^{1/2}) = 0$. For small Bi, one has

$$Bo_{\text{sf}} = \frac{k^2 + (n - 1/2)^2\pi^2}{1 + Pr^{1/2}} + \frac{2Bi}{(1 + Pr^{1/2})^2} + O(Bi^2) \quad \text{for } Bi \ll 1 \tag{78}$$

and more generally, $(n - 1/2)\pi \leq x_n \leq n\pi$.

These asymptotic results are represented in Fig. 10, in the plane (Bo, k^2), for the lowest values of the vertical wavenumber $n = 1, 2, \dots$. Note that for given k and n, $Bo_{\text{sf}} < Bo_{\text{res}} < Bo_{\text{int}}$, for all Pr and Bi.

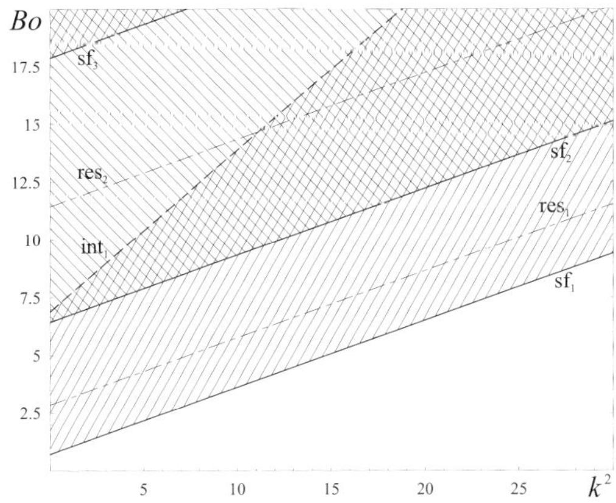

Figure 10. Asymptotic results in the limit Ra, $Ma \to -\infty$, for $Bi = 0$ and $Pr = 6$: marginal stability thresholds of surface ($\mathrm{sf_n}$) and internal ($\mathrm{int_n}$) waves, together with loci of resonances ($\mathrm{res_n}$), for most dangerous modes $n = 1, 2, \ldots$. The reference state is unstable to one wave mode in the hatched regions, to two modes in the crossed-hatched regions, ...

The above results and the following discussions must be understood in the asymptotic limit Ra, $Ma \to -\infty$, $Bo = O(1)$. Considering Bo as a free parameter progressively increased from 0 (pure Marangoni effect), instability first occurs to surface waves at the minimal value of Bo_{sf}. According to Eq. (77) and Fig. 10, this is $Bo_{\mathrm{min}} = x_1^2/(1 + Pr^{1/2})$ at $k = 0$, $n = 1$. For $Bi = 0$, this gives $Bo_{\mathrm{min}} = \pi^2/4(1 + Pr^{1/2}) = 0.715$ for $Pr = 6$, in agreement with fully numerical results (e.g. in Fig. 9(left), the right asymptote of the marginal stability curve appears at $k = 0$, precisely at this value of Bo).

Then, considering fixed values of k and $n = 1$ in Fig. 10 for simplicity, the growth constant of surface waves eventually keeps increasing with Bo, while that of internal waves remains negative. In the vicinity of the resonance point $Bo = Bo_{\mathrm{res}}$, it is shown in (Rednikov et al., 2000) that a continuous transition occurs between the surface wave mode and the internal wave mode (and vice-versa), a phenomenon called *mode-mixing*. There, a higher degree of amplification is predicted for the unstable mixed mode, while the stable mixed mode is more damped (however the real parts of both modes still remain small compared with the frequency). This mode mixing results in the fact that above Bo_{res}, the mode with positive growth rate turns out to be the internal wave mode. Increasing Bo again (still at fixed k and $n = 1$), the amplification rate of the internal wave mode decreases, and passes through zero at $Bo_{\rm int}(n=1)$. This does not mean that the reference state is stable above this value however, because another wave mode with $n=2$ has become amplified at $Bo_{\rm sf}(n=2) < Bo_{\rm int}(n=1)$ (see Fig. 10). In fact, it can be seen that the number of unstable modes globally grows with Bo, since $Bo_{\rm int}$ grows faster than $Bo_{\rm sf}$ when n increases. Thus when Bo is increased, the system remains unstable (in the limit $Ra \to -\infty$ and $Ma = Ra/Bo$).

Another (more physical) representation is obtained at given Bond number Bo. At given n, the resolution of $Bo = Bo_{\rm int}$, $Bo = Bo_{\rm res}$ and $Bo = Bo_{\rm sf}$ with respect to the horizontal wavenumber k leads to, say, $k_{\rm int}$, $k_{\rm res}$ and $k_{\rm sf}$ respectively. For $n=1$, a solution $k_{\rm sf} > 0$ exists if $Bo > x_1^2/(1+Pr^{1/2})$, while $k_{\rm int} > 0$ exists if $Bo > \pi^2(1+2Pr^{1/2})/(Pr+Pr^{1/2})$. In this case, it is readily shown that $k_{\rm sf} > k_{\rm res} > k_{\rm int}$, as also seen in Fig. 10. Values $k_{\rm int}$ and $k_{\rm sf}$ of the wavenumber indeed correspond to the left and right asymptotes of marginal stability curves of Fig. 9(left) (limits of the stability boundary for $Ma \to -\infty$, defining the interval of unstable wavenumbers at fixed n).

Note that the effect of Bi (hence, of possible evaporation/condensation processes at the interface) turns out to affect the surface wave only at this order (as formulas (75-77) show), the growth rate of which is always decreasing with Bi. Actually, even the interval of unstable wavenumbers $[k_{\rm int}, k_{\rm sf}]$ shrinks, $k_{\rm sf}$ getting closer to $k_{\rm res}$ when the effective Biot number Bi increases.

To conclude this section, it is worth emphasizing the generality of the mode-mixing process described above, by pointing out another example where very similar asymptotic features are observed (with some subtle differences, however). As mentionned in section 2.1, the capillary-gravity wave is yet another high-frequency wave mode which can resonantly interact and mix with the surface wave mode described in this section. In the gravitational time scale $\tau_{\rm grav} = (d/g)^{1/2}$, the lowest order dispersion relation of ideal capillary-gravity waves indeed reads $\omega_{\rm c\text{-}g}^2 = k(1+B^{-1}k^2)\tanh[k]$ where $B = \rho g d^2/\sigma$ is the *static* Bond number (σ here denotes the surface tension). This is indeed the result found in the limit $\tau_{\rm grav} \ll \tau_{\rm th} \sim \tau_{\rm visc}$ with $B = O(1)$ (see for instance Rednikov et al., 2001).

Using the thermal time scale $\tau_{\rm th}$ for all three waves (internal, surface, and capillary-gravity), the corresponding dispersion relations read

$$\omega_{\rm int}^2 = \frac{k^2 Pr\,(-Ra)}{k^2+n^2\pi^2} \qquad n = 1, 2, ... \tag{79}$$

$$\omega_{\rm sf}^2 = \frac{k^2 Pr\,(-Ma)}{1+Pr^{1/2}} \tag{80}$$

$$\omega_{\rm c\text{-}g}^2 = PrGak(1+B^{-1}k^2)\tanh[k] \tag{81}$$

where $Ga = gd^3/\nu\kappa$ is the Galileo number.

The possibility of resonance between either two of these wave modes can be analyzed with the help of a diagram of ω^2 versus k^2. For instance, inspection of the slopes of the dispersion relations at the origin readily shows that $\omega_{\text{int}} = \omega_{\text{sf}}$ at some non-zero k provided $Bo > n^2\pi^2/(1 + Pr^{1/2})$, which turns out to be achievable for usual liquids such as water with depths in the centimeter range, at least for $n = 1$ (as discussed earlier).

Similarly, resonant interaction between longitudinal surface waves and transverse capillary-gravity waves will certainly occur at some finite k provided $Ga < -Ma/(1 + Pr^{1/2})$. Even though resonance can in principle occur for larger values of Ga (depending on B), this criterion turns out to be quite accurate in practice, and can be rearranged as $B < \gamma|\theta|/\sigma(1+Pr^{1/2})$. Hence, for usual liquids, the static Bond number should be rather small (because $\gamma|\theta| \ll \sigma$), i.e. the depth d has to be submillimetric to reach resonance between these waves. In mass transfer systems, these waves might be more easily observable since the variation of surface tension with concentration may be extremely large. Their detailed asymptotic analysis, complemented by fully numerical results, has been achieved in (Rednikov et al., 2001).

Finally, resonance between internal and capillary-gravity waves turns out to be impossible, independently of the liquid depth. Indeed, taking into account that $Ra = \alpha\theta Ga$, it is easy to show that $\omega^2_{\text{c-g}}/\omega^2_{\text{int}} > (\alpha|\theta|)^{-1}(k^2 + n^2\pi^2)\tanh[k]/k > (\alpha|\theta|)^{-1}$. Hence, resonance would require quite unachievable temperature drops in usual liquids (in addition to the fact that the Boussinesq approximation underlying our analysis relies on the hypothesis $\alpha|\theta| \ll 1$).

3.5 Direct numerical simulations

Let us briefly come back to a more general perspective. In the previous sections, a general mathematical framework has been described, which allows studying instabilities in a single layer of liquid, either simply heated from below or above, or undergoing evaporation at its upper free surface (with the possible complication of heat conduction in the gas and the presence of an inert component).

Still, apart for simple reference states (§3.3) and their linear stability analyses (§3.4), it is generally impossible to solve the full system of equations (7), (8) and (9) analytically, even with the simplest boundary conditions (10), (11), (12) and $\partial_z T + Bi(T - T_\infty) = 0$ (i.e. constant Biot number).

A first way to proceed in the analysis of such thermohydrodynamic instabilities would be to undertake direct numerical simulations, e.g. using the FLUENT software (see Fig. 11) or home-made codes.

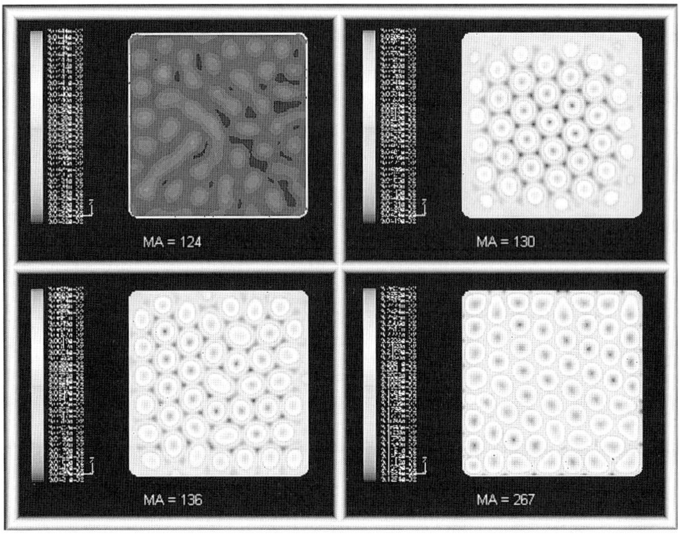

Figure 11. Direct 3D numerical simulations of purely surface-tension-driven convection in a square container with rigid insulating sidewalls. The different simulations represent the free surface temperature field obtained after a certain time, for various imposed temperature differences (the calculated Marangoni number is indicated).

While this clearly shows that the above system of equations and boundary conditions is quite suitable to describe experiments on layers heated from below (and therefore also in more complex cases including evaporation, for instance), such direct three-dimensional simulations are particularly time-consuming, not only due to the fine mesh needed to capture all details of the flow, but also to the differences in time scales inherent to the physical system considered. Indeed, near the instability threshold the evolution is quite slow (for large Prandtl number fluids like silicone oils, it may be shown that instability develops after a time of order $\epsilon^{-1}\tau_{\rm th}$, where $\epsilon = (Ma - Ma_c)/Ma_c$ is a supercriticality parameter), a phenomenon analogous to the so-called "critical slowing down" of equilibrium phase transitions. Another time scale is associated with pattern organization on the lateral scale L of the container, which is about $(L/d)^2$ times larger than the time it takes for convection cells to develop.

Note finally that Fig. 11 also shows that near threshold, the hexagonal pattern is quite regular, with few defects, while their number increases at increasing ϵ. In addition, the distance at which the pattern "feels" the presence of boundaries also seems to diverge near the critical point, which again shows an analogy with the correlation length of phase transitions. Such analogies have been studied in details, though rather from models of Bénard convection and other patterned structures, to which the rest of these lecture notes is dedicated.

4 Simplified models for non-equilibrium patterns

Rather than attempting direct numerical simulations of the basic equations and boundary conditions applying to a specific Bénard set-up, alternative approaches exist, which allow to understand most fundamental aspects of pattern and wave formation in systems such as those considered in these lectures. Note that most of the convection models presented hereafter can be systematically derived from equations such as presented in §3, at least in some asymptotic limit (small supercriticality ϵ, large-scale flows compared to liquid depth, ...). However, the presentation made hereafter will remain at a phenomenological level, in order to avoid too technical calculations, and to emphasize the role of symmetries enjoyed by the physical set-up considered.

4.1 The Swift-Hohenberg equation and its variants

Consider first the equation

$$\partial_t \phi = \epsilon\phi - (\Delta + 1)^2\phi + \delta\phi^2 - \phi^3 \tag{82}$$

for an order-parameter $\phi(\vec{r} = x\vec{1}_x + y\vec{1}_y, t)$ which may for instance represent the two-dimensional free surface temperature field in a Marangoni-Bénard problem (or the vertical liquid velocity at mid-depth of the layer in a Rayleigh-Bénard set-up). When $\delta = 0$, this equation is the celebrated Swift-Hohenberg (SH) equation proposed by these authors (Swift and Hohenberg, 1977), as a phenomenological model capturing most of the essential features of pattern formation in Rayleigh-Bénard convection. For $\delta \neq 0$, Eq. (82) is a modified SH equation, which rather seems to qualitatively apply to Marangoni-Bénard convection, as seen in the following.

First, remark that $\phi = 0$ is a particular steady solution of Eq. (82), which actually corresponds to the uniform state of heat conduction prevailing at small ΔT. In order to determine whether this reference state is stable, we allow small perturbations in the form of normal (i.e. Fourier) modes

$$\phi(\vec{r}, t) = \eta \exp[\sigma t] \exp[i\vec{k}.\vec{r}] \qquad \text{with } \eta \ll 1 \tag{83}$$

where σ is the growth rate and $\vec{k} = k_x\vec{1}_x + k_y\vec{1}_y$ is a two-dimensional wavevector. Inserting this expression in Eq. (82) and linearizing with re-

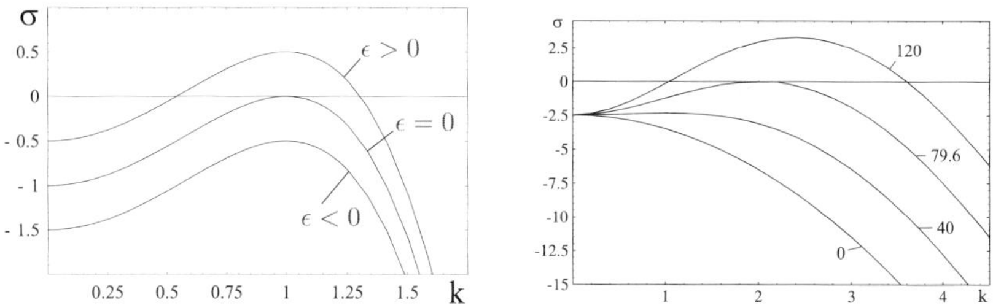

Figure 12. Left: dispersion relation of the Swift-Hohenberg equation (82). Right: dispersion relation for perturbations of the conductive state in the Marangoni-Bénard problem, with $Ra = Bi = 0$ and $Pr \to \infty$, for several values of Ma.

spect to the small parameter η, it is seen that the equation is indeed satisfied provided

$$\sigma = \epsilon - \left(1 - k^2\right)^2 \tag{84}$$

which is the dispersion relation for perturbations, i.e. the relation between their growth rate σ and their wavenumber $k = |\vec{k}|$. This relation is represented in Fig. 12, and compared with the exact dispersion relation for the Marangoni-Bénard problem described in §3.

It is seen from Fig. 12 that both dispersion relations show similar behavior near the critical point ($\epsilon = 0$ or $Ma = Ma_c = 79.6$). Namely, all Fourier modes have negative growth rate below this critical point (hence, the reference state is stable), while a band of wavenumbers lead to a positive growth rate as soon as it is exceeded (unstable reference state). At criticality, only one wavenumber k_c has a zero growth rate (it is then said that the reference state is neutrally – or marginally – stable).

For $\epsilon > 0$, some Fourier modes will therefore be amplified exponentially, and the assumption of small η in Eq. (83) eventually fails after some time. In this case, nonlinear effects (terms ϕ^2 and ϕ^3) start to couple the different Fourier modes, in addition to saturating the initial exponential growth of fluctuations. Some direct numerical simulations of the modified SH equation are shown in Fig. 13. At small values of δ and ε, the system shows a coexistence of hexagons and rolls for a long time, before the hexagonal pattern invades the whole simulation domain. At larger values of δ, the hexagons are strongly favored even after a much shorter time, although the final pattern seems to contain more defects (actually, some pentagon-heptagon pairs can be seen, and there are several domains of regular hexagons with different orientation separated by "grain boundaries", quite similarly to what is observed experimentally).

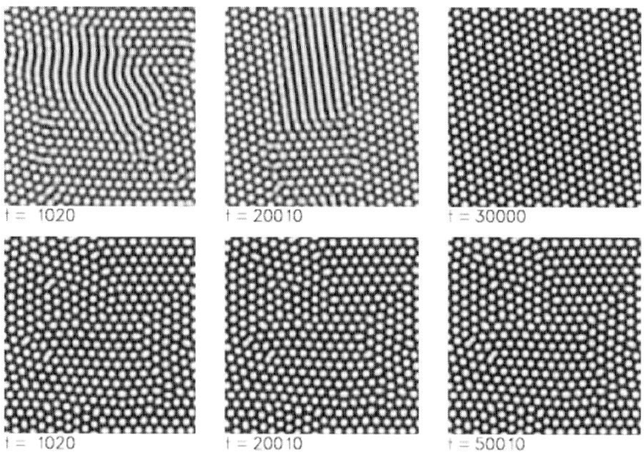

Figure 13. Direct numerical simulations of the modified Swift-Hohenberg equation (82). Top row : $\epsilon = 0.1$, $\delta = 0.26$. Bottom row : $\epsilon = 0.1$, $\delta = 1.3$. Courtesy of M. Bestehorn.

However, despite this nice qualitative agreement with Marangoni-Bénard convection, there are also important differences. First, it can be shown that the dynamics described by Eq. (82) is always relaxational (monotonic evolution to a steady state, minimizing some energy functional). However, several authors have later shown that this is not the case for Marangoni-Bénard convection, with important consequences on the dynamics of defects (see e.g. Colinet et al., 2002, and references therein). Second, Eq. (82) does not show a transition to square patterns at increasing ϵ, even though this seems to be a robust feature of Marangoni-Bénard convection (Eckert et al., 1998; Eckert and Thess, 1999). Variants of the Swift-Hohenberg equation have therefore been constructed (Bestehorn and Friedrich, 1999) in order to better match particular physical set-ups. As an example, adding terms such as $\phi\Delta^2(\phi^2)$ in Eq. (82) indeed produces squares provided some conditions on the coefficients of the various terms are satisfied (see also lectures of M. Bestehorn).

4.2 Basic symmetries of Bénard set-ups

The SH equation and its variants quite nicely model Bénard convection, which can be attributed to the fact that the dispersion relations have a similar behavior near the critical point (see Fig. 12), in addition to the fact that their symmetry properties are identical to those obeyed by full systems of equations and boundary conditions such as given in §3. To simplify the presentation, we however restrict the following considerations to Eq. (82) in this section.

Given a particular solution $\phi_0(\vec{r}, t)$ of the modified Swift-Hohenberg equation, consider first the translated solution $\phi(\vec{r}, t) = \phi_0(\vec{r} - \vec{r}_0, t)$, where $\vec{r}_0$ is an arbitrary horizontal vector. Clearly, the equation describing the evolution of the translated solution ϕ is identical to that describing the evolution of ϕ_0. Indeed, as ϕ_0 is a solution of Eq. (82), we have

$$\partial_t \phi_0(\vec{r}, t) = \epsilon \phi_0(\vec{r}, t) - (\Delta_{\vec{r}} + 1)^2 \phi_0(\vec{r}, t) + \delta \phi_0(\vec{r}, t)^2 - \phi_0(\vec{r}, t)^3 \tag{85}$$

where $\Delta_{\vec{r}}$ is the horizontal Laplacian with respect to $\vec{r}$-variables. Substituting $\vec{r}$ by $\vec{r} - \vec{r}_0$ in Eq. (85) directly leads to Eq. (82) for ϕ, as $\Delta_{\vec{r}-\vec{r}_0} = \Delta_{\vec{r}}$, indicating that the (modified) SH equation is invariant by translation. For a *laterally infinite* Bénard set-up, it is indeed natural to expect that any translation of a particular solution of the problem is itself a solution.

Consider now the rotation symmetry, by transforming the particular solution $\phi_0(\vec{r}, t)$ onto a rotated solution $\phi(\vec{r}, t) = \phi_0(R_\theta \vec{r}, t)$, where R_θ is a rotation operator (with angle θ). Substituting $\vec{r}$ by $R_\theta \vec{r}$ in Eq. (85) leads to

$$\partial_t \phi(\vec{r}, t) = \epsilon \phi(\vec{r}, t) - (\Delta_{R_\theta \vec{r}} + 1)^2 \phi(\vec{r}, t) + \delta \phi(\vec{r}, t)^2 - \phi(\vec{r}, t)^3 \tag{86}$$

Now, using the matrix representation

$$R_\theta = \begin{pmatrix} \cos\theta & \sin\theta \\ -\sin\theta & \cos\theta \end{pmatrix} \tag{87}$$

and defining $\vec{r}' = R_\theta \vec{r}$, we have

$$\begin{aligned} \Delta_{\vec{r}'} = \frac{\partial^2}{\partial x'^2} + \frac{\partial^2}{\partial y'^2} &= \left(\cos\theta \frac{\partial}{\partial x} + \sin\theta \frac{\partial}{\partial y}\right)^2 + \left(-\sin\theta \frac{\partial}{\partial x} + \cos\theta \frac{\partial}{\partial y}\right)^2 \\ &= \frac{\partial^2}{\partial x^2} + \frac{\partial^2}{\partial y^2} = \Delta_{\vec{r}} \end{aligned} \tag{88}$$

and hence the equation satisfied by the rotated pattern ϕ reduces to the original equation, which is thus invariant by rotation. From both cases of translation and rotation, it is seen that the condition of invariance of Eq. (82) with respect to a symmetry operator S acting on $\vec{r}$, is that the Laplacian operator is invariant with respect to this transformation, i.e. $\Delta_{S\vec{r}} = \Delta_{\vec{r}}$.

It is straightforward to check that this is verified for the parity operator P (defined by $P\vec{r} = -\vec{r}$, in fact $R_\pi = P$), and for the mirror reflection

$$M_\theta = \begin{pmatrix} \cos 2\theta & \sin 2\theta \\ \sin 2\theta & -\cos 2\theta \end{pmatrix} \tag{89}$$

with respect to a line passing through the origin and forming an angle θ with the x-axis. Note that any mirror reflection with angle θ can be obtained as the composition of a mirror reflection with respect to x, followed by a rotation of angle -2θ (i.e. 2θ in the counter-clockwise direction).

For autonomous systems, the origin of time is arbitrary, so that another invariance to be satisfied is the time-translation $t \to t+\Delta t$, which is trivially satisfied by Eq. (82).

To summarize, Eq. (82) has been found to be invariant to translations, rotations (including parity), mirror reflections and time-translations, which is also the case for the full system of equations and boundary conditions presented in §3. These basic symmetries will be sufficient for our purposes, and it will be shown in the next section that they indeed strongly determine the nature and stability of new solutions appearing above instability threshold.

Importantly, for systems in uniform rotation around a vertical axis, some symmetries do not hold anymore. Indeed, in this case centrifugal and Coriolis forces (Chandrasekhar, 1961) have to be incorporated in the Navier-Stokes equations (written in the rotating frame), which breaks both translation and reflection symmetries (although invariance by rotation continues to hold). This will have important consequences, as seen in the following sections. Note finally that other more specific symmetries will be encountered in the following.

4.3 Symmetries and amplitude equations

Near instability threshold, i.e. for sufficiently small supercriticality ϵ, it is possible to describe the weakly nonlinear evolution of several types of patterns using equations describing the evolution of amplitudes of fluctuations around the reference state. While this can be done rigorously using perturbation expansions in powers of ϵ, e.g. from systems of PDE's such as presented in §3 (basic techniques are described in Colinet et al., 2001), we will rather use the symmetry considerations presented in the last section to obtain the form of such amplitude equations with much less calculations. Again, we will continue to use the modified SH equation as a model, qualitatively equivalent to the original full system of equations and boundary conditions.

Bifurcation of rolls Let us first attempt to solve Eq. (82) by

$$\phi = A_1(t)\exp[i\vec{k}_1.\vec{r}] + A_1^*(t)\exp[-i\vec{k}_1.\vec{r}] \tag{90}$$

where $\vec{k}_1$ is an arbitrary wavevector, and A_1 a complex amplitude in general (an asterisk denotes the complex conjugate). For small $|A_1|$, Eq. (82) may be linearized and is then found to be satisfied if

$$\dot{A}_1 = \sigma A_1 \tag{91}$$

where the dot means the time derivative, and σ is given by Eq. (84) with $k = |\vec{k}_1|$. Hereafter, we select $k = 1$ (critical wavenumber) for simplicity, such as $\sigma = \epsilon$. Clearly, Eq. (91) is solved by an exponential, such as in Eq. (83). Hence, the question is to determine how to complete Eq. (91) such as to avoid this unphysical exponential growth for $\epsilon > 0$.

Assuming a more general evolution equation of the form

$$\dot{A}_1 = f(A_1, A_1^*) \tag{92}$$

where f is an analytical function of its arguments, we will expand this amplitude equation in Taylor series as

$$\dot{A}_1 = \sigma A_1 + \delta_1 A_1^2 + \delta_2 |A_1|^2 + \delta_3 A_1^{*2} + \beta_1 A_1^3 + \beta_2 A_1^2 A_1^* + \beta_3 A_1 A_1^{*2} + \beta_4 A_1^{*3} \tag{93}$$

where the expansion has been limited to cubic order (to be justified a posteriori). Now, this form may be greatly simplified by making use of the symmetry considerations. Consider the translational invariance $\vec{r} \to \vec{r} - \vec{r}_0$. If $\phi_0(\vec{r})$ is a solution, $\phi_0(\vec{r} - \vec{r}_0)$ must also be a solution. As seen from (90), this translated solution may also be obtained by the transformation $A_1 \to B_1 = A_1 \exp[-i\vec{k}_1.\vec{r}_0]$ acting on the amplitude A_1. Thus, B_1 should satisfy the amplitude equation (93) if A_1 does. Substituting A_1 by B_1 in this equation, the result should be valid for all $\vec{r}_0$ and we see that all terms not proportional to $\exp[-i\vec{k}_1.\vec{r}_0]$, after substitution, should cancel, i.e. we get $\delta_1 = \delta_2 = \delta_3 = \beta_1 = \beta_3 = \beta_4 = 0$. Redefining $\beta_2 = -\beta$, the equation (90) reduces to the Landau equation

$$\dot{A}_1 = \sigma A_1 - \beta A_1 |A_1|^2 \tag{94}$$

The only terms remaining are those multiplied by $\exp[i\varphi]$ under the transformation $A_1 \to A_1 \exp[i\varphi]$ (equivariant terms). Note that the equivariant function $A_1(\sigma - \beta|A_1|^2)$ is the product of the equivariant function A_1 and the invariant function $\sigma - \beta|A_1|^2$ (i.e. it remains invariant under the transformation).

Now, although translational invariance has already greatly simplified the amplitude equation, we have to consider the effect of other symmetries. The rotational invariance $\vec{r} \to \vec{r}' = R_\theta \vec{r}$ will not be useful here. According to Eq. (90), its action on the amplitude is $A_1 \to A_1 \exp[i\vec{k}_1.(R_\theta\vec{r}-\vec{r})]$, i.e. the phase generally depends on $\vec{r}$, a situation excluded in the present analysis. In the case of parity $P\vec{r} = R_\pi \vec{r} = -\vec{r}$, the corresponding action on the amplitude is $A_1 \to A_1^*$. Thus, the equation (94) should be satisfied under this change, which is only possible if $\sigma = \sigma^*$ and $\beta = \beta^*$ (real coefficients). Finally, it can be seen that the mirror reflection symmetry faces the same problems as rotation. Strictly speaking, the fact that rotation and reflection invariances cannot be satisfied is a consequence of our arbitrary choice of a particular direction $\vec{k}_1$ in the horizontal plane, which clearly breaks these symmetries.

The amplitude equation (94) for rolls (or stripes) may be solved by first separating modulus and phase $A = r\exp[i\varphi]$. This leads to $\dot{\varphi} = 0$ and to $\dot{r} = \sigma r - \beta r^3$ (pitchfork bifurcation), which indeed saturates the exponential growth provided $\beta > 0$ (supercritical pitchfork bifurcation, see Fig. 14). Otherwise (subcritical bifurcation), higher-order terms have to be included in Eqs (93) and (94).

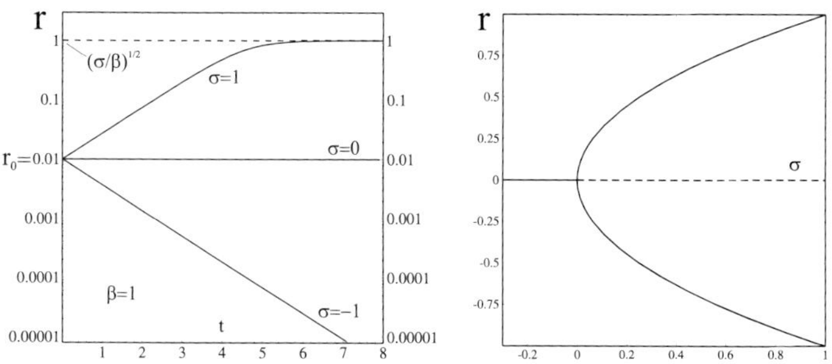

Figure 14. Supercritical pitchfork bifurcation. (a): temporal evolution of the amplitude r starting from an initial fluctuation $r_0 = 0.01$, for several values of the linear growth rate σ. (b): steady states and their stability (full lines: stable solutions, dashed lines: unstable solutions).

Bifurcation of hexagonal patterns We now consider the important case of the interaction of three Fourier modes with wavevectors $\vec{k}_1$, $\vec{k}_2$ and $\vec{k}_3$ oriented at 120° apart on the critical circle (of radius $k = 1$). Note that $\vec{k}_1 + \vec{k}_2 + \vec{k}_3 = 0$, i.e. the basic wavevectors form a "resonant triad", to employ the language of waves.

The planform function is therefore written

$$\phi = A_1 \exp[i\vec{k}_1.\vec{r}] + A_2 \exp[i\vec{k}_2.\vec{r}] + A_3 \exp[i\vec{k}_3.\vec{r}] + c.c. \quad (95)$$

and the amplitudes are assumed to satisfy evolution equations of the form

$$\dot{A}_p = f_p(A_1, A_1^*, A_2, A_2^*, A_3, A_3^*) \quad p = 1, 2, 3 \quad (96)$$

which should be Taylor-expanded up to cubic order, just as for roll patterns. Actually, most of the terms in the resulting long expressions can readily be eliminated by considering the translation invariance $\vec{r} \to \vec{r} - \vec{r}_0$, which is equivalent to the action $A_p \to A_p \exp[-i\vec{k}_p.\vec{r}_0]$, $p = 1, 2, 3$. Thus, in the equation for A_1 for example, the linear term A_1 remains as for rolls, and the quadratic term $A_2^* A_3^*$ is equivariant as well (it is multiplied by $\exp[i(\vec{k}_2 + \vec{k}_3).\vec{r}_0] = \exp[-i\vec{k}_1.\vec{r}_0]$ under the translation, just as $\dot{A}_1$). It is easily seen that these are the only equivariant linear and quadratic terms, and going on to cubic order finally leads to

$$\dot{A}_1 = \sigma A_1 + \delta A_2^* A_3^* - A_1 \left[\beta |A_1|^2 + (\gamma + R)|A_2|^2 + (\gamma - R)|A_3|^2\right] \quad (97)$$

where all the coefficients σ, δ, β, γ and R are real, due to the parity invariance $\vec{r} \to -\vec{r}$ (just as for rolls). The equations for the other amplitudes are obtained from cyclic permutations $A_1 \to A_2$, $A_2 \to A_3$ and $A_3 \to A_1$, which results from the rotational invariance (here restricted to angles $\theta = 2\pi/3$ due to our particular choice of wavevectors which breaks the continuous symmetry group of rotation, as was the case for roll patterns). Note that the coefficients of cubic terms in Eq. (97) have been written in such a form for reasons that will become apparent hereafter.

Finally, it remains to express the possible invariance to mirror reflections which, as mentioned at the end of §4.2, can be broken by Coriolis forces in case the Bénard layer is made to rotate uniformly around a vertical axis (still neglecting the possible breaking of translational invariance by centrifugal forces, expected to be small).

Consider a non-rotating system for the moment : still, not every reflection plane is permitted, because of our particular selection of wavevectors. We may for instance select the bissector of the angle between wavevectors $\vec{k}_2$ and $\vec{k}_3$, and denote this particular reflection by the matrix M. Then, as $\vec{k}_2.M\vec{r} = M\vec{k}_2.\vec{r} = \vec{k}_3.\vec{r}$, the reflection operation $\vec{r} \to M\vec{r}$ applied to Eq. (95) is seen to correspond to the action $A_2 \to A_3$ and $A_3 \to A_2$. In case reflection symmetry holds, amplitude equations should be invariant under this action, which leads to $R = 0$ in Eq. (97).

As a conclusion, R is the parameter which measures the breaking of reflection (or chiral) symmetry by Coriolis forces, and is therefore expected to increase with the angular speed. We now briefly examine some consequences this has on the dynamics of the patterns above instability threshold.

For $R = 0$, the function

$$\begin{aligned} V =& -\sigma\left(|A_1|^2+|A_2|^2+|A_3|^2\right)-\delta\left(A_1A_2A_3+A_1^*A_2^*A_3^*\right) \\ &+\frac{\beta}{2}\left(|A_1|^4+|A_2|^4+|A_3|^4\right)+\gamma\left(|A_1|^2|A_2|^2+|A_1|^2|A_3|^2+|A_2|^2|A_3|^2\right) \end{aligned} \tag{98}$$

is a Lyapunov function for the system of amplitude equations. Indeed, it is seen that $\dot{A}_p = -\partial V/\partial A_p^*$, $p = 1, 2, 3$ and $\dot{V} = -2\sum_{p=1}^{3} |\partial V/\partial A_p|^2 \leq 0$, thus implying a monotonic decrease of V up to one of its local minima. In the non-rotating case $R = 0$, the behavior of Eq. (97) – and its cyclic permutations – is thus severely restricted by this property : in particular, it excludes all non-monotonic behaviors such as oscillations, or chaotic dynamics. Note that for $R \neq 0$, it is not possible to find a function V satisfying the same conditions, and we may therefore expect more complex behaviors in case rotation occurs.

Separating moduli and phases as $A_p = r_p \exp[i\varphi_p]$, $p = 1, 2, 3$, and summing up the imaginary parts of amplitude equations, we get an equation for the sum of the phases $\varphi = \varphi_1 + \varphi_2 + \varphi_3$ as

$$\dot{\varphi} = -\delta\left(\frac{r_2r_3}{r_1}+\frac{r_1r_2}{r_3}+\frac{r_1r_3}{r_2}\right)\sin\varphi \tag{99}$$

which is actually valid for $R \neq 0$ as well, and shows that the only steady solutions for φ are $\varphi = 0$ and $\varphi = \pi$ (as the quantity between parentheses is always positive, the evolution is monotonic towards $\varphi = 0$ if $\delta > 0$ and $\varphi = \pi$ if $\delta < 0$). This phenomenon is called phase synchronization, which is particular to the case of resonant triads, i.e. it is directly linked with the quadratic terms of amplitude equations (97).

Collecting the real parts of amplitude equations, we get relations allowing to determine not only the steady state solutions, but also their stability (see for instance Colinet et al., 2001). In particular, relevant steady solutions are rolls

$$r_1 = r_2 = 0, r_3 = (\sigma/\beta)^{1/2} \qquad \rightarrow \textbf{Rolls}\ \text{(Rl)} \tag{100}$$

and hexagons

$$r_1 = r_2 = r_3 = \frac{\delta \pm \sqrt{\delta^2+4\sigma(\beta+2\gamma)}}{2(\beta+2\gamma)} \qquad \rightarrow \textbf{Hexagons}\ \text{(H)} \tag{101}$$

In addition to the reference state $r_{1,2,3} = 0$ (unstable for $\sigma > 0$, as usual), there are also mixed solutions, not given here for conciseness (and because they turn out to be always unstable). A typical bifurcation diagram is given in Fig. 15, for the non-rotating case $R = 0$. Note that hexagons with $\varphi = 0$ ($\delta > 0$) are denoted by H+ (and correspond to hexagonal structures with upflow at the center of an hexagon, which are usually observed for high Prandtl number fluids), while those with $\varphi = \pi$ ($\delta < 0$) are denoted by H− (downflow at the center).

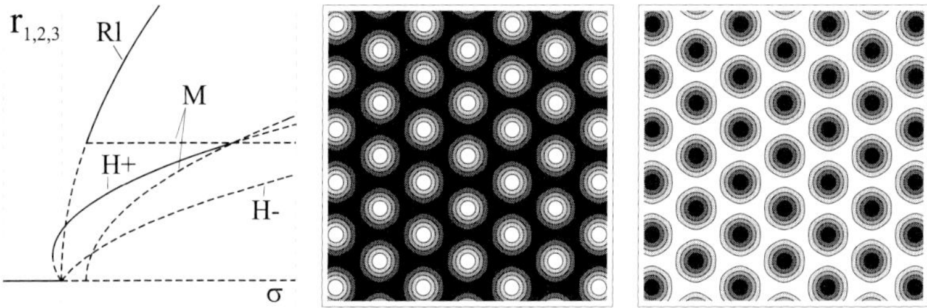

Figure 15. Left : bifurcation diagram for the competition between rolls (Rl) and hexagons (H+ and H−) in the non-rotating case $R = 0$, and for $\gamma > \beta > 0$ (typical case in Marangoni-Bénard convection). Stable solutions are represented as plain lines, while unstable branches are dashed. The branches labeled M correspond to an unstable mixed solution. Center: reconstruction of the up-hexagon solution H+ from Eq. (95). Right: reconstruction of the down-hexagon solution H− from Eq. (95).

It is seen in Fig. 15 (left) that the only stable solution just above the instability threshold $\sigma = 0$ is the hexagonal pattern with $r_{H+} = r_{1,2,3}$ given by Eq. (101). Note that, decreasing σ, this pattern remains stable even for $\sigma < 0$, until a transition to the reference state $r_{1,2,3} = 0$ takes place. Hence, the primary bifurcation from the reference state to hexagonal patterns is associated with an *hysteresis loop*, whose width depends on the problem studied (it is small in practice for Marangoni-Bénard convection).

For increasing σ, the hexagonal solution becomes unstable when $r_{H+} > \delta/(\gamma - \beta)$ (for $\gamma > \beta > 0$), and the only stable solution above this secondary bifurcation turns out to be the roll pattern. As remarked at the end of §4.1, this behavior is not realistic for Marangoni-Bénard convection, since a square pattern is rather observed at increasing supercriticality. A more complete analysis using a system of six amplitude equations is then necessary, in order to allow combinations of plane waves with orthogonal wavevectors, indeed representing square patterns. This is not done here for conciseness (see e.g. Colinet et al., 2001).

To conclude this section, allowing the Bénard set-up to rotate around a vertical axis, i.e. for $R \neq 0$, the secondary instability of hexagonal patterns still occurs at $r_{H+} > \delta/(\gamma - \beta)$ (for $\gamma > \beta > 0$), but is now found to be oscillatory (see e.g. Echebarria and Riecke, 2000). The so-called Hopf bifurcation then leads to a limit cycle, i.e. r_1, r_2 and r_3 oscillate periodically, after some transient period (see Fig. 16).

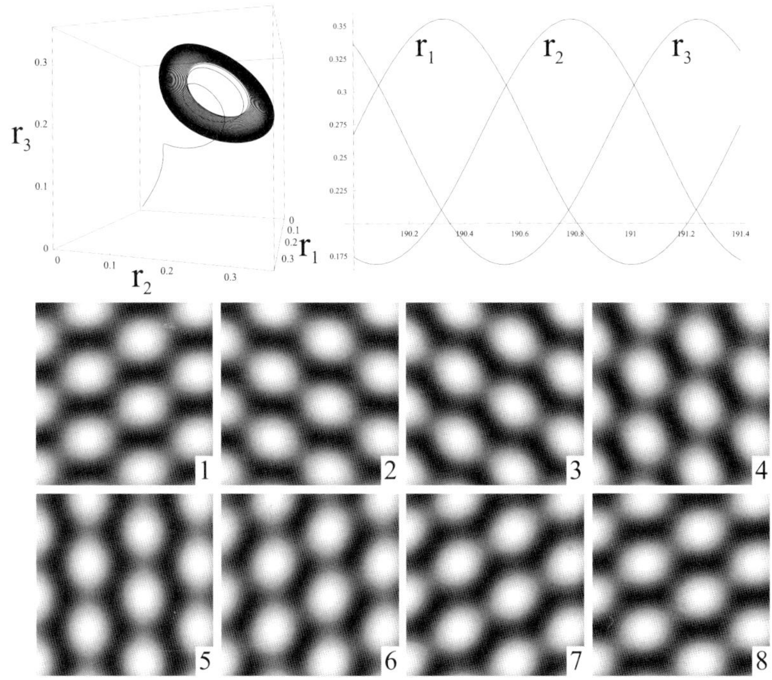

Figure 16. Oscillating hexagons in systems with broken chiral symmetry, obtained by direct numerical integration of ODE's (97) with $\sigma = 0.5$, $\delta = 1$, $\beta = 1$, $\gamma = 5$, and $R = 20$. Top left : trajectory in the phase space spanned by $r_p = |A_p|, p = 1, 2, 3$. Top right : evolution of amplitudes during one oscillation period. Bottom : 8 snapshots of the reconstructed pattern, sampled during one oscillation period.

4.4 Long-wave order-parameter equations for patterns

Another class of models applying to various types of Bénard instabilities (as well as to other physical situations) can be derived when the primary instability has a large spatial scale (i.e. a small critical wavenumber) compared to some other characteristic dimension of the problem. In this case, multiscale asymtotic expansions can generally be applied to the basic systems of equations and boundary conditions, which typically leads to models of lower dimensionality than the original one. Rather than to follow this route, we will attempt in this section to derive the form of such long-wave instability models from symmetry considerations, as already done for amplitude equations of the Landau type considered in §4.3.

Long-wave instability modes are often linked with a particular conservation law (or symmetry property). For instance, when both boundaries of a Bénard set-up are heat-insulating, a uniform shift of the temperature is possible, and there is no force which acts in restoring the initial temperature[8]. Hence, the system is neutrally stable to such a uniform shift, and the linear stability problem should possess at least one zero eigenvalue. Note that for long-wave surface deformational modes described in §2.4, the corresponding symmetry is an an invariance with respect to a shift of the height of the interface, which is also neutrally stable as long as the shift is uniform (see Fig. 17).

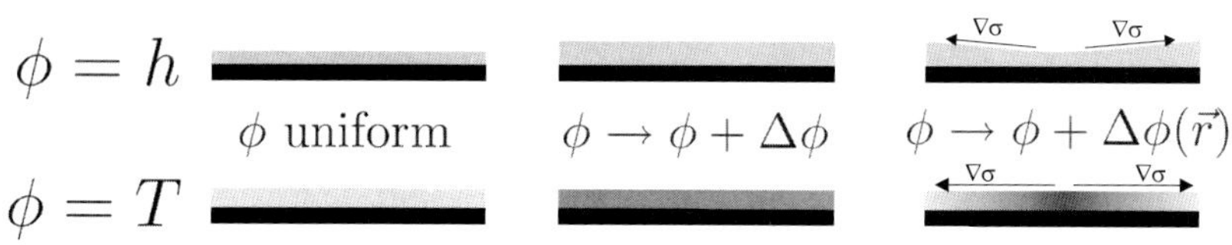

Figure 17. Sketch illustrating how neutral stability with respect to a uniform shift $\Delta\phi$ of the order parameter ϕ is weakly broken by large-scale non-uniformities of this shift. Top row : surface deformational mode. Bottom row : thermocapillary mode.

Now, when the shift is non-uniform, i.e. it is slowly modulated along the horizontal coordinates, a slow dynamics can be expected, which may couple to convection flows and possibly lead to instability of the horizontally uniform state. Thus, denoting the control parameter by μ, in each case there exists an eigenvalue $\sigma(k, \mu)$ such that $\sigma(0, \mu) = 0$ for all μ. We here restrict attention to systems for which σ is real, and can be expanded as

$$\sigma(k, \mu) = k^2\mu - k^4 + O(k^6) \tag{102}$$

[8]Such symmetry property would hold for the system considered in §3, provided we rather use adiabatic boundary conditions $\partial_z T = 0$ in place of Eqs (10) and (20). In this case, the full system would indeed be invariant to a shift $T \to T + \Theta$, up to a redefinition of the pressure in the Navier-Stokes equations (8).

where positive constants in front of the two terms have been set to unity without loss of generality (suitably rescaling time and μ).

Now, there exist many systems in which the symmetry responsible for the zero eigenvalue at $k = 0$ is only approximate. This happens for instance when top and/or bottom boundaries have a finite but small thermal conductivity, such that the homogeneous mode $k = 0$ itself has a slow evolution (assumed to be a slow exponential damping here). In such case, Eq. (102) can be completed by a small term $-\alpha$ with $\alpha > 0$.

If ϕ is the conserved (or quasi-conserved) scalar quantity (i.e. the temperature T, or the fluid depth h), the linear part of the order-parameter equation for ϕ may be written, consistently with Eq. (102) complemented by $-\alpha$, as

$$\partial_t \phi = -\alpha\phi - \mu\Delta\phi - \Delta^2\phi \tag{103}$$

hence formally equivalent to the linear part of the modified Swift-Hohenberg equation (82), even though the control parameter μ here appears to affect the "diffusion coefficient" of the quantity ϕ, i.e. the coefficient of the Laplacian term. This is the reason why systems described by such kind of linear behavior are sometimes called "negative viscosity" systems when $\mu > 0$. A rather complete discussion of such systems has been established (Nepomnyashchy, 1995).

Above the instability threshold, Eq. (103) should be completed by nonlinear terms, whose form is restricted by symmetry considerations, in the same spirit as in §4.2 and §4.3. Therefore, using again Eqs (87) and (89), any term appearing in the equation should be unaffected by rotations $\vec{r} \to R_\theta \vec{r}$ (including parity $\theta = \pi$), reflections $\vec{r} \to M_\theta \vec{r}$, in addition to translations $\vec{r} \to \vec{r} - \vec{r}_0$ and time shifts $t \to t + \Delta t$. From §4.2, it is clear that the linear part (103) indeed satisfies these properties, since the Laplacian operator is unaffected by these transformations. However, while the operator $\vec{\nabla}$ is also unaffected by space (and time) translations, it is useful for the following to note that it transforms according to $\vec{\nabla} \to R_\theta \vec{\nabla}$ and $\vec{\nabla} \to M_\theta \vec{\nabla}$ under rotation and reflection, respectively.

In addition, possible nonlinear terms are also restricted by the scalings $\partial_t \sim \alpha \sim \mu\Delta \sim \Delta^2$, as all terms must be of the same order in Eq. (103). Therefore, if we assume $\mu \sim \varepsilon^2 \ll 1$, we have $\vec{\nabla} \sim \varepsilon$, and $\alpha \sim \partial_t \sim \varepsilon^4$, which indeed shows that the order-parameter evolves on slow time and length scales. Note that no scaling is imposed yet on the amplitude of ϕ.

Consider first systems in which there is a strict invariance with respect to uniform shifts[9] $\phi \to \phi + \tilde{\phi}$. As all the terms must be invariant to this transformation, we have $\alpha = 0$ and nonlinear terms such as ϕ^2 or ϕ^3 cannot

[9] The shift is now denoted by $\tilde{\phi}$ in order to avoid confusion with the Laplacian operator.

appear in the evolution equation. Assuming ϕ to remain small, it can be checked that $(\vec{\nabla}\phi)^2$ indeed respects all symmetries, and that no other term can be found at this order. This yields the celebrated Kuramoto-Sivashinsky equation

$$\partial_t\phi = -\mu\Delta\phi - \Delta^2\phi + (\vec{\nabla}\phi)^2 \tag{104}$$

which was originally derived in the context of front propagation in combustion or solidification (Sivashinsky, 1977, 1983), and has since then been found in a number of other physical systems (see also the lectures of A.A. Nepomnyashchy in this volume). Nowadays, this equation is still intensively investigated as a prototype PDE displaying rich spatio-temporal dynamics not only in two, but also in one dimension (see Fig. 18).

Considering now that the invariance $\phi \to \phi + \tilde{\phi}$ is weakly broken by the linear term, we rather get the damped KS equation

$$\partial_t\phi = -\alpha\phi - \mu\Delta\phi - \Delta^2\phi + (\vec{\nabla}\phi)^2 \tag{105}$$

which has been used as a model to describe transitions from steady hexagonal patterns to oscillating and even chaotic structures (see e.g. Paniconi and Elder, 1997).

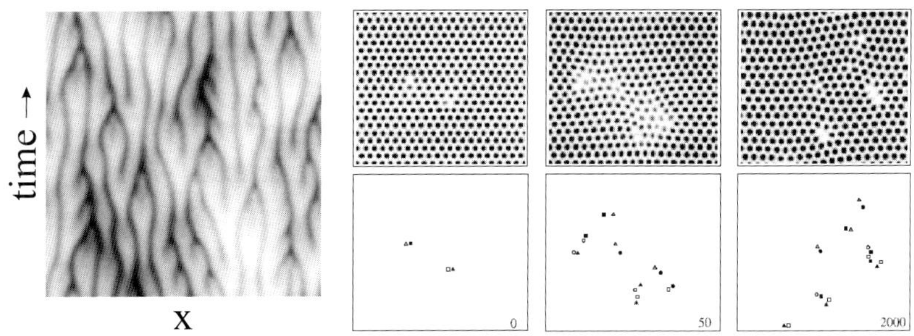

Figure 18. Direct numerical simulations of the 1D KS equation $\dot{\phi} = -\phi'' - \phi'''' + (\phi')^2$ (left) and of the 2D damped KS equation (105) (right). Left : space-time plot of $\phi - \langle\phi\rangle$, in a typical state of (so-called) phase turbulence. Right : evolution of an hexagonal pattern initially containing two pentagon-heptagon defects, showing the nucleation of new defects (indicated by symbols in bottom plots), and wavelength adjustment by defect motion after sufficiently long time (Colinet et al., 2002).

As the term $(\vec{\nabla}\phi)^2$ is sufficient to saturate the exponential growth of ϕ, the scaling of the order-parameter is $\phi \sim \varepsilon^2$, in order for all terms to be of the same order in the evolution equation. Note that another term of the same order satisfying all symmetries but the invariance to a uniform shift, is the term $\phi\Delta\phi$. However, both these quadratic terms do not appear in general for Bénard problems, unless some non-Boussinesq effects or surface deformation are considered. In the latter case, one gets the equation

$$\partial_t\phi = -\alpha\phi - \mu\Delta\phi - \Delta^2\phi - \gamma\vec{\nabla}.\left[\phi\vec{\nabla}\phi\right] + \gamma'\left\{(\vec{\nabla}\phi)^2 - \left\langle(\vec{\nabla}\phi)^2\right\rangle\right\} \quad (106)$$

the 1D version of which has been found by Garcia-Ybarra, Castillo and Velarde (Garcia-Ybarra et al., 1987a,b). Note that $\vec{\nabla}.[\phi\vec{\nabla}\phi] = (\vec{\nabla}\phi)^2 + \phi\Delta\phi$. Within some ranges of values of the coefficients γ and γ', the instability is saturated and very complex chaotic regimes occur, similarly to the Kuramoto-Sivashinsky equation.

At the contrary, for non-Boussinesq cases, but undeformable interfaces (Knobloch, 1990), γ' vanishes and only the non-saturating term $\vec{\nabla}.(\phi\vec{\nabla}\phi)$ remains at this order. Higher-order nonlinear terms should then be considered. By symmetry, next-order quadratic terms should contain at least four gradient operators, such that the balance with linear terms leads to $\phi \sim O(1)$. Hence, there is no longer any reason to consider quadratic terms before cubic ones, and there are many terms in general which are compatible with symmetry requirements. For a particular problem, several of them may vanish, such that we will not attempt here to pursue the discussion in general terms, but rather focus on some particular cases.

For Bénard convection between poorly conducting undeformable boundaries, and in the limit of large Prandtl number, it has been shown (Knobloch, 1990) that the general form of the order-parameter equation reads

$$\begin{aligned}\partial_t\phi =& -\alpha\phi - \mu\Delta\phi - \Delta^2\phi + \kappa\vec{\nabla}.\left[(\vec{\nabla}\phi)^2\vec{\nabla}\phi\right] + \beta\vec{\nabla}.\left[\Delta\phi\vec{\nabla}\phi\right] + \\ & \delta\Delta\left[(\vec{\nabla}\phi)^2\right] - \gamma\vec{\nabla}.\left[\phi\vec{\nabla}\phi\right]\end{aligned} \quad (107)$$

where γ quantifies non-Boussinesq effects (e.g. variation of fluid properties with temperature), while other coefficients can be calculated for particular problems (buoyancy-driven or surface-tension-driven convection, one-layer or two-layer system, ...). In particular, for values of parameters corresponding to usual Marangoni-Bénard convection, direct numerical simulations (see Fig. 19) show a very nice agreement with recent experiments (Schatz and Neitzel, 2001) showing a strong coarsening effect (increase of the mean wavelength of convection structures with supercriticality), in addition with the development of sharp gradients at the cold boundaries of polygonal convection cells (see also Thess and Orszag, 1995; Thess et al., 1995; Colinet et al., 2001). This shows that long-wave models such as the Knobloch equation (107) are quite useful to understand some important features of the transitions to highly supercritical Marangoni convection, even though they might not be able to describe transitions to complex time-dependent regimes characteristic of interfacial turbulence. This remains an open problem (see also Nepomnyashchy et al., 2002).

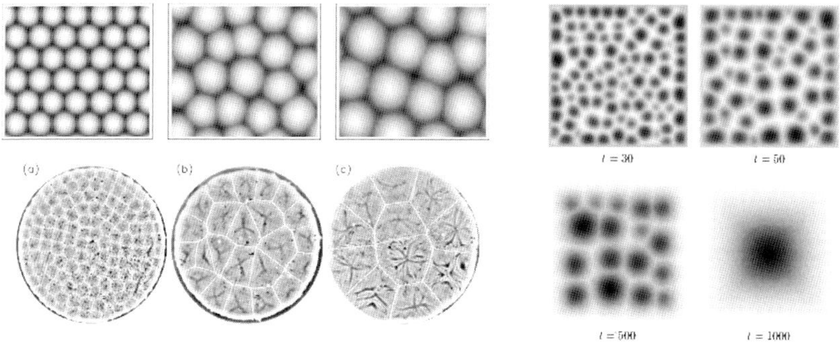

Figure 19. Left : pattern coarsening at increasing values of the supercriticality in Marangoni-Bénard convection. Top : direct numerical simulations of the Knobloch equation (107). Bottom : experiments in thin layers of silicon oil (courtesy of M. Schatz and G.P. Neitzel). Right : time evolution of the pattern predicted from the Knobloch equation at a fixed high supercriticality (Pontes et al., 1996, 1999), showing pattern coarsening ending up with one convection cell in the simulation domain (courtesy of M.G. Velarde).

To conclude, we will briefly reconsider the effect of uniform rotation of the layer around a vertical axis, which was treated by amplitude equations of the Landau type in the previous section. Considering here again that only the reflection (or chiral) symmetry is broken by the rotation, we will consider the model

$$\begin{aligned}\partial_t \phi =& -\alpha\phi - \mu\Delta\phi - \Delta^2\phi + \kappa\vec{\nabla}.\left[(\vec{\nabla}\phi)^2\vec{\nabla}\phi\right] + \beta\vec{\nabla}.\left[\Delta\phi\vec{\nabla}\phi\right] + \\ & \delta\Delta\left[(\vec{\nabla}\phi)^2\right] - \gamma\vec{\nabla}.\left[\phi\vec{\nabla}\phi\right] + R\vec{1}_z.\left[\vec{\nabla}\phi \times \vec{\nabla}\Delta\phi\right]\end{aligned} \tag{108}$$

the last term of which being indeed the lowest-order term breaking this invariance property (its sign is changed by the transformation $\vec{\nabla} \to M_\theta \vec{\nabla}$). Note that the same term has also been used in (Young and Riecke, 2003), though rather in combination with modified Swift-Hohenberg equations. These authors have identified a regime of spatiotemporal chaos which is induced, in hexagonal structures, by a persistent mechanism of nucleation of dislocations by pentagon-heptagon defects. Note that this effect is quite similar to that shown in Fig. 18, although no permanent chaotic regimes are observed there, probably due to the absence of rotation.

Direct numerical simulation of the modified Knobloch equation (108) can be performed efficiently in large-scale periodic domains, using spectral techniques based on fast Fourier transforms. A typical result is presented in Fig. 20, which shows chaotic regimes reminiscent of the periodic structures of Fig. 16, even though many defects are present, coexisting with islands of relatively ordered hexagonal structures. Note finally that other variants of the Knobloch model exist, including couplings with mean flow effects at low Prandtl numbers (Shtilman and Sivashinsky, 1991), or free surface deformation for two-layer systems (Golovin et al., 1995).

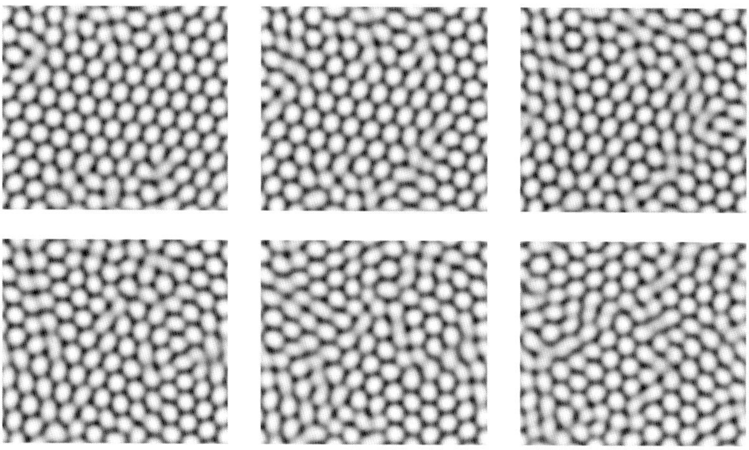

Figure 20. Snapshots of a chaotic state developing on the background of an hexagonal pattern, obtained by direct numerical simulations of the Knobloch equation modified by a term breaking the chiral symmetry (mimicking rotation-induced Coriolis forces).

5 Acknowledgments

The author would like to thank J. Bush, A. Hosoi, K. Eckert, A. Wierschem, M.G. Velarde, C.S. Iorio, M. Bestehorn, G.P. Neitzel and M. Schatz for allowing reproduction of some of their research results or drawings in these lecture notes. The author gratefully acknowledges financial support of the Fonds de la Recherche Scientifique - F.N.R.S., of the Communauté Française de Belgique through the ARCHIMEDES project (ARC 04-09/308), of the Belgian Federal Science Policy Office and of the European Space Agency through the CIMEX (Convection and Interfacial Mass Exchange) research program.

Bibliography

H. Bénard. Les tourbillons cellulaires dans une nappe liquide transportent de la chaleur par convection en régime permanent. *Ann. Chim. Phys.*, 23:62, 1901.

M. Bestehorn and R. Friedrich. Rotationally invariant order parameter equations for natural patterns in nonequilibrium systems. *Phys. Rev. E*, 59:2642–2652, 1999.

R.B. Bird, W.E. Stewart, and E.N. Lightfoot. *Transport Phenomena*. Wiley, 1960.

M.J. Block. Surface tension as the cause of Bénard cells and surface deformation in a liquid film. *Nature*, 178:650–651, 1956.

J.P. Burelbach, S.G. Bankoff, and S.H. Davis. Nonlinear stability of evaporating/condensing liquid films. *J. Fluid Mech.*, 195:463–494, 1988.

S. Chandrasekhar. *Hydrodynamic and Hydromagnetic Stability*. Dover, 1961.

C.I. Christov and M.G. Velarde. Dissipative solitons. *Physica D*, 86:323–347, 1995.

P. Colinet, J.C. Legros, and M.G. Velarde. *Nonlinear Dynamics of Surface-Tension-Driven Instabilities*. Wiley-VCH, 2001.

P. Colinet, A.A. Nepomnyashchy, and J.C. Legros. Multiplication of defects in hexagonal patterns. *Europhys. Lett.*, 57:480–486, 2002.

M.C. Cross and P.C. Hohenberg. Pattern formation outside of equilibrium. *Rev. Mod. Phys.*, 65:851–1112, 1993.

B. Echebarria and H. Riecke. Instabilities of hexagonal patterns with broken chiral symmetry. *Physica D*, 139:97–108, 2000.

K. Eckert and A. Grahn. Plumes and finger regimes driven by an exothermic interfacial reaction. *Phys. Rev. Lett.*, 82:4436, 1999.

K. Eckert and A. Thess. Nonbound dislocations in hexagonal patterns: pentagon lines in surface-tension-driven Bénard convection. *Phys. Rev. E*, 60:4117–4124, 1999.

K. Eckert, M. Bestehorn, and A. Thess. Square cells in surface-tension-driven Bénard convection: experiment and theory. *J. Fluid Mech.*, 356: 155–197, 1998.

K. Eckert, M. Acker, and Y. Shi. Chemical pattern formation driven by a neutralization reaction. Part I : Mechanism and basic features. *Phys. Fluids*, 16:385–399, 2004.

P.L. Garcia-Ybarra, J.L. Castillo, and M.G. Velarde. A nonlinear evolution equation for Bénard–Marangoni convection with deformable boundary. *Phys. Lett. A*, 122:107–110, 1987a.

P.L. Garcia-Ybarra, J.L. Castillo, and M.G. Velarde. Bénard–Marangoni convection with a deformable interface and poorly conducting boundaries. *Phys. Fluids*, 30:2655–2661, 1987b.

A.A. Golovin, A.A. Nepomnyashchy, and L.M. Pismen. Pattern formation in large-scale Marangoni convection with deformable interface. *Physica D*, 81:117–147, 1995.

E. Guyon, J.-P. Hulin, and L. Petit. *Hydrodynamique Physique*. Savoirs Actuels, Editions du CNRS, 1991.

B. Haut and P. Colinet. Surface-tension-driven instabilities of a pure liquid layer evaporating into an inert gas. *J. Colloid Interface Sci.*, 285:296–305, 2005.

J. Israelachvili. *Intermolecular and surface forces*. Academic Press, 1992.

E. Knobloch. Pattern selection in long-wavelength convection. *Physica D*, 41:450–479, 1990.

M. Maillard, L. Motte, A.T. Ngo, and M.P. Pileni. Rings and hexagons made of nanocrystals : a Marangoni effect. *J. Phys. Chem. B*, 104: 11871–11877, 2000.

T. Molenkamp. *Marangoni Convection, Mass Transfer and Microgravity*. Rijksuniversiteit Groningen, 1998.

A.A. Nepomnyashchy. Order parameter equations for long-wavelength instabilities. *Physica D*, 86:90–95, 1995.

A.A. Nepomnyashchy, M.G. Velarde, and P. Colinet. *Interfacial Phenomena and Convection*. Chapman & Hall/CRC, 2002.

A. Oron, S.H. Davis, and S.G. Bankhoff. Long-scale evolution of thin liquid films. *Rev. Mod. Phys.*, 69:931–980, 1997.

M. Paniconi and K.R. Elder. Stationary, dynamical and chaotic states of the two-dimensional damped Kuramoto–Sivashinsky equation. *Phys. Rev. E*, 56:2713–2721, 1997.

J.R.A. Pearson. On convection cells induced by surface tension. *J. Fluid Mech.*, 4:489, 1958.

J. Pontes, C.I. Christov, and M.G. Velarde. Numerical study of patterns and their evolution in finite geometries. *Int. J. Bif. Chaos*, 6:1883–1890, 1996.

J. Pontes, C.I. Christov, and M.G. Velarde. Numerical approach to pattern selection in a model problem for Bénard convection in finite fluid layer. *Ann. Univ. Sofia*, 93:157–175, 1999.

A.Ye. Rednikov and M.G. Velarde. Nonlinear waves and (dissipative) solitons in thin liquid layers. In H.-C. Chang, editor, *Nonlinear Waves in Multi-Phase Flow*. Kluwer, 2000.

A.Ye. Rednikov, P. Colinet, M.G. Velarde, and J.C. Legros. Rayleigh–Marangoni oscillatory instability in a horizontal liquid layer heated from above: coupling and mode-mixing of internal and surface dilational waves. *J. Fluid Mech.*, 405:57–77, 2000.

A.Ye. Rednikov, P. Colinet, M.G. Velarde, and J.C. Legros. Oscillatory thermocapillary instability in a liquid layer with deformable open surface: capillary-gravity waves, longitudinal waves and mode-mixing. *J. Non-Eq. Thermodyn.*, 25:381–405, 2001.

M. F. Schatz and G.P. Neitzel. Experiments on thermocapillary instabilities. *Annu. Rev. Fluid Mech.*, 33:93–127, 2001.

L. Shtilman and G. Sivashinsky. Hexagonal structure of large-scale Marangoni convection. *Physica D*, 52:477–488, 1991.

G.I. Sivashinsky. Nonlinear analysis of hydrodynamic instability of laminar flames. part 1. derivation of basic equations. *Acta Astronautica*, 4:1177–1206, 1977.

G.I. Sivashinsky. On cellular instability in the solidification of a dilute binary alloy. *Physica D*, 8:243–248, 1983.

P. Stephan and C.A. Busse. Analysis of the heat transfer coefficient of grooved heat pipe evaporator walls. *Int. J. Heat Mass Transfer*, 35: 383–391, 1992.

J.B. Swift and P.C. Hohenberg. Hydrodynamic fluctuations at the convective instability. *Phys. Rev. A*, 15:319, 1977.

A. Thess and S.A. Orszag. Surface-tension-driven Bénard convection at infinite Prandtl number. *J. Fluid Mech.*, 283:201–230, 1995.

A. Thess, D. Spirn, and B. Jüttner. Viscous flow at infinite Marangoni number. *Phys. Rev. Lett.*, 75:4614–4617, 1995.

J. Thomson. On certain curious motions observable on the surfaces of wine and other alcoholic liquours. *Philosophical Magazine*, 10:330, 1855.

J.S. Turner. *Buoyancy Effects in Fluids.* Cambridge University Press, 1973.

S.J. VanHook, M.F. Schatz, J.B. Swift, W.D. McCormick, and H.L. Swinney. Long-wavelength surface-tension-driven Bénard convection: experiment and theory. *J. Fluid Mech.*, 345:45, 1997.

L. Weh and H. Linde. Marangoni-stress-driven 'solitonic' (periodic) wave trains rotating in an annular container during heat transfer. *J. Colloid Interface Sci.*, 187:159–165, 1997.

A. Wierschem, H. Linde, and M.G. Velarde. Internal waves excited by the Marangoni effect. *Phys. Rev. E*, 62:6522–6530, 2000.

Y.-N. Young and H. Riecke. Penta-hepta defect chaos in a model for rotating hexagonal convection. *Phys. Rev. Lett.*, 90:134502, 2003.

Nonlinear dynamics of fronts

Alexander A. Nepomnyashchy
Department of Mathematics
Technion - Israel Institute of Technology, Israel

Abstract The subject of the present chapter is the front propagation in reaction-diffusion systems. We start with the consideration of the basic model, Fisher-Kolmogorov equation, and discuss the linear and nonlinear criteria of the velocity selection for fronts between a stable and an unstable stationary states. Then we consider the dynamics of plane and curved fronts between stable stationary states. The last section of this chapter contains the description of combustion fronts and their stability.

Contents

1 Introduction

This lecture is about waves, but unusual ones. The waves that propagate on the water surface, electromagnetic waves, sound waves etc. are governed by *hyperbolic* equations. A typical example of such an equation is the classical *wave equation*,

$$u_{tt} - c^2 u_{xx} = 0, \ -\infty < x < \infty, \ t > 0; \ u(x,0) = u_0(x), \ u_t(x,0) = v_0(x). \tag{1}$$

Its general solution has the form

$$u(x,t) = u_+(x - ct) + u_-(x + ct).$$

The physical system governed by (1) is *conservative*: the energy integral

$$E = \int_{-\infty}^{\infty} \left(\frac{1}{2} u_t^2 + \frac{1}{2} c^2 u_x^2 \right) dx$$

is conserved. The energy conservation is the physical reason of the wave propagation without a decay. In reality, waves in water decay because of the viscosity, but that decay may be rather slow.

A typical example of a *parabolic* equation is the diffusion equation,

$$u_t = D u_{xx}, \ -\infty < x < \infty, \ t > 0; \ u(x,0) = u_0(x); \ D > 0.$$

The solution of this equation is constructed with the help of the Green function

$$G(x - x', t - t') = \frac{1}{\sqrt{4\pi D(t - t')}} \exp\left[-\frac{(x - x')^2}{4D(t - t')} \right].$$

This equation describes a decay of the initial disturbance rather than waves. The energy

$$E = \int_{-\infty}^{\infty} \frac{1}{2} u^2 dx$$

is not conserved but decreases:

$$\frac{dE}{dt} = -D \int_{-\infty}^{\infty} u_x^2 dx < 0,$$

which is characteristic for a non-conservative, *dissipative*, system.

In order to get a persistent wave in a dissipative system, we have to introduce an *energy source* that will compensate the energy dissipation. This source can be related to an intrinsic *instability* of the system.

First, let us discuss simpler examples governed by ordinary differential equations. An example of a conservative system is the harmonic oscillator:

$$\frac{d^2x}{dt^2} + \omega^2 x = 0.$$

The energy

$$E = \frac{1}{2}\left(\frac{dx}{dt}\right)^2 + \frac{1}{2}\omega^2 x^2$$

is conserved, all the solutions

$$x(t) = A\cos(\omega t + \phi)$$

describe oscillations with a constant amplitude A.

If the dissipation is taken into account, the oscillator is described by the equation

$$\frac{d^2x}{dt^2} + 2\gamma\frac{dx}{dt} + \omega^2 x = 0.$$

The energy decreases:

$$\frac{dE}{dt} = -2\gamma\left(\frac{dx}{dt}\right)^2 \leq 0.$$

The solution

$$x(t) = Ae^{-\gamma t}\cos(\sqrt{\omega^2 - \gamma^2}t + \phi)$$

describes decay of oscillations ($\gamma < \omega$ is assumed).

In order to get stationary oscillations in the presence of dissipation, one needs an energy source. An example is provided by the *nonlinear oscillator*

$$\frac{d^2x}{dt^2} + 2\gamma(-1 + x^2)\frac{dx}{dt} + \omega^2 x = 0.$$

The solution $x = 0$ is *linearly unstable*. Indeed, solutions of the linearized equation,

$$\frac{d^2x}{dt^2} - 2\gamma\frac{dx}{dt} + \omega^2 x = 0,$$

grow with time:

$$x(t) = Ae^{\gamma t}\cos(\sqrt{\omega^2 - \gamma^2}t + \phi).$$

The evolution of energy is governed by the equation

$$\frac{dE}{dt} = \gamma(1 - x^2)\left(\frac{dx}{dt}\right)^2.$$

Thus, the energy grows when $|x| < 1$ and decreases when $|x| > 1$. One can expect that there is a closed orbit on the phase plane $(x, dx/dt)$ which is situated in both regions $|x| < 1$ and $|x| > 1$ in such a way that the gain and the loss of energy are fully compensated. Such an orbit (*limit cycle*) will describe sustained periodic oscillations in the non-conservative system.

Now we will add a diffusion to the dynamics.

2 Reaction-diffusion systems

Let us consider a simple dynamic system which describes the multiplication of animals:

$$\frac{dC}{dt} = bC - dC,$$

where C is the number of animals per unit area, b is the multiplication rate and d is the death rate. If $b > d$, the animals multiply in an exponential way:

$$C(t) = C(0)e^{rt},$$

where $r = b - d$ is the constant growth rate. The model described above is not realistic: it is natural to assume that the growth rate actually decreases with the growth of the number of animals, due to lack of resources, and the multiplication is completely stopped when the density of animals C reaches a certain value C_∞. Assume that the dependence of the growth rate $r(C)$ on C is linear:

$$\frac{dC}{dt} = r(C)C, \tag{2}$$

where

$$r(C) = r_0\left(1 - \frac{C}{C_\infty}\right) \tag{3}$$

(*Verhulst model*;Verhulst (1838)). The solution of the equation (2), (3) with the initial condition $C(0) = C_0$ is

$$C(t) = \frac{C_\infty}{1 + e^{-r_0 t}(C_\infty - C_0)/C_0}.$$

Equation (2), (3) has two equilibrium points, unstable point $C = 0$ and stable point $C = C_\infty$. For any initial condition such that $C_0 \neq 0$, $C_0 \neq C_\infty$, solution $C(t)$ tends to C_∞ as $t \to \infty$.

An obvious rescaling of variables C and t allows to rewrite the system (2), (3) in the form

$$\frac{du}{dt} = u(1 - u), \tag{4}$$

with equilibrium points $u = 0$ (unstable) and $u = 1$ (stable).

Assume now that the density of the population u is not uniform and depends also on the coordinate x. Assume also that the animals perform a random walk along the coordinate x, which can be described by the additional diffusion term in the right-hand side of equation (4):

$$u_t = u_{xx} + u(1-u). \tag{5}$$

Equation (5) is called *Fisher-Kolmogorov* equation (Fisher (1937), Kolmogoroff e.a. (1937)). This equation describes both multiplication of animals ("reaction") and their migration ("diffusion"), and hence it is an example of a *reaction-diffusion system.*

A general reaction-diffusion system can be constructed in the framework of chemical kinetics. Assume that we have a multicomponent mixture with the concentrations of components C_i, $i = 1, \dots, n$. If the components are distributed uniformly in space, the reactions between them are described by some kinetic equations

$$\frac{dC_i}{dt} = f_i(C_1, \dots, C_n), \; i = 1, \dots, n.$$

If the distributions of the components are non-uniform, one has to take into account nonzero fluxes $\mathbf{q}_i(\mathbf{x}, t)$ of components:

$$\frac{\partial C_i}{\partial t} = -\nabla \cdot \mathbf{q}_i + f_i(C_1, \dots, C_n), \; i = 1, \dots, n.$$

Typically,

$$\mathbf{q}_i = -\sum_{j=1}^{n} D_{ij} \nabla C_j,$$

hence

$$\frac{\partial C_i}{\partial t} = \sum_{j=1}^{n} \nabla \cdot (D_{ij} \nabla C_j) + f_i(C_1, \dots, C_n), \; i = 1, \dots, n.$$

If one can disregard the dependence of D_{ij} on $C_1, \dots, C_n$, then

$$\frac{\partial C_i}{\partial t} = \sum_{j=1}^{n} D_{ij} \Delta C_j + f_i, \; i = 1, \dots, n.$$

The non-diagonal elements of the diffusion matrix are usually smaller than the diagonal ones. If we assume that $D_{ij} = D_i \delta_{ij}$, we get

$$\frac{\partial C_i}{\partial t} = D_i \Delta C_i + f_i(C_1, \dots, C_n), \; i = 1, \dots, n.$$

Usually, the rates of reactions strongly depend on the temperature, therefore one has to take into account the heat production caused by the reaction:

$$\frac{\partial C_i}{\partial t} = D_i \Delta C_i + f_i(C_1, \ldots, C_n, T), \ i = 1, \ldots, n; \tag{6}$$

$$\frac{\partial T}{\partial t} = \chi \Delta T + q(C_1, \ldots, C_n, T) \tag{7}$$

(the dependence of the heat diffusion coefficient χ on the variables $C_1, \ldots, C_n$, T is ignored).

The reaction-diffusion equations may have solutions in the form of propagating fronts. As an example, let us consider the system (6), (7) with $n = 2$, where C_1 is the concentration of the fuel and C_2 is the concentration of the combustion product (we assume that the oxygen is abundant and hence its consumption is irrelevant). The system of equations can be written as follows:

$$\frac{\partial C_1}{\partial t} = D_1 \Delta C_1 - f(T) C_1, \tag{8}$$

$$\frac{\partial C_2}{\partial t} = D_2 \Delta C_2 + f(T) C_1, \tag{9}$$

$$\frac{\partial T}{\partial t} = \chi \Delta T + q f(T) C_1. \tag{10}$$

We can assume that a pure fuel is located at $x \to -\infty$ and a pure product is located at $x \to \infty$ (behind the combustion front); that leads to the boundary conditions:

$$x \to -\infty : \ C_1 \to 1, \ C_2 \to 0, \ T \to T_0;$$

$$x \to \infty : \ C_1 \to 0, \ C_2 \to 1, \ T \to T_0 + q.$$

Note that we have to postulate that $f(T_0) = 0$. Later we will study the combustion front solutions in more detail.

One more example of a reaction-diffusion system: phase-field description of a first-order phase transition, i.e., solidification. Let us characterize the liquid/solid system by two variables, temperature T and phase-order parameter ϕ. The phase transition is determined by the thermodynamical potential $F(\phi, T)$ which has two minima as a function of ϕ for a fixed value of T, at the point $\phi = \phi_l$ (liquid) and $\phi = \phi_s$ (solid). In the phase transition point, $T = T_c$, both minima have the same depth: $F(\phi_l, T_c) = F(\phi_s, T_c)$. For $T > T_c$, $F(\phi_l, T) < F(\phi_s, T)$; for $T < T_c$, $F(\phi_s, T) < F(\phi_l, T)$. Of course, there should be a maximum of F in a certain point ϕ_c between two minima, $\phi_l < \phi_c < \phi_s$, which corresponds to an unstable intermediate state. The kinetic equation that describes the phase transition, looks as follows:

$$\frac{\partial \phi}{\partial t} = D\Delta\phi - \kappa \frac{\partial F}{\partial \phi}(\phi, T), \tag{11}$$

where D and κ are positive constants. The heat transfer is modelled by the equation

$$\frac{\partial T}{\partial t} = \chi \Delta T + q \frac{\partial \phi}{\partial t} \tag{12}$$

(the last term in the right-hand side is caused by the latent heat of the phase transition). The phase transition front corresponds to the following boundary conditions for the order parameter ϕ,

$$x \to -\infty : \ \phi \to \phi_l; \ x \to \infty : \ \phi \to \phi_s,$$

and appropriate boundary conditions for temperature T.

Biological systems provide many examples of reaction-diffusion models, besides the Kolmogorov-Fisher equation mentioned above. Let us mention a few of them (for details, see Murray (2002)).

Calcium waves in eggs of fish and amphibiae. Target-like and spiral waves of Ca^{++} ions are observed. A simple model has been suggested:

$$u_t = D\Delta u + A(u - u_1)(u_2 - u)(u - u_3). \tag{13}$$

Chemotaxis. Animals are attracted by some substances that they can secrete themselves. In the absence of water, some amoebae form slimes that propagate through wavy motions (targets and spirals).

Animals' disease. The population of foxes can be described by two variables, $S(\mathbf{x}, t)$ (the density of healthy animals) and $I(\mathbf{x}, t)$ (the density of ill animals). The governing equations are:

$$\frac{\partial I}{\partial t} = rIS - aI + D_I \Delta I$$

(the terms in the right-hand side correspond to the infection, death and random motion of the ill animals),

$$\frac{\partial S}{\partial t} = -rIS + D_S \Delta S$$

(infection). The model is rather similar to combustion.

Now we will start investigating the basic features of the reaction-diffusion equation, with an emphasis on the front solutions.

3 The Fisher-Kolmogorov equation

We start with the simplest problem,

$$u_t = u_{xx} + u(1-u), \ -\infty < x < \infty, \ t > 0; \tag{14}$$

$$u(0,x) = u_0(x) \geq 0, \ u_0 < \infty \text{ as } x \to \pm\infty. \tag{15}$$

Note that negative values of u are meaningless (because u is the rescaled density of the population). It can be shown that if $u_0(x) \geq 0$, $u(x,t) \geq 0$ for any $t > 0$.

3.1 Stationary solutions

First, let us discuss the stationary solutions of equation (14) that satisfy the following ordinary differential equation:

$$u_{xx} + u(1-u) = 0. \tag{16}$$

Formally, equation (16) is equivalent to the equation of motion of a particle in the potential (see Fig.1(a))

$$U(u) = \frac{1}{2}u^2 - \frac{1}{3}u^3 \tag{17}$$

(variable x serves as the time variable for that fictitious particle) so that the "energy"

$$E = \frac{1}{2}u_x^2 + U(x)$$

is conserved. Obviously, all the solutions that are finite at $x \to \pm\infty$ have alternating signs and hence physically meaningless, except two constant solutions, $u = 0$ (with $E = 0$) and $u = 1$ (with $E = 1/6$); see Fig.2.

Linearizing equation (14) around a constant solution $u = u_0$ (u_0 is either 0 or 1), we obtain:

$$\tilde{u}_t = \tilde{u}_{xx} + f'(u_0)\tilde{u}, \ -\infty < x < \infty;$$

$$|\tilde{u}| < \infty, \ x \to \pm\infty.$$

Here $f(u_0) = u_0(1-u_0)$, hence $f'(u_0) = 1 - 2u_0$. Using the Fourier transformation, we find:

$$\tilde{u}(x,t) = \int_{-\infty}^{\infty} \hat{u}(k,t)e^{ikx}\frac{dk}{2\pi},$$

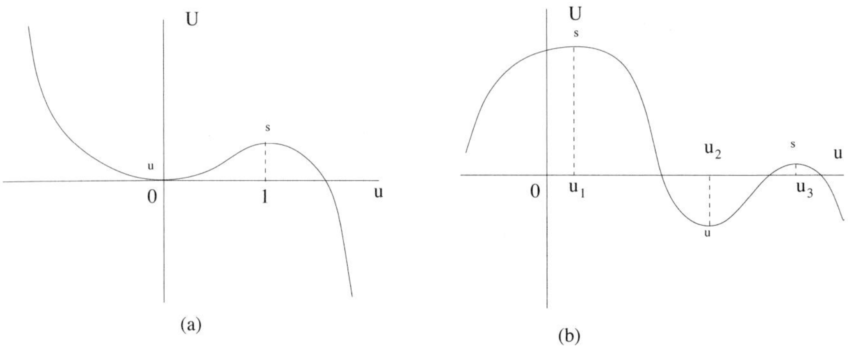

Figure 1. The shapes of the potential (a) for a system governed by the Fisher-Kolmogorov equation; (b) for a multistable system.

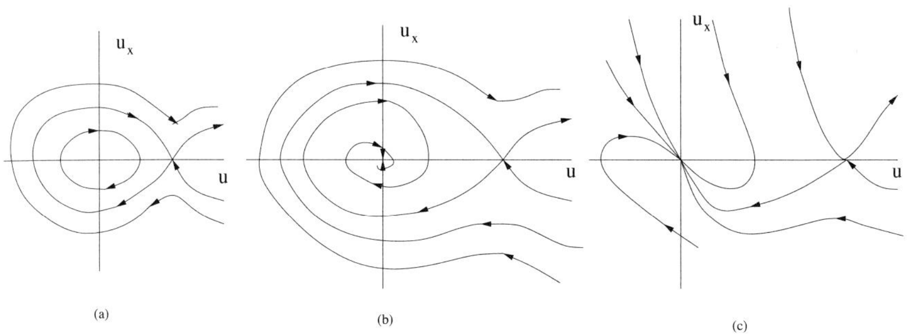

Figure 2. Orbits in the phase plane for traveling wave solutions governed by the Fisher-Kolmogorov equation; (a) $c = 0$; (b) $0 < c < 2$; (c) $c > 2$.

$$\frac{d\hat{u}(k,t)}{dt} = \sigma(k)\hat{u}(k,t),$$

where the growth rate of the component with the wavenumber k,

$$\sigma(k) = f'(u_0) - k^2.$$

Thus, for the solution $u_0 = 0$, $\sigma(k) = 1 - k^2$; therefore, that solution is unstable with respect to disturbances with $|k| < 1$, while for the solution $u_0 = 1$, $\sigma(k) = -1 - k^2 < 0$ for all k, hence that solution is stable.

3.2 Front solutions

Let us consider now traveling wave (front) solutions of equation (14) in the form $u(x,t) = u(z)$, $z = x - ct$ (c is the front velocity). Assume $c > 0$ (solutions with $c < 0$ can be found by means of the reflection transformation $x \to -x$ from those with $c > 0$). We obtain equation

$$u'' + cu' + u(1-u) = 0 \tag{18}$$

that describes the motion of a fictitious particle in the same potential $U(u)$ defined by (17) but this time with the friction c. Using the mechanical analogy, we come to the conclusion that all the solutions which are finite as $z \to \pm\infty$ are the following ones: (i) uniform unstable solution $u = 0$ (the particle is in the stationary point corresponding to the minimum of the potential $U(z)$) (ii) uniform stable solution $u = 1$ (the particle is in the stationary point corresponding to the maximum of the potential $U(z)$); (iii) front solution $u(z)$ with $u(-\infty) = 1$ and $u(\infty) = 0$ (the particle starts moving in the point $u = 1$ at $z = -\infty$ to the left and stops in the point $u = 0$ at $z = \infty$). Because only solutions with $u > 0$ for all z are permitted, function $u(z)$ should be monotonically decreasing rather than oscillating. Linearizing equation (18) around the solution $u = 0$ and looking for solutions $\tilde{u}$ of the linearized equation in the form $\tilde{u} \sim \exp(-az)$, we find that a satisfies the relation

$$a^2 - ca + 1 = 0, \tag{19}$$

hence

$$a_{\pm} = \frac{c}{2} \pm \sqrt{\frac{c^2}{4} - 1}. \tag{20}$$

Therefore, if $c < 2$, the tail of the front solution at large z oscillates changing its sign, and such solutions should be discarded (see Fig.2(b)). For $c > 2$, the asymptotics of the solution at large z is

$$u(z) \sim A\exp(-a_{-}z) + B\exp(-a_{+}z). \tag{21}$$

In the case $c = 2$,

$$u(z) \sim (A + Bz)\exp\left(-\frac{c}{2}\right)z.$$

For a qualitative investigation of solutions, it is convenient to define $v(z) = u'(z)$ and rewrite equation (18) as the system

$$\frac{dv}{dz} = -cv - u(1-u), \quad \frac{du}{dz} = v$$

or as a first-order ordinary differential equation,

$$\frac{dv}{du} = -c - \frac{u(1-u)}{v}.$$

The analysis shows that all the front solutions with $c > 2$ are non-negative and monotonically decreasing (see Fig.2(c)); A in (21) never vanishes, hence the solution tends to zero along the *generic*, relatively slow direction:

$$u \sim \exp(-a_- z). \tag{22}$$

For $c > 2$, $a_- < 1$. The solution with $c = 2$ is also non-negative and monotonically decreasing.

Thus, we have obtained a family of front solutions moving with different velocities $c \geq 2$ and having different tail asymptotics. It is interesting to note that the front velocity c is determined solely by the shape of its edge described by the parameter a, as if it was pulled by the edge; such a front is called *pulled front.* According to (19),

$$c = a + \frac{1}{a}.$$

Because in the region of the edge $u \ll 1$, the problem is actually governed by a linearized equation. Thus, in the case of a pulled front we can expect getting important information while considering a linearized equation. Below we will discuss the selection of the front solution which serves indeed as the large-time asymptotic for the solution of a Cauchy problem (14), (15) with a generic initial condition $u_0(x)$.

3.3 Motion of the front edge

Assume that $u_0(x)$ is Fourier-transformable:

$$u_0(x) = \int_{-\infty}^{\infty} \hat{u}_0(k) e^{ikx} \frac{dk}{2\pi}, \; \hat{u}_0(k) = \int_{-\infty}^{\infty} u_0(x) e^{-ikx} dx.$$

Specifically, that means that $u_0(x)$ has a finite support or tends to zero sufficiently fast as $z \to \pm\infty$. Assume that $u_0(x)$ is very small so that we can use the linearized problem,

$$u_t = u_{xx} + u, \; -\infty < x < \infty, \; t > 0; \; u(x,0) = u_0(x).$$

Using the Fourier transformation, we find that

$$u(x,t) = \int_{-\infty}^{\infty} \hat{u}_0(k) e^{ikx + (1-k^2)t} \frac{dk}{2\pi}. \tag{23}$$

Expression (23) determines the propagation of the front on the linear stage of its development. Let us define the front velocity in the following way. Consider the evolution of the field $u(x, t)$ in a reference frame moving with the velocity v, i.e. investigate the solution's behavior along the rays $x = X + vt$, X =const. We can expect that if we take v sufficiently large, we will outstrip the front, thus u will decay. If we take v sufficiently small, we will lag behind the front, so that u will grow. Assume that there exists such a value c that for the expression (23)

$$\lim_{t\to\infty} u(X + vt, t) = 0, \; v > c$$

and

$$\lim_{t\to\infty} u(X + vt, t) \to \infty, \; v < c.$$

Then we will call c the *linear front velocity*.

In order to estimate the large time asymptotics of the expression

$$u(X + vt, t) = \int_{-\infty}^{\infty} \hat{u}_0(k) e^{ikX} e^{\Phi(k,t)} \frac{dk}{2\pi},$$

$$\Phi(k, t) = (ikv + 1 - k^2)t,$$

we can use the method of the stationary phase. The saddle point corresponding to the zero of the function $d\Phi/k$ in the complex plane of k, is located at $k = iv/2$. Let us shift the integration path in such a way that it will pass through the saddle point: $k = K + iv/2$. Then we get:

$$u(X+vt, t) = \exp\left(-\frac{vX}{2}\right) \exp\left[\left(1 - \frac{v^2}{4}\right)t\right] \int_{-\infty}^{\infty} \frac{dK}{2\pi} \hat{u}_0\left(\frac{iv}{2} + K\right) e^{iKX} e^{-K^2 t}.$$

Obviously, for large t only the interval of K with $|K| \sim 1/\sqrt{t}$ essentially contributes to the integral. Hence the integral is equal approximately to

$$\hat{u}_0\left(\frac{iv}{2}\right) \int_{-\infty}^{\infty} \frac{dK}{2\pi} e^{iKX} e^{-K^2 t} \sim \frac{C}{\sqrt{t}},$$

where C is a constant. Thus, we find that

$$u(X + vt, t) \sim \exp\left(-\frac{vX}{2}\right) \exp\left[\left(1 - \frac{v^2}{4}\right)t - \frac{1}{2}\ln t\right].$$

Using the definition of the front velocity given above, we find that the velocity of the front that is created in the natural way from a generic initial condition is $c = 2$.

Note that this is the lowest value of the front velocity which is allowed by the condition of the non-oscillatory edge behavior. At the same time, the selected front is the steepest one among all the fronts on the *generic* branch $a = a_-(c)$. Let us mention also the rigorous result of Kolmogorov, Petrovsky and Piscounov on the selection of the front with $c = 2$ in the case of the initial condition with a finite support (Kolmogoroff e.a. (1937)).

A similar analysis can be done for a general reaction-diffusion equation

$$u_t = u_{xx} + f(u).$$

The linear criterion selects the front with the velocity $c_l = 2\sqrt{f'(0)}$.

3.4 Non-generic fronts

For $f(u)$ different from $u(1-u)$, it may happen that the coefficient A in (21) vanishes for a certain value of c. In this case the separatrix on the phase plane (u, v) which leaves the saddle point $(1, 0)$, tends to the stable node $(0, 0)$ along the *non-generic* ("fast") direction, $u \sim \exp(-a_+ z)$. If such a non-generic front exists, it is steeper than the front selected by the linear criterion described in the previous subsection. It turns out that the non-generic front wins the competition with the family of generic fronts.

As an example demonstrating the possibility of a non-generic front, let us consider the modification of the Fisher-Kolmogorov equation with $f(u) = u(1-u)(1+u/b)$, $0 < b < 1$.

Define $v = du/dz$ and rewrite the problem

$$u_{zz} + cu_z + u(1-u)(1+u/b),\ u(-\infty) = 1,\ u(\infty) = 0 \tag{24}$$

in the form

$$\frac{dv}{du} = -c - \frac{u(1-u)(1+u/b)}{v},\ v(0) = v(1) = 0. \tag{25}$$

Let us find an exact solution $v(u)$ in the form of a polynomial. Taking into account the boundary conditions for $v(u)$ and the assumption that $u(z)$ is a monotonically decreasing function, we come to the conclusion that the simplest trial function is

$$v = -au(1-u),\ a > 0. \tag{26}$$

Substituting (26) into (25), we find that the latter equation is satisfied if

$$a = \frac{1}{\sqrt{2b}},\ c = \sqrt{2b} + \frac{1}{\sqrt{2b}}.$$

The solution of the original problem (24) in this case is

$$u(z) = \frac{1}{2}\left[1 - \tanh\frac{a}{2}(z - z_0)\right], \; z_0 = \text{const.} \tag{27}$$

If $b > 1/2$, then $a < 1$, thus the obtained exact solution belongs to the family of generic fronts, and the corresponding front is less steep than the front with $a = 1$ selected by the linear criterion. However, if $b < 1/2$, then $a > 1$, hence the solution (27) is a non-generic front which is steeper than all the generic fronts. The steepest front will win the competition.

There exists a theorem (Aronson and Weinberger (1978)) that generalizes the results of Kolmogorov, Petrovsky and Piscounov to the case of a function $f(u)$ which is not convex. If $f(0) = f(1) = 0$, the function $u_0(x)$ has a finite support, and $0 \le u_0(x) \le 1$, the solution $u(x,t)$ tends as $t \to \infty$ to a traveling wave with the phase velocity c_* such that $c^- \le c_* \le c^+$, $c^{\pm} = 2\sqrt{d^{\pm}}$, where

$$d^- = f'(0), \; d^+ = \sup_{0<u<1} \frac{f(u)}{u}.$$

4 Multistability

4.1 Fronts between locally stable phases

In the previous section we considered fronts governed by the reaction-diffusion equation

$$u_t = u_{xx} + f(u) \tag{28}$$

in the case where the function $f(u) = u(1-u)$ had two zeros, $u_1 = 0$ and $u_2 = 1$, such that $f'(u_1) > 0$ (hence the solution $u = u_1$ was unstable) and $f'(u_2) < 0$ (hence the solution $u = u_2$ was stable). The fronts separated two domains, one with u around u_1 and another one with u around u_2. When the front propagated, the unstable state was ousted by the stable one. As we have seen, there existed a family of front solutions $u = u(z)$, $z = x - ct$ with different values of c. This feature of stable-unstable "pulled" fronts is clear from the following mechanical interpretation of the problem

$$u_{zz} + cu_z + f(u) = 0, \; u(-\infty) = u_2, \; u(\infty) = u_1.$$

A particle starts its motion in the point $u = u_2$ of the maximum of the potential $U(u)$ ($f(u) = U'(u)$) and moves to the point $u = u_1$ of the minimum of the potential $U(u)$ in the presence of the friction force proportional to $c > 0$. Because the energy of the particle $E = u_z^2/2 + U(u)$ decreases due to the friction, it is obvious that the particle moving to the left from the point $u = u_2 = 1$ finishes its way at the bottom of the well, in the point $u = u_1 = 0$, for any $c > 0$. If $0 < c < 2$, the point $u = u_1, v = u' = 0$ is an attracting focus, thus the particle performs oscillations. For $c > 2$, that point is an attracting node, and u decreases monotonically.

Let us consider now the case where the potential $U(u)$ has several alternating maxima and minima in the points $u = u_i$ (see Fig.1(b)). In the points of maxima, $U'(u_i) = f(u_i) = 0$ and $U''(u_i) = f'(u_i) < 0$, while in the points of minima $U'(u_i) = f(u_i) = 0$ and $U''(u_i) = f'(u_i) > 0$. Let us linearize equation (28) around the equilibrium point $u = u_i$: $u = u_i + \tilde{u}$, $|\tilde{u}| \ll 1$. We obtain:

$$\tilde{u}_t = \tilde{u}_{xx} + f'(u_i)\tilde{u}, \ -\infty < x < \infty; \ |\tilde{u}| < \infty, \ x \to \pm\infty.$$

We find that for the normal disturbance $\tilde{u} \sim \exp[ikx + \sigma(k)t]$, $\sigma(k) = f'(u_i) - k^2$. Therefore, the maxima of the potential $U(u)$ correspond to stable equilibrium states ($\sigma(k) < 0$ for all k), while the minima correspond to unstable equilibrium states ($\sigma(k) > 0$ for some k). Thus, the existence of several maxima of the potential $U(u)$ leads to a multistability of the physical system.

Consider the situation where the potential has two maxima $u = u_1$ and $u = u_3$ and a minimum $u = u_2$ between them, $u_1 < u_2 < u_3$. Equation (28) can be considered a simplest model of an (isotermic) phase transition between two locally stable phases, $u = u_1$ or $u = u_3$. Indeed, both *uniform* solutions $u = u_1$ and $u = u_3$ are stable with respect to small disturbances. Of course, there exist families of stable/unstable fronts corresponding to particle trajectories starting in points $u = u_1$ or $u = u_3$ and finishing in the point $u = u_2$, which describe ousting the unstable state by a stable one.

Let us discuss now fronts between stable phases. Assume that we have an initial distribution of $u(x, 0) = u_0(x)$ such that

$$\lim_{x \to -\infty} = u_1, \ \lim_{x \to \infty} = u_3. \tag{29}$$

Which phase will win when they are in contact? Will $u(x, t)$ tend to u_1 or u_3 as $t \to \infty$?

Assume that $U(u_1) > U(u_3)$, and let us discuss the existence of a traveling wave solution, $u(x, t) = u(z)$, $z = x - ct$, which satisfies the problem

$$u_{zz} + cu_z + U'(u) = 0, \ u(-\infty) = u_1, \ u(\infty) = u_3.$$

This problem describes the motion of the particle which starts on the top of the hill at $u = u_1$ and finishes exactly on the top of another hill at $u = u_3$. Is such a motion possible?

At $z \to -\infty$, the energy of the particle $E = U(u_1)$. During the motion, it decreases, and this decrease depends on the friction magnitude c (Fig.3). If the friction is sufficiently small, the particle will arrive in the point $u = u_3$ with the energy larger than $U(u_3)$, and it will pass through the point $u = u_3$. If the friction is sufficiently large, the energy of particle will become less than $U(u_3)$ in the course of motion, and the particle will never reach the hill top at $u = u_3$ but will eventually tend to the valley bottom u_2. However, one can expect, that due to the continuous dependence of the solution on the parameter c, there will be such a value of c that for this value $E \to U(u_3)$ as $u \to u_3$, so that the particle will move from one hill top to another one. Such an exceptional trajectory, which leads from one saddle point $(u_1, 0)$ to another saddle point $(u_3, 0)$ on the phase plane $(u, v = du/dz)$, will exist for an exceptional value of the front velocity c. Thus, in the contradistinction to stable/unstable fronts, the stable/stable front will be characterized by a definite value of velocity c. Because $c > 0$, the phase $u = u_1$ with a larger value of the potential $V(u)$ will oust the phase $u = u_3$ with a smaller value of the potential.

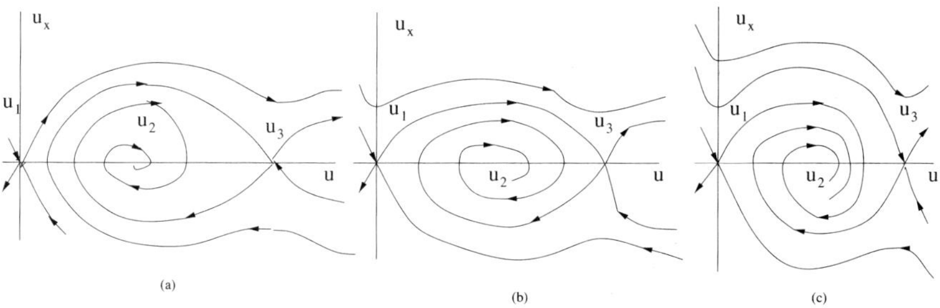

Figure 3. Front velocity selection for a multistable system. Orbits in the phase plane: (a) $c < c_*$; (b) $c = c_*$; (c) $c > c_*$.

If $U(u_1) = U(u_3)$, the particle will move from $u = u_1$ to $u = u_3$ if the energy is conserved, i.e. in the case $c = 0$. Thus, the front between two phases with equal values of the potential will be motionless. Obviously, if $U(u_1) < U(u_3)$, then $c < 0$ (we can just perform the reflection of the coordinate x and repeat the arguments presented above).

4.2 Lyapunov functional

The rule that determines which phase is truly stable and which one is only metastable (i.e. it is stable with respect to small disturbances but it is ousted by the truly stable phase) can be formulated in a clear way using the notion of the *Lyapunov functional*. In order to avoid the divergences, let us consider the problem in a large but finite region:

$$u_t = u_{xx} + f(u), \ -L < x < L, \ t > 0; \ u(x,0) = u_0(x), \ -L < x < L; \quad (30)$$

$$u_x(-L,t) = u_x(L,t) = 0, \ t > 0. \quad (31)$$

Let us define the Lyapunov functional

$$F[u(x)] \equiv \int_{-L}^{L} \left[\frac{1}{2}u_x^2 - U(u)\right] dx, \ f(u) = U'(u).$$

Actually, that is the action of a frictionless particle moving according to the law

$$u_{xx} + U'(u) = 0.$$

The variation of the Lyapunov functional on the class of functions satisfying the boundary conditions (31) can be transformed to

$$\delta F = -\int_{-L}^{L} dx[u_{xx} + U'(u)]\delta u(x)$$

by integration by parts, hence (30) can be written as

$$u_t = -\frac{\delta F}{\delta u}. \quad (32)$$

For any solution of (32), (31), $F = F[u(x,t)]$ is a function of t. Let us calculate the derivative of this function:

$$\frac{dF}{dt} = \int_{-L}^{L} dx[u_x u_{xt} - U'(u)u_t] = -\int_{-L}^{L} u_t[u_{xx} + U'(u)]dx = -\int_{-L}^{L} u_t^2 dx \leq 0.$$

If function $U(u)$ is bounded from above in the region of its definition, then the Lyapunov functional $F(t)$ is bounded from below. That means that $F(t)$ is either constant (at the equilibrium states) or a decreasing function tending to a constant as $t \to \infty$. In physical terms, the system tends to an equilibrium state as $t \to \infty$.

The Lyapunov functional corresponding to a uniform equilibrium state $u = u_i$ =const is equal to $F[u_i] = -2LU(u_i)$. Because the Lyapunov functional decreases with time, the uniform state with higher value of U ousts that with lower value of U.

4.3 Allen-Cahn equation

A basic model of a multistable system, suggested in the context of the phase transition, is the *Allen-Cahn equation*,

$$u_t = u_{xx} + u - u^3 \tag{33}$$

(which is actually equation (24) with $b = 1$). The potential $U(u)$ can be defined as

$$U(u) = -\frac{1}{4}(1-u^2)^2$$

(see Fig.4(a)) and the Lyapunov functional in the infinite region $-\infty < x < \infty$ is

$$F[u(x)] = \int_{-\infty}^{\infty} dx \left[\frac{1}{2}u_x^2 + \frac{1}{4}(1-u^2)^2\right].$$

Let us discuss stationary solutions of equation (33) which satisfy the equation

$$u_{xx} + u - u^3 = 0; \tag{34}$$

the first integration gives

$$\frac{u_x^2}{2} + U(u) = E = \text{const}.$$

The potential has two maximum points, $u = \pm 1$, with $U(-1) = U(1) = 0$, and one minimum point, $u = 0$, with $U(0) = -1/4$; these points correspond to uniform stationary solutions of equation (33). Obviously, solutions $u = \pm 1$ are stable and solution $u = 0$ is unstable.

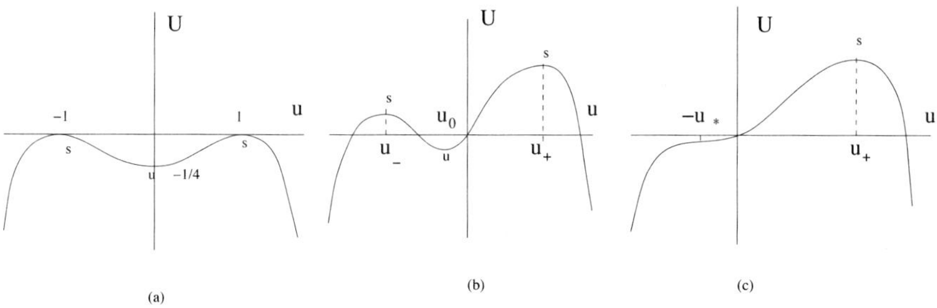

Figure 4. The shapes of the potential for the Allen-Cahn equation; (a) $\mu = 0$; (b) $0 < \mu < \mu_*$; (c) $\mu = \mu_*$.

The physical interpretation of the model described above can be as follows: u is the spontaneous magnetization directed along the definite axis (due to the crystal anisotropy). Solutions $u = 1$ and $u = -1$ correspond to the opposite orientations of the magnetization in a ferromagnet, while solution $u = 0$ corresponds to the unstable paramagnetic phase which has no spontaneous magnetization in the absence of the external magnetic field.

Two more families of solutions, which correspond to the energy $E = u_x^2/2 + U(u) = 0$, are

$$u_k(x) = \tanh\frac{x - x_0}{\sqrt{2}} \text{ and } u_{ak}(x) = -\tanh\frac{x - x_0}{\sqrt{2}}, \ x_0 = \text{const}.$$

These solutions, which are called *kink* and *antikink* solutions, respectively, describe the motionless front (domain wall) between two stable states $u = \pm 1$. The Lyapunov functional for uniform stable solutions $u = \pm 1$ is equal to 0, while the Lyapunov functional for kink and antikink solutions is

$$F_k = \int_{-\infty}^{\infty} dx \frac{1}{2} \cosh^{-4}\frac{x - x_0}{\sqrt{2}} = \frac{1}{\sqrt{2}} \int_{-1}^{1} du(1 - u^2) = \frac{2\sqrt{2}}{3} > 0 \qquad (35)$$

(the variable change $u = \tanh(x - x_0)/\sqrt{2}$ is used).

Though the kink solution has a larger value of the Lyapunov functional than the uniform state, it is stable with respect to small disturbances. To show that, let us linearize equation (33) around the kink solution and formulate the eigenvalue problem for the disturbance growth rate:

$$\tilde{u}_{xx} + \left(1 - 3\tanh^2\frac{x - x_0}{\sqrt{2}}\right)\tilde{u} = \sigma\tilde{u}. \qquad (36)$$

The continuous spectrum of the Schrödinger equation (36), which corresponds to the eigenfunctions that do not decay as $x \to \pm\infty$, is $\sigma(k) = -2 - k^2 < 0$. In order to find the discrete spectrum of (36), let us define $z = \tanh(x - x_0)/\sqrt{2}$ and rewrite (36) as the Legendre equation

$$(1 - z^2)\frac{d^2\tilde{u}}{dz^2} - 2z\frac{d\tilde{u}}{dz} + \left[s(s + 1) - \frac{\epsilon^2}{1 - z^2}\right]\tilde{u} = 0,$$

where $s = 2$, $\epsilon^2 = 4 + 2\sigma$. According to general theory of the Legendre equation, the discrete spectrum includes s eigenvalues, corresponding to $\epsilon = 1, \ldots, s$. In our case, we get two eigenvalues:

$$\epsilon = 2, \ \sigma = 0, \ \tilde{u} = 1 - z^2 = \cosh^{-2}\frac{x - x_0}{\sqrt{2}}$$

(this neutral disturbance corresponds to an infinitesimal spatial translation of the kink) and

$$\epsilon = 1, \ \sigma = -\frac{3}{2}, \ \tilde{u} = z(1 - z^2)^{1/2} = \frac{\tanh[(x - x_0)/\sqrt{2}]}{\cosh[(x - x_0)/\sqrt{2}]}.$$

Thus, there are no positive eigenvalues for σ.

All other stationary solutions with $-1/4 < E < 0$, which correspond to oscillations of the particle, can be found analytically by means of elliptic Jacobi functions; however, all of them are unstable.

4.4 Interaction between kinks

If the system governed by (33) evolves starting with a given $u(x,0) = u_0(x)$, such that $u_0(x)$ does not have the same sign everywhere, the initial evolution leads to the creation of domains with $u \approx \pm 1$ separated by kink-like domain walls. Because each domain wall gives a positive contribution (35) to the system's Lyapunov functional, the system tends to diminish its number. Indeed, adjacent kink and antikink attract, approach each other and eventually annihilate, which leads to the decrease of the Lyapunov functional by $2F_k$. The decrease of the number of domain walls leads to the growth of the domains' sizes, which is called *coarsening*.

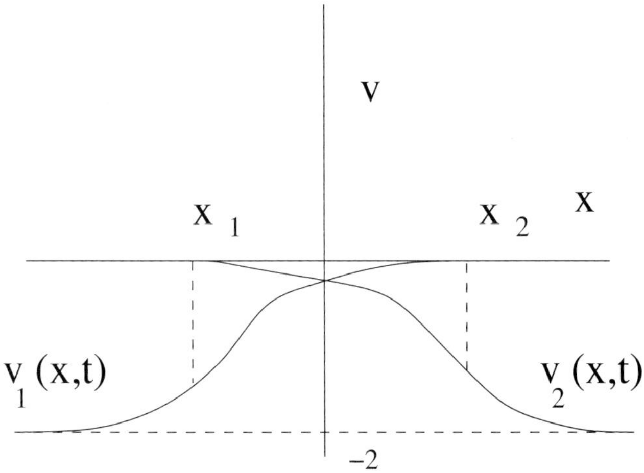

Figure 5. Interaction of distant kinks.

In order to estimate the rate of coarsening, let us calculate the interaction between a kink with the center located at $x = x_1$, and an antikink with the center located at $x = x_2$; assume $x_1 < x_2$, $x_2 - x_1 \gg 1$. It is convenient to define new variable v such that $u = 1 + v$. Equation (33) is rewritten as

$$v_t = v_{xx} - 2v - 3v^2 - v^3. \tag{37}$$

The kink solution is

$$v_1(x,t) = \tanh\frac{x-x_1}{\sqrt{2}} - 1, \tag{38}$$

the antikink solution is

$$v_2(x,t) = -\tanh\frac{x-x_2}{\sqrt{2}} - 1.$$

Let us construct solution to (37) which consists of both kinks (see Fig.5). Because equation (37) is nonlinear, we cannot just add kink and antikink solutions. However, near the center of the kink $x \approx x_1$, the antikink solution is exponentially small,

$$v_2(x,t) \approx -2\exp[(x-x_2)\sqrt{2}], \tag{39}$$

and it can be considered as a weak disturbance characterized by the small parameter

$$\delta = \exp[(x_1 - x_2)\sqrt{2}].$$

One can expect that the interaction of the kinks can lead to their slow motion, with the velocity of order δ, and to some distortions of the same order. So, we construct the solution in the form

$$v(x,t) = v_1(x,t) + v_2(x,t) + \sum_{n=1}^{\infty} v^{(n)}(x,t), \tag{40}$$

where $v^{(n)} = O(\delta^n)$ and $d/dt = O(\delta)$. Substituting (40) into (37), at the leading order we obtain the following equation for $v^{(1)}$:

$$\frac{\partial^2 v^{(1)}}{\partial x^2} - (2 + 6v_1 + 3v_1^2)v^{(1)} = -\frac{\partial v_1}{\partial t} - 6v_1 v_2 - 3v_1^2 v_2. \tag{41}$$

The solution $v^{(1)}$ is bounded if the right-hand side of equation (41) is orthogonal to the solution $\partial v_1/\partial x$ of the homogeneous (self-adjoint) problem; thus, we obtain:

$$\frac{dx_1}{dt}\int_{-\infty}^{\infty} dx \left(\frac{\partial v_1}{\partial x}\right)^2 = 6\int_{-\infty}^{\infty} dx v_1 \frac{\partial v_1}{\partial x} v_2 + 3\int_{-\infty}^{\infty} dx v_1^2 \frac{\partial v_1}{\partial x} v_2. \tag{42}$$

Substituting expressions (38), (39) into (42), we find:

$$\frac{2\sqrt{2}}{3} \cdot \frac{dx_1}{dt} = 16e^{-(x_2-x_1)\sqrt{2}}. \tag{43}$$

Similarly, we can obtain:

$$\frac{2\sqrt{2}}{3} \cdot \frac{dx_2}{dt} = -16e^{-(x_2-x_1)\sqrt{2}}. \tag{44}$$

Thus, the motion of two domain walls is equivalent to the motion of two attracting particles with inverse mobilities $2\sqrt{2}/3$. Their interaction energy is

$$w(x_2 - x_1) = -8\sqrt{2}\exp[-\sqrt{2}(x_2 - x_1)]. \tag{45}$$

It can be shown that this is just the Lyapunov functional corresponding the solution (40) minus the energy $2F_k$ of isolated kink and antikink.

In the general case the equations of motion for the domain wall centers are:

$$\frac{2\sqrt{2}}{3} \cdot \frac{dx_i}{dt} = -\frac{\partial W}{\partial x_i}, \tag{46}$$

where

$$W = \sum_i w(x_i - x_{i-1}),$$

$$w(x_i - x_{i-1}) = -8\sqrt{2}\exp[-\sqrt{2}(x_i - x_{i-1})]. \tag{47}$$

Note that $W = F - NF_k$, where N is the number of domain walls.

Because of the exponential interaction law, only the attraction to the nearest neighbour is important for every "particle". Let us consider the interaction of a pair of domain walls governed by the system of equations (43), (44). Setting $l = x_2 - x_1$, we find that the distance between domain walls is determined by the equation

$$\frac{2\sqrt{2}}{3}\frac{dl}{dt} = -32e^{-l\sqrt{2}}. \tag{48}$$

If $l(0) = l_0 \gg 1$, the solution is

$$l = l_0 + \frac{1}{\sqrt{2}}\ln\left(1 - 48te^{-l_0\sqrt{2}}\right).$$

The distance between two domain walls becomes $O(1)$ at

$$t \sim t_0 = \frac{1}{48}e^{l_0\sqrt{2}}. \tag{49}$$

Equation (48) does not describe the last stages of interaction on short distances of order $O(1)$ and the annihilation of domain walls, but it is clear that the duration of these stages is of $O(1)$, hence it is much less than

t_0. Equation (49) gives an estimate of the time necessary for annihilation of domain walls originally separated by the distance l_0. Vice versa, for any $t \gg 1$ only the domain walls with the original separation greater than $\ln(48t)/\sqrt{2}$ can survive. In a large but finite system with the length l_0, formula (49) gives the evolution time necessary for reaching a final uniform steady state (with $u = 1$ or $u = -1$).

Thus, late stages of the system evolution towards a stable stationary state can be described as an exponentially slow motion of domain walls and their annihilation.

4.5 Phase transition in an external field

Let us consider now the situation when two locally stable phases are energetically non-equivalent. Let us take a system with the Lyapunov functional

$$F[u(x)] = \int \left[\frac{1}{2} u_x^2 + \frac{1}{4}(1 - u^2)^2 - \mu u \right] dx \tag{50}$$

which corresponds to the dynamic equation

$$u_t = u_{xx} + u - u^3 + \mu. \tag{51}$$

The last term in the expression (50) describes the influence of an external magnetic field, which make the orientation of the magnetization in the direction of the field preferable. The uniform stationary states satisfy the equation

$$u - u^3 + \mu = 0. \tag{52}$$

The multistability takes place in the interval $-\mu_* < \mu < \mu_*$, where $\mu_* = 2/(3\sqrt{3})$ (see Fig.4(b) and Fig.4(c)). In this interval, equation (50) has three solutions: stable solution $u = u_+ > 1/\sqrt{3}$, another stable solution $u = u_- < -1/\sqrt{3}$, and an intermediate unstable solution u_0, $-1/\sqrt{3} < u_0 < 1/\sqrt{3}$. Obviously, the potential

$$U(u) = -\frac{1}{4}(1 - u^2)^2 + \mu u$$

is higher, and the Lyapunov functional is lower for the solution u_+ than for the solution u_-, if $\mu > 0$. Therefore, we can expect that the front between stable states u_+ and u_- will move (for $\mu > 0$) in such a way that the state u_+ will oust the state u_-.

Indeed, consider the traveling wave solution $u = u(z)$, $z = x - ct$, which satisfies equation

$$u'' + cu' + dU(u)/du = 0 \tag{53}$$

with the boundary conditions:

$$u(-\infty) = u_+, \; u(\infty) = u_-.$$

Multiplying equation (53) by u' and integrating over z from $-\infty$ to ∞, we find:

$$c\int_{-\infty}^{\infty} (u')^2 dx + U(u_-) - U(u_+),$$

hence

$$c = \frac{U(u_+) - U(u_-)}{\int_{-\infty}^{\infty}(u')^2}. \qquad (54)$$

One can see that the stable/stable front is pushed by the difference $U(u_+) - U(u_-)$ between the potentials of phases. This kind of fronts is called *pushed fronts*.

In the case $0 < \mu \ll 1$ the front velocity can be found analytically. Let us construct a kink or antikink solution in the form

$$u = u_0[x - x_0(\tau)] + \mu u_1 + \ldots, \; \tau = \mu t. \qquad (55)$$

Substitute (55) into (51) and collect terms of the same order in μ. At the leading order, we find:

$$(u_0)_{xx} + u_0 - u_0^3 = 0,$$

thus

$$u_0 = \pm \tanh \frac{x - x_0(\tau)}{\sqrt{2}}.$$

At the first order, we obtain the following non-homogeneous equation:

$$(u_1)_{xx} + u_1 - 3u_0^2 u_1 = -1 - (u_0)_x \frac{dx_0}{d\tau}.$$

The solvability condition (orthogonality of the right-hand side to the solution of the homogeneous problem, $(u_0)_x$) gives

$$\frac{2\sqrt{2}}{3}\frac{dx_0}{d\tau} = \mp 2,$$

or

$$\frac{2\sqrt{2}}{3}\frac{dx_0}{dt} = \mp 2\mu$$

(the upper sign is for a kink and the lower sign is for an antikink).

Thus, for $\mu > 0$ the phase with $u = u_+ > 0$ is preferable. However, the phase with $u = u_- < 0$ is metastable and will exist in the absence of the phase $u = u_+$. Moreover, only a sufficiently large nucleus of the truly stable phase can initiate the phase transition. Indeed, let is consider such a nucleus; its boundaries are a kink at $x = x_1$ and an antikink at $x = x_2$. The dynamics of nucleus boundaries is governed by the system of equations which take into account both "pushing" the fronts outwards by the difference of potentials and the attractive interaction between the kink and the antikink:

$$\frac{2\sqrt{2}}{3}\frac{dx_1}{dt} = -2\mu + 16e^{-(x_2-x_1)\sqrt{2}},$$

$$\frac{2\sqrt{2}}{3}\frac{dx_2}{dt} = 2\mu - 16e^{-(x_2-x_1)\sqrt{2}}.$$

If the distance between the kinks $x_2 - x_1 = l_*$,

$$l_* = \frac{1}{\sqrt{2}}\ln\frac{8}{\mu},$$

the nucleus boundaries are motionless (critical nucleus). Obviously, the critical nucleus is unstable: if the distance between the kink and antikink $l < l_*$, the attracting force prevails, and the nucleus disappears. In the opposite case $l > l_*$ the nucleus grows with the velocity tending to a constant value proportional to μ (see Fig.6).

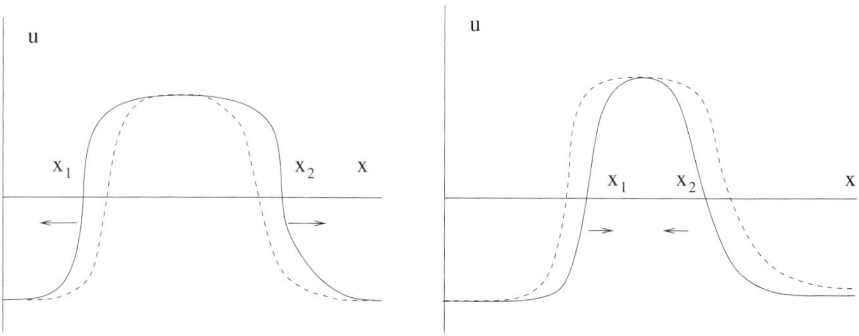

Figure 6. Interaction of domain walls. Left:$l > l_*$; right:$l < l_*$.

4.6 Domain wall pinned by an inhomogeneity

Let us consider now the case when the front between two stable states propagates in a weakly inhomogeneous medium, i.e. the governing equation is

$$u_t = u_{xx} + f(u,x),\ f(u,x) = f_0(u) + \epsilon f_1(u,x),\ 0 < \epsilon \ll 1.$$

In the context of the phase transition, the inhomogeneity can be caused by a spatial non-uniformity of the temperature field. Let us consider the following perturbed Allen-Cahn equation:

$$u_t = u_{xx} + [1 + \epsilon f_1(x)]u - u^3,$$

with the Lyapunov functional

$$F = \int dx \left\{ \frac{u_x^2}{2} - \frac{(1+\epsilon f_1(x)]u^2}{2} + \frac{u^4}{4} + \frac{1}{4} \right\}.$$

We search the solution in the form

$$u = u_0(x - x(\tau)) + \epsilon u_1 + \ldots,\ \tau = \epsilon t.$$

At the first order in ϵ,

$$(u_0)_{xx} + u_0 - u_0^3 = 0,$$

hence

$$u_0 = \pm \tanh \frac{x - x_0(\tau)}{\sqrt{2}}.$$

At the second order in ϵ we obtain:

$$(u_1)_{xx} + (1 - 3u_0^2)u_1 = -(u_0)_x \frac{dx_0}{d\tau} - f_1(x)u_0.$$

The solvability condition gives:

$$\frac{2\sqrt{2}}{3}\frac{dx_0}{dt} = -\frac{d}{dx_0}V_{ih}(x_0),$$

where the potential of the interaction between the domain wall and the inhomogeneity is

$$V_{ih}(x_0) = -\frac{1}{2}\int_{-\infty}^{\infty} f_1(x_0 + y)u_0^2(y)dy. \tag{56}$$

Specifically, if the inhomogeneity has a δ-like shape,

$$f_1(x) = -2V_0\delta(x - x_*),$$

the interaction potential is

$$V_{ih}(x_0) = -V_0 \cosh^{-2} \frac{x_0 - x_*}{\sqrt{2}}.$$

If there are many domain walls and many inhomogeneities, the equations of motion of domain walls are (cf. (46))

$$\frac{2\sqrt{2}}{3}\frac{dx_i}{dt} = -\frac{\partial W}{\partial x_i},$$

$$W = \sum_i w(x_i - x_{i-1}) + \sum_i V_{ih}(x_i),$$

where $w(x_i - x_{i-1})$ is determined by the expression (47), and $V_{ih}(x_i)$ corresponds to (56).

As an example let us consider two domain walls with coordinates x_1 and x_2 which are near the distant δ-shaped attracting inhomogeneities:

$$f_1(x) = -2V_0\delta(x - x_{1*}) - 2V_0\delta(x - x_{2*}),\ V_0 > 0,\ x_{2*} > x_{1*}.$$

$x_{2*} - x_{1*} = l_* \gg 1$, $|x_1 - x_{1*}| = O(1)$, $|x_2 - x_{2*}| = O(1)$. The equation of motion for the left domain wall is:

$$\frac{2\sqrt{2}}{3}\frac{dx_1}{dt} = 16e^{-(x_2-x_1)\sqrt{2}} - \sqrt{2}V_0 \sinh\frac{x_1 - x_{1*}}{\sqrt{2}} \cosh^{-3}\frac{x_1 - x_{1*}}{\sqrt{2}} \tag{57}$$

(one can disregard the interaction of between the left domain wall and the right inhomogeneity). The first term in the right-hand side of equation can be estimated as $16\exp(-l_*\sqrt{2})$. The minimum of the second term in the right-hand side of the equation is equal to $-2\sqrt{2}V_0/3\sqrt{3}$. Thus, we come to the conclusion that if

$$\frac{2\sqrt{2}}{3\sqrt{3}}V_0 > 16e^{-l_*\sqrt{2}}, \tag{58}$$

the domain wall will not be able to escape from the potential well created by the inhomogeneity. Hence, the coarsening will be stopped when the distances between the neighbor domain walls satisfy the inequality (58).

Similarly, in the case of domain walls pushed by the asymmetry of phases (see the previous subsection, we find the criterion of pinning:

$$V_0 > \frac{3\sqrt{3}}{\sqrt{2}}\mu.$$

4.7 Curved fronts of the phase transition

In the previous sections we considered fronts depending on only one spatial variable. Now we will consider fronts in two spatial variables. We shall deal only with stable-stable fronts in this subsection.

Let us consider the reaction-diffusion equation

$$u_t = \Delta_{\mathbf{x}} u + f(u,\mu),\ \Delta_{\mathbf{x}} = \frac{\partial^2}{\partial x^2} + \frac{\partial^2}{\partial y^2},\ f(u,\mu) = u - u^3 + \mu \tag{59}$$

in the interval of μ corresponding to the bistability. Let $u = u_+(\mu)$ and $u = u_-(\mu)$ be stable solutions of the equation $f(u,\mu) = 0$. Of course, equation (59) has solutions of the type $u = u(\mathbf{x}\cdot\mathbf{n}, t)$ (n is a unit two-dimensional vector), which correspond to plane fronts with the normal vector $\mathbf{n}$. These solutions are identical to those considered in the previous section. They are motionless as $\mu = 0$ and move with the definite velocity $c(\mu)$ as $\mu \neq 0$.

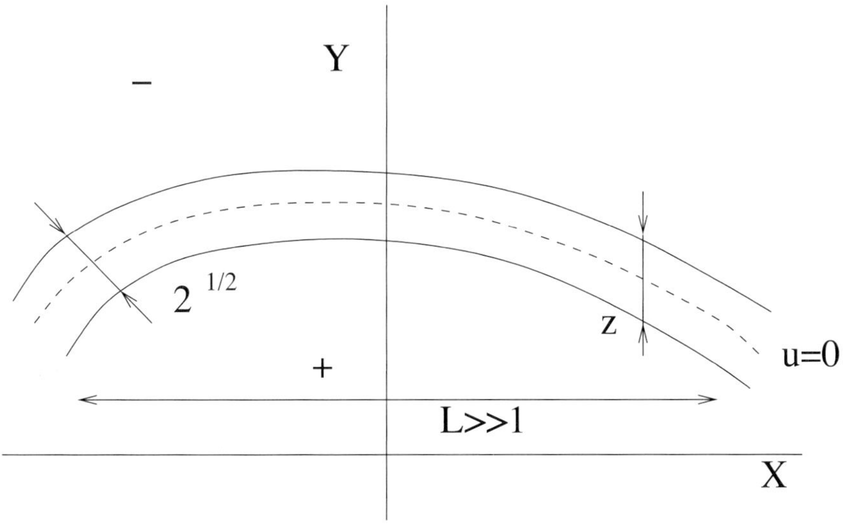

Figure 7. Curved front.

Let us consider now a curved front separating two regions with $u \approx u_+$ and $u \approx u_-$ (see Fig.7). We shall assume that the characteristic size of each region and the curvature radius of the front are large as compared to the width of the front. Introduce new "slow" variables $X = \epsilon x$, $Y = \epsilon y$. In the framework of these variables, the front, which has the thickness $O(1)$ in the original variables, is a thin stripe of the thickness $O(\epsilon)$ around a certain line

$Y = h(X)$. Across this thin stripe, the function u changes rapidly. Thus, we have an *internal layer* which we will describe using the "fast" variable

$$z = (Y - h(X))/\epsilon.$$

For sake of simplicity, we will assume that the parameter μ is also $O(\epsilon)$: $\mu = \mu_1\epsilon$. It is reasonable to expect that the velocity of the front will be $O(\epsilon)$ in original variables and hence $O(\epsilon^2)$ in new variables. Therefore, we introduce a "slow" time variable $T = \epsilon^2 t$. Finally, we search solution of (59) in the form

$$u = u(z, X, Y, T);\ X = \epsilon x;\ Y = \epsilon y;\ T = \epsilon^2 t;\ z = \frac{Y - h(X,T)}{\epsilon}.$$

The transformation of the partial derivatives in (59) is as follows:

$$\frac{\partial}{\partial t} = -\epsilon h_T \frac{\partial}{\partial z} + \epsilon^2 \frac{\partial}{\partial T};\ \frac{\partial}{\partial x} = -h_X \frac{\partial}{\partial z} + \epsilon \frac{\partial}{\partial X};\ \frac{\partial}{\partial y} = \frac{\partial}{\partial z} + \epsilon \frac{\partial}{\partial Y}.$$

Equation (59) is rewritten as

$$-\epsilon h_T u_z + \epsilon^2 u_T = \left(-h_X \frac{\partial}{\partial z} + \epsilon \frac{\partial}{\partial X}\right)\left(-h_X \frac{\partial}{\partial z} + \epsilon \frac{\partial}{\partial X}\right) u +$$

$$\left(\frac{\partial}{\partial z} + \epsilon \frac{\partial}{\partial Y}\right)^2 u + u - u^3 + \mu_1 \epsilon.$$

The solution is constructed in the form

$$u = u_0 + \epsilon u_1 + \ldots,\ h = h_0 + \ldots$$

At the leading order, we obtain:

$$[1 + (h_0)_X^2](u_0)_{zz} + u_0 - u_0^3 = 0.$$

The front solution is

$$u_0 = -\tanh \frac{z}{\sqrt{2[1 + (h_0)_X^2]}} \tag{60}$$

(we assume that the phase $u \approx u_+ = -1 + O(\epsilon)$ is located above the front, and the phase $u \approx u_- = 1 + O(\epsilon)$ is located below the front).

At the next order, we find:

$$[1 + (h_0)_X^2](u_1)_{zz} + (1 - 3u_0^2)u_1 =$$

$$-(h_0)_T(u_0)_z + (h_0)_{XX}(u_0)_z + 2(h_0)_X(u_0)_{zX} - 2(h_0)_X(h_1)_X(u_0)_{zz} - \mu_1.$$

The condition of the solvability on the class of bounded functions is the orthogonality of the right-hand side to the solution of the homogeneous equation:

$$[-(h_0)_T + (h_0)_{XX}] \int_{-\infty}^{\infty} (u_0)_z^2 dz + (h_0)_X \frac{\partial}{\partial X} \int_{-\infty}^{\infty} (u_0)_z^2 dz + 2\mu_1 = 0.$$

Substituting (60) and evaluating the integrals, we obtain the following equation that determines the motion of the curved stable/stable front (the subscript 0 in h_0 is omitted):

$$\frac{h_T}{\sqrt{1+h_X^2}} = \frac{h_{XX}}{(1+h_X^2)^{3/2}} + \frac{3\sqrt{2}}{2}\mu_1.$$

The term in the left-hand side is the normal velocity of the front v_n, while the first term in the right-hand side is the curvature κ of the front. Thus, the law of the front motion can be written in the following invariant form independent on the choice of the system of coordinate and the shape of the front:

$$v_n = \kappa + \frac{3\sqrt{2}}{2}\mu_1. \tag{61}$$

Specifically, in the case $\mu_1 = 0$ we get just the relation $v_n = \kappa$, which is characteristic for a *curvature flow*. The physical origin of the curvature contribution into the normal velocity of the curved domain wall is the positive contribution of domain wall into the Lyapunov functional (see (35)): the system tends to diminish the length of the domain wall in order to diminish the value of the Lyapunov functional.

As an example, let us consider a round droplet of the phase u_+ in the infinite sea of the phase u_-. According to (61), the evolution of the droplet radius $R(t)$ is governed by the equation

$$\frac{dR}{dt} = -\frac{1}{R} + m, \; m = \frac{3\sqrt{2}}{2}\mu_1$$

with a certain initial condition $R(0) = R_0$.

In the case $m = 0$ (both phases are equivalent), the solution is

$$R = \sqrt{R_0^2 - 2t}.$$

The droplet collapses during the finite time $t_* \approx R_0^2/2$ (the final stage of the evolution when the radius of the droplet becomes of the order of the domain wall width is not described by the asymptotic theory described above).

In the case $m < 0$ (the outer phase is truly stable), the solution can be presented in an implicit form as $g_+(R) = g_+(R_0) + t$, where

$$g_+(R) = R_*^2[-\rho + \ln(1+\rho)],\ \rho = R/R_*,\ R_* = 1/|m|.$$

The function $g_+(R)$ is negative and monotonically decreasing for all R. Hence, with the growth of t the radius $R(t)$ decreases monotonically and reaches zero at $t = t_* = -g(R_0)$.

In the case $m > 0$ (the inner phase is truly stable), the solution can be presented in an implicit form as $g_-(R) = g_-(R_0) + t$, where

$$g_-(R) = R_*^2[\rho + \ln|1-\rho|],\ \rho = R/R_*,\ R_* = 1/m.$$

The function $g_-(R)$ is decreasing for $\rho < 1$ and increasing for $\rho > 1$. Hence, if the initial radius of the droplet $R_0 < R_*$, $R(t)$ decreases and reaches zero at $t = t_* = -g(R_0)$. But if the initial radius of the droplet $R_0 > R_*$, $R(t)$ increases and tends to ∞ as $t \to \infty$. There is also a special stationary solution $R(t) = R_*$ which describes the critical nucleus of the truly stable phase in the metastable sea.

5 Combustion fronts

5.1 Formulation of the problem

In the previous section we considered fronts described by one variable. Now we will consider an example when the problem is described by two variables, temperature T and fuel concentration C:

$$\frac{\partial C}{\partial t} = D\Delta C - f(T)C, \tag{62}$$

$$\frac{\partial T}{\partial t} = \chi\Delta T + qf(T)C \tag{63}$$

(cf. equation (8) and (10); we will not consider the product concentration, because it does not influence the evolution of the temperature and of the fuel concentration). We shall consider the front solution with the boundary conditions

$$x \to -\infty :\ T = T_-,\ C = C_-;\ x \to \infty :\ T = T_+,\ C = 0$$

(it is assumed that the fuel is fully consumed by the reaction). T_+ is not an independent parameter: it is determined by the heat balance; the corresponding expression will be given below (equation (65)). The reaction rate is assumed to be

$$f(T) = A\exp(-E/RT) \tag{64}$$

(*Arrhenius law*), where R is the universal constant, and E is the activation energy. Formally, the law (64) is incompatible with the boundary condition $T(x \to -\infty) = T_-$, because $f(T_-) \neq 0$. However, if $RT_- \ll E$, then $f(T_-)$ is exponentially small and its nonzero value can be disregarded.

In the analysis presented below we follow the works by Sivashinsky, Matkowsky and Margolis (see, e.g., Sivashinsky (1977), Matkowsky and Sivashinsky (1979), Margolis and Matkowsky (1985)).

As a first step, we shall consider the one-dimensional plane front solution

$$T = T(z),\ C = C(z),\ z = x + vt$$

moving to the left ($v > 0$). We find:

$$vC_z = DC_{zz} - f(T)C;\ vT_z = \chi T_{zz} + qf(T)C.$$

Integrating both equations from $z = -\infty$ to $z = \infty$ and using the boundary conditions, we find:

$$-vC_- = -\int_{-\infty}^{\infty} f(T)Cdx,\ v(T_+ - T_-) = q\int_{-\infty}^{\infty} f(T)Cdx.$$

Thus,

$$T_+ = T_- + qC_-. \tag{65}$$

Let us transform the one-dimensional version of equations (62), (63) to a non-dimensional form:

$$T = T_- + \Theta T_+(1-\sigma);\ \sigma = T_-/T_+ < 1;\ C = YC_-;\ z = \xi\chi/v;\ t = \tau\chi/v^2.$$

We obtain:

$$\frac{\partial Y}{\partial \tau} + \frac{\partial Y}{\partial \xi} = L^{-1}\frac{\partial^2 Y}{\partial \xi^2} - \Lambda Y \exp\left[\frac{Z(\Theta - 1)}{\sigma + (1-\sigma)\Theta}\right], \tag{66}$$

$$\frac{\partial \Theta}{\partial \tau} + \frac{\partial \Theta}{\partial \xi} = \frac{\partial^2 \Theta}{\partial \xi^2} + \Lambda Y \exp\left[\frac{Z(\Theta - 1)}{\sigma + (1-\sigma)\Theta}\right]. \tag{67}$$

Here

$$L = \frac{\chi}{D}(\text{ Lewis number }),\ Z = \frac{E(1-\sigma)}{RT_+}(\text{ Zeldovich number }),$$

$$\Lambda = A\exp\left(-\frac{E}{RT_+}\right)\frac{\chi}{v^2}.$$

The boundary conditions:

$$Y(-\infty) = 1,\ \Theta(-\infty) = 0;\ Y(\infty) = 0,\ \Theta(\infty) = 1. \tag{68}$$

Later on, we shall assume that $Z \gg 1$, which is indeed a typical situation.

As we mentioned above, the boundary condition $\Theta(-\infty) = 0$ cannot be actually satisfied, because according to the equation

$$\frac{\partial \Theta}{\partial \tau}(-\infty) \sim \exp\left(-\frac{Z}{\sigma}\right) \neq 0.$$

However, the violation of this boundary condition is exponentially small for $Z \gg 1$, and we shall disregard it.

5.2 Plane stationary front

The plane stationary front is described by the problem (66) - (68) with the time derivative $\partial Y/\partial \tau$ and $\partial \Theta/\partial \tau$ omitted.

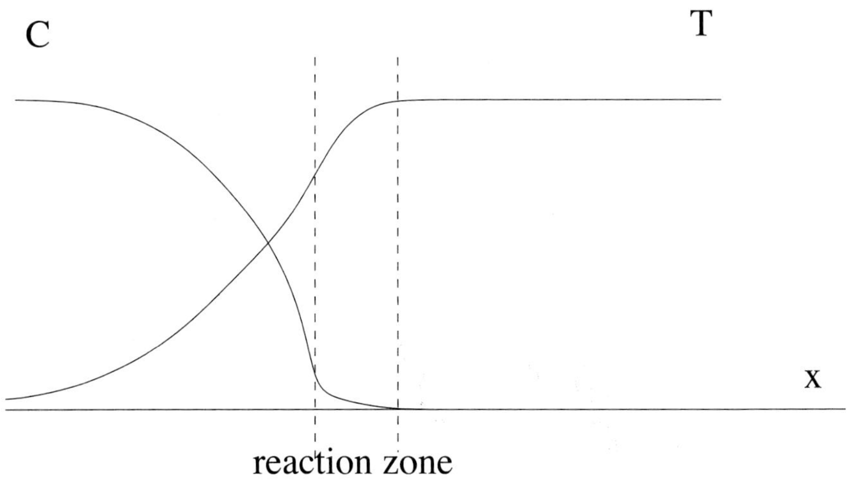

Figure 8. The structure of the combustion front.

For large Z, the reaction rate grows very fast with the growth of temperature. Because of that, the reaction takes place in a thin layer called *combustion front* where the temperature is already close to the maximum one, $\Theta \approx 1$, while the concentration of the fuel C is still different from zero (see Fig.8). The thickness of the combustion front is $O(\epsilon)$, where $\epsilon \equiv Z^{-1}$. Below we will develop the asymptotic theory of the combustion front propagation. We will find solutions outside the front (*outer solutions*), where there is no reaction, and the solution inside the front (*inner solutions*), where the reaction takes place. Matching of these solutions will give us the value of Λ (which can be called "eigenvalue" of the nonlinear problem), i.e. the front velocity.

In the outer regions the reaction term can be neglected: in the region $\xi < 0$, because the temperature is too low, and in the region $\xi > 0$, because the fuel concentration is too low. The solutions can be presented in the form of a series in ϵ,

$$Y = Y_0 + \epsilon Y_1 + \ldots, \ \Theta = \Theta_0 + \epsilon\Theta_1 + \ldots$$

Actually, we are interested only in the leading-order terms governed by the equations

$$\frac{dY_0}{d\xi} = L^{-1}\frac{d^2Y_0}{d\xi^2}, \ \frac{d\Theta_0}{d\xi} = \frac{d^2\Theta_0}{d\xi^2}. \tag{69}$$

In the region $\xi < 0$, the solution has to satisfy the boundary conditions

$$Y_0(-\infty) = 1, \ \Theta_0(-\infty) = 0.$$

The solution is

$$Y_0(\xi) = 1 + A_- e^{L\xi}, \ \Theta_0(\xi) = B_- e^{\xi}, \tag{70}$$

where A_- and B_- are constants. In the region $\xi > 0$, equations (69) are solved with the boundary conditions

$$Y_0(\infty) = 0, \ \Theta_0(\infty) = 1.$$

The solution is just

$$Y_0(\xi) = 0, \ \Theta_0(\xi) = 1. \tag{71}$$

The intensive fuel consumption and heat release lead to jumps of concentration and heat *fluxes* across the front, i.e. discontinuity of the spatial derivatives of the concentration and temperature fields. However, the diffusion and heat diffusion equations do not allow jumps of the concentration and temperature themselves. Hence, solutions (70) and (71) should be matched in the point $\xi = 0$:

$$Y_0(0^-) = Y_0(0^+) = 0, \ \Theta_0(0^-) = \Theta_0(0^+) = 1. \tag{72}$$

That gives the following expressions for the outer solution in the region $\xi < 0$:

$$Y_0(\xi) = 1 - e^{L\xi}, \ \Theta_0(\xi) = e^{\xi}. \tag{73}$$

Note that from the point of view of the outer expansion, the reaction takes place in the infinitely thin layer, i.e. the outer solution can be considered as the solution of the equations system

$$\frac{dY_0}{d\xi} = L^{-1}\frac{d^2Y_0}{d\xi^2} - H_0\delta(\xi), \quad \frac{d\Theta_0}{d\xi} = \frac{d^2\Theta_0}{d\xi^2} + H_0\delta(\xi), \tag{74}$$

where H_0 is the integral of the reaction term across the front which determines the jumps of the fluxes:

$$H_0 = L^{-1}\left[\frac{dY_0}{d\xi}\right] = -\left[\frac{d\Theta_0}{d\xi}\right],$$

where

$$[f] \equiv f(0^+) - f(0^-). \tag{75}$$

Solution (73) is obtained with $H_0 = 1$.

Let us consider now the fine structure of the combustion front in the region $\xi = O(\epsilon)$. For this goal, let us introduce the "fast" variable $X = \epsilon^{-1}\xi$, and assume that the solution is the function of the variable X:

$$Y = y_0(X) + \epsilon y_1(X) + \epsilon^2 y_2(X) + \dots, \quad \Theta = \theta_0(X) + \epsilon\theta_1(X) + \epsilon^2\theta_2(X) + \dots$$

Because of the relations (72), $y_0(X) = 0$, $\theta_0(X) = 1$. The exponential expression in the reaction rate is just $\exp(\theta_1)$ at the leading order. The balance between the reaction term and the diffusion term is reached if $\Lambda = O(\epsilon^{-2})$; therefore, we will use the expansion

$$\Lambda = \epsilon^{-2}\Lambda_{-2} + \epsilon^{-1}\Lambda_{-1} + \dots$$

Finally, we get the following system of equations for $y_1(X)$, $\theta_1(X)$:

$$L^{-1}\frac{d^2y_1}{dX^2} - \Lambda_{-2}y_1\exp(\theta_1) = 0, \tag{76}$$

$$\frac{d^2\theta_1}{dX^2} + \Lambda_{-2}y_1\exp(\theta_1) = 0. \tag{77}$$

The boundary conditions for functions $y_1(X)$, $\theta_1(X)$ are obtain from the matching with the solution (73), which can be written for small $\xi = \epsilon X$ as

$$Y_0 = -\epsilon LX + \dots, \quad \Theta_0 = 1 + \epsilon X + \dots,$$

and with the solution (71). We find:

$$X \to -\infty : \; y_1(X) \sim -LX, \; \theta_1(X) \sim X; \; X \to \infty : \; y_1(X) \to 0, \; \theta_1(X) \to 0.$$

Adding equations (76) and (77), we find

$$\frac{d^2}{dX^2}(\theta_1 + L^{-1}y_1) = 0.$$

Taking into account the boundary conditions we find:

$$y_1 = -L\theta_1. \tag{78}$$

Substituting (78) into (77), we obtain the following nonlinear eigenvalue problem:

$$\frac{d^2\theta_1}{dX^2} - L\Lambda_{-2}\theta_1 \exp(\theta_1) = 0, \tag{79}$$

$$X \to -\infty : \; \theta_1(X) \sim X; \; X \to \infty : \; \theta_1(X) \to 0.$$

The first integral of the equation (79) is:

$$\frac{1}{2}(\theta_1')^2 - \Lambda_{-2}L(\theta_1 - 1)\exp(\theta_1) = C, \tag{80}$$

where C is a constant which is obtained from the boundary condition at $X \to \infty$:

$$C = \Lambda_{-2}L. \tag{81}$$

Evaluating both sides of (80) at $X \to -\infty$, we find that

$$\Lambda_{-2} = \frac{1}{2L}.$$

Returning to the original variables, we find the following formula for the combustion front velocity:

$$v = \sqrt{\frac{2A}{D}\frac{\chi R T_+^2}{E(T_+ - T_-)}} \exp\left(-\frac{E}{2RT_+}\right).$$

5.3 Dynamics of curved fronts

Let us generalize now the theory described in the previous subsections. Let fuel concentration and temperature be functions of three spatial variables, $C(x_1, x_2, x_3, t)$ and $T(x_1, x_2, x_3, t)$. Assume that the combustion front is a two-dimensional surface, $x_1 = \Phi(x_2, x_3, t)$. Define the new system of variables, $x = x_1 - \Phi(x_2, x_3, t)$, $y = x_2$, $z = x_3$, $\tau = t$. Using the same transformation to non-dimensional variables, as in the previous subsections (i.e., using the velocity v of the plane front for non-dimensionalization), we obtain the following system of equations:

$$\frac{\partial Y}{\partial \tau} - \Phi_\tau \frac{\partial Y}{\partial x} = L^{-1}\Delta Y - w, \quad \frac{\partial \Theta}{\partial \tau} - \Phi_\tau \frac{\partial \Theta}{\partial x} = \Delta\Theta + w,$$

$$w = \Lambda Y \exp\left[\epsilon^{-1} \frac{\Theta - 1}{\sigma + (1-\sigma)\Theta}\right].$$

The expression for the Laplacian in the coordinates (x, y, z) is

$$\Delta f = (1+\Phi_y^2+\Phi_z^2)\frac{\partial^2 f}{\partial x^2}+\frac{\partial^2 f}{\partial y^2}+\frac{\partial^2 f}{\partial z^2}-(\Phi_{yy}+\Phi_{zz})\frac{\partial f}{\partial x}-2\left(\Phi_y\frac{\partial^2 f}{\partial x \partial y}+\Phi_z\frac{\partial^2 f}{\partial x \partial z}\right).$$

When considering the outer solution, we can replace w by $H\delta(x)$, where H is still unknown function which is equal to the integral of w over the combustion front.

For sake of simplicity, we shall develop the theory in the case where the Lewis number is close to L, i.e., $L = 1 + \epsilon\beta$, hence, $L^{-1} = 1 - \epsilon\beta + \ldots$. We construct the outer solution in the form

$$Y = Y_0(x,y,z,\tau)+\epsilon Y_1(x,y,z,\tau)+\ldots;\ \Theta = \Theta_0(x,y,z,\tau)+\epsilon\Theta_1(x,y,z,\tau)+\ldots.$$

Note that according to our assumption of the full consumption of the fuel at the reaction front, the outer solution for Y is zero in all the order in the region $x > 0$. The inner solution is presented in the form

$$Y = y_0(X,y,z,\tau)+\epsilon y_1(X,y,z,\tau)+\ldots;\ \Theta = \theta_0(X,y,z,\tau)+\epsilon\theta_1(X,y,z,\tau)+\ldots,$$

where $X = \epsilon^{-1}x$. Also, we use the expansions

$$\Phi = \Phi_0 + \epsilon\Phi_1 + \ldots,\ H = H_0 + \epsilon H_1 + \ldots.$$

For Y_0 and Θ_0, we obtain:

$$\frac{\partial Y_0}{\partial \tau} - (\Phi_0)_\tau \frac{\partial Y_0}{\partial x} = \Delta Y_0 - H_0\delta(x), \tag{82}$$

$$\frac{\partial \Theta_0}{\partial \tau} - (\Phi_0)_\tau \frac{\partial \Theta_0}{\partial x} = \Delta \Theta_0 + H_0\delta(x); \tag{83}$$

$$x = -\infty:\ Y_0 = 1,\ \Theta_0 = 0;\ x = \infty:\ Y_0 = 0,\ \Theta_0 = 1.$$

Let us add equations (82) and (83). We find that the quantity $S_0 = Y_0 + \Theta_0$ satisfy the equation

$$\frac{\partial S_0}{\partial \tau} - (\Phi_0)_\tau \frac{\partial S_0}{\partial x} = \Delta S_0$$

with boundary conditions

$$x = -\infty:\ S_0 = 1;\ x = \infty:\ S_0 = 1.$$

Choose solution $S_0 = 1$ (assuming that it is compatible with the initial conditions). Then $\Theta_0 = 1$ in the region $x > 0$.

Let us consider now the inner solution. Because of the continuity of the fields of Y_0, Θ_0, $Y_0 = 0$ and $\Theta_0 = 1$ as $x = 0$. Therefore, the inner expansion can be written as

$$Y = \epsilon y_1(X, y, z, \tau) + \ldots; \ \Theta = 1 + \epsilon\theta_1(x, y, z, \tau) + \ldots.$$

Because we used the velocity of the plane front for non-dimensionalization, the expansion $\Lambda = \Lambda_{-2}\epsilon^{-2} + \ldots$ coincides with that used in the previous section. Thus, in our case $\Lambda_{-2} = 1/2$. We obtain the following system of equations in the leading order:

$$[1+(\Phi_0)_y^2+(\Phi_0)_z^2]\frac{\partial^2 y_1}{\partial X^2}-\frac{1}{2}y_1\exp(\theta_1) = 0,\ [1+(\Phi_0)_y^2+(\Phi_0)_z^2]\frac{\partial^2 \theta_1}{\partial X^2}+\frac{1}{2}y_1\exp(\theta_1) = 0.$$

Adding both equations, we find that for the quantity $s_1 = y_1 + \theta_1$,

$$\frac{\partial^2 s_1}{\partial X^2} = 0.$$

From the matching with the outer solution for $x > 0$ we find that $s_1 = \theta_1(\infty)$, $y_1 = \theta_1(\infty) - \theta_1$, hence

$$[1 + (\Phi_0)_y^2 + (\Phi_0)_z^2](\theta_1)_{XX} + \frac{1}{2}[\theta_1(\infty) - \theta_1]\exp(\theta_1) = 0 \tag{84}$$

(here and below $f(a)$ means $f(a, y, z, \tau)$). Multiplying both sides of equation (84) by $(\theta_1)_X$ and integrating from $-\infty$ to ∞, we find:

$$(\theta_1)_X(-\infty) = \frac{\exp\left(\frac{1}{2}\theta_1(\infty)\right)}{\sqrt{1 + (\Phi_0)_y^2 + (\Phi_0)_z^2}}.$$

Because $(\theta_1)_X(\infty) = 0$, we find that the change of the derivative $(\theta_1)_X$ across the reaction zone,

$$(\theta_1)_X(\infty) - (\theta_1)_X(-\infty) = -\frac{\exp\left(\frac{1}{2}\theta_1(\infty)\right)}{\sqrt{1 + (\Phi_0)_y^2 + (\Phi_0)_z^2}}. \tag{85}$$

The matching conditions

$$\theta_1(\infty) = \Theta_1(0^+),\ \frac{\partial\theta_1}{\partial X}(\infty) = \frac{\partial\Theta_0}{\partial x}(0^+),\ \frac{\partial\theta_1}{\partial X}(-\infty) = \frac{\partial\Theta_0}{\partial x}(0^-)$$

allow us to represent the relation (85) as the jump of the derivative for the outer solution:

$$[(\Theta_0)_x] = -\frac{\exp(\Theta_1(0^+)/2)}{\sqrt{1 + (\Phi_0)_y^2 + (\Phi_0)_z^2}}$$

(notation (75) is used).

Now we can determine the function H_0 in (82), (83). Let us integrate (83) over x from $x = -\delta$ to $x = \delta$ and take the limit $\delta \to 0$. We find:

$$(1 + (\Phi_0)_y^2 + (\Phi_0)_z^2)[(\Theta_0^+)_x] + H_0 = 0.$$

Thus,

$$H_0 = \sqrt{1 + (\Phi_0)_y^2 + (\Phi_0)_z^2}\exp(\Theta_1(0^+)/2),$$

so that equation (83) can be rewritten as

$$\frac{\partial \Theta_0}{\partial \tau} - (\Phi_0)_\tau \frac{\partial \Theta_0}{\partial x} = \Delta\Theta_0 + \sqrt{1 + (\Phi_0)_y^2 + (\Phi_0)_z^2}\exp(\Theta_1(0^+)/2)\delta(x). \quad (86)$$

We come to the conclusion that the leading order equation do not allow to get a closed formulation of the outer problem: equation for Θ_0 contains $\Theta_1(0^+)$. Thus, we have to consider the next order of the outer problem.

In the next order, we obtain equations:

$$\frac{\partial Y_1}{\partial \tau} - (\Phi_0)_\tau \frac{\partial Y_1}{\partial x} - (\Phi_1)_\tau \frac{\partial Y_0}{\partial x} = \Delta Y_1 - \beta\Delta Y_0 - w_1,$$

$$\frac{\partial \Theta_1}{\partial \tau} - (\Phi_0)_\tau \frac{\partial \Theta_1}{\partial x} - (\Phi_1)_\tau \frac{\partial \Theta_0}{\partial x} = \Delta\Theta_1 + w_1,$$

where w_1 is the next order reaction term. Adding both equations and taking into account that $S_0 = Y_0 + \Theta_0 = 1$, we find the following equation for S_1:

$$\frac{\partial S_1}{\partial \tau} - (\Phi_0)_\tau \frac{\partial S_1}{\partial x} = \Delta S_1 + \beta\Delta\Theta_0. \quad (87)$$

In the region $x > 0$, $Y_1 = 0$, hence $\Theta_1(0^+) = S_1(0^+)$. Thus, equation (86) can be written as

$$\frac{\partial \Theta_0}{\partial \tau} - (\Phi_0)_\tau \frac{\partial \Theta_0}{\partial x} = \Delta\Theta_0 + \sqrt{1 + (\Phi_0)_y^2 + (\Phi_0)_z^2}\exp(S_1(0^+)/2)\delta(x) \quad (88)$$

(because S_1 is a continuous function, we can write just $S_1(0)$ instead of $S_1(0^+)$). Equations (87) and (88) with boundary conditions

$$x \to -\infty: \ \Theta_0 = 0, \ S_1 = 0; \ x \to \infty: \ \Theta_0 = 1, \ S_1 = 0$$

give a closed description of the problem.

Instead of using δ-function, we can consider non-smooth solutions of homogeneous equations

$$\frac{\partial \Theta_0}{\partial \tau} - (\Phi_0)_\tau \frac{\partial \Theta_0}{\partial x} = \Delta\Theta_0, \ \frac{\partial S_1}{\partial \tau} - (\Phi_0)_\tau \frac{\partial S_1}{\partial x} = \Delta S_1 + \beta\Delta\Theta_0 \quad (89)$$

Relation (101) is satified in two cases. The first case is $r_- = 0$, i.e. $k = 0$. Indeed, there always exist a neutral disturbance corresponding to a shift of the position of the front. Another case is

$$\beta = -2(r_+ - r_-)^2 = -2(1 + 4k^2). \tag{102}$$

Relation (102) determines the *monotonic instability boundary* of the planar front (see Fig.9). When $\beta > -2(1 + 4k^2)$, there is no monotonic instability. When $\beta < -2(1 + 4k^2)$, the front is unstable with respect to disturbances with wavenumbers $0 < k^2 < k_m^2$, where

$$k_m^2 = -\frac{\beta + 2}{8}.$$

This kind of instability is called *longwave instability*.

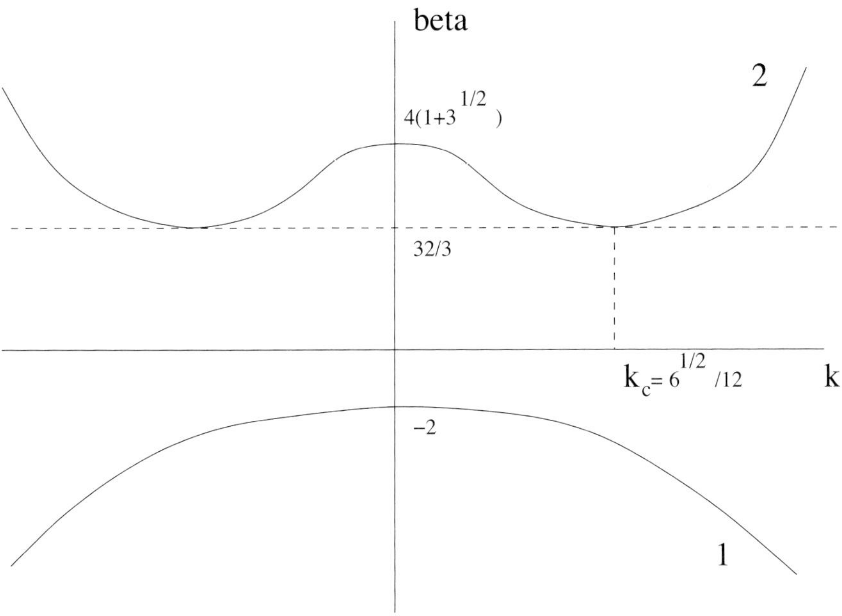

Figure 9. Instabilities of combustion front. Line 1: monotonic instability; line 2: oscillatory instability.

Oscillatory instability Let us consider now the oscillatory instability, $\sigma = i\omega$. The eigenfunctions are:

$$f(x) = -e^{r_+ x}, \ x < 0; f(x) = 0, \ x > 0,$$

$$g(x) = (2r_- + \beta)e^{r_+ x} - \frac{\beta(k^2 - r_+^2)}{2r_+ - 1} x e^{r_+ x}, \ x < 0; g(x) = 2r_- e^{r_- x}, \ x > 0,$$

where

$$r_\pm = \frac{1}{2} \pm \sqrt{\frac{1}{4} + k^2 + i\omega}.$$

The dispersion relation is

$$64(i\omega)^3 + (i\omega)^2[8(\beta + 2 + 8k^2) + 16(1 + 8k^2) - \beta^2] + \\ 2i\omega(\beta + 2 + 8k^2)(1 + 12k^2) + k^2(\beta + 2 + 8k^2)^2 = 0. \tag{103}$$

Taking the real and imaginary parts of equation (103), we find:

$$\omega^2 = \frac{1}{32}(\beta + 2 + 8k^2)(1 + 12k^2);$$

$$\beta^2(1 + 12k^2) - 8\beta(1 + 8k^2) - 32(1 + 8k^2)^2 = 0. \tag{104}$$

Equation (104) determines the *oscillatory instability boundary*. In the contradistinction to the monotonic instability boundary, which has a threshold $\beta_c = -2$ corresponding to the wavenumber $k = 0$, the minimum of the curve $\beta(k^2)$ determined by (104) is located at finite value of k^2, $k^2 = k_c^2 = 1/24$. The threshold value of β, corresponding to the oscillatory instability, is $\beta_c = \beta(k_c^2) = 32/3$. Note that the longwave limit if $\beta(k^2)$, $\beta(0) = 4(1 + \sqrt{3})$, is higher than β_c. Thus, for $L > 1$ a *shortwave oscillatory instability* takes place.

5.5 Nonlinear development of front instabilities

Monotonic longwave instability Let us consider the nonlinear evolution of an unstable monotonic mode near the instability threshold, i.e. for $\beta = -2(1 + \delta^2)$, $\delta \ll 1$.

The analysis of the linear dispersion relation shows that for such value of β the wavenumbers of growing disturbances have $k = O(\delta)$, while the growth rate is $O(\delta^4)$. That allows us to expect that the the curved front is characterized by the followinf scales: $Y = \delta y$, $Z = \delta z$, $T = \delta^4 \tau$. We construct solutions in the following form:

$$\Phi = \bar{\Phi} + \delta^2 \phi(Y, Z, T), \ \Theta = \bar{\Theta} + \delta^4 \theta(x, Y, Z, T), \ S = \bar{S} + \delta^4 s(x, Y, Z, T).$$

The disturbances are expanded into series:

$$\phi = \phi_0 + \delta^2 \phi_1 + \ldots; \ \theta = \theta_0 + \delta^2 \theta_1 + \ldots; \ s = s0 + \delta^2 s_1 + \ldots.$$

The leading order problem can be presented in the following form:

$$\frac{\partial \theta_0}{\partial x} = \frac{\partial^2 \theta_0}{\partial x^2} - \left(\frac{\partial^2 \phi_0}{\partial Y^2} + \frac{\partial^2 \phi_0}{\partial Z^2}\right) \cdot \left(\begin{array}{c} e^x,\ x<0 \\ 0,\ x>0 \end{array}\right) + \frac{1}{2} s_0 \delta(x),$$

$$\frac{\partial s_0}{\partial x} = \frac{\partial^2 s_0}{\partial x^2} - 2\frac{\partial^2 \theta_0}{\partial x^2} - 2\left(\frac{\partial^2 \phi_0}{\partial Y^2} + \frac{\partial^2 \phi_0}{\partial Z^2}\right) \cdot \left(\begin{array}{c} xe^x,\ x<0 \\ 0,\ x>0 \end{array}\right).$$

The solution is:

$$\theta_0 = \left(\begin{array}{c} \left(\frac{\partial^2 \phi_0}{\partial Y^2} + \frac{\partial^2 \phi_0}{\partial Z^2}\right) xe^x,\ x<0 \\ 0,\ x>0 \end{array}\right).$$

$$s_0 = \left(\begin{array}{c} 2\left(\frac{\partial^2 \phi_0}{\partial Y^2} + \frac{\partial^2 \phi_0}{\partial Z^2}\right)(1+x^2)e^x,\ x<0 \\ 2\left(\frac{\partial^2 \phi_0}{\partial Y^2} + \frac{\partial^2 \phi_0}{\partial Z^2}\right),\ x>0 \end{array}\right).$$

Note the obtained solution does not satisfy the boundary condition $s(x = \infty, Y, Z, T) = 0$. Actually, because of the longwave nature of the disturbance, it decays on a large spatial scale, $X = \delta x$. The transition between the value of s_0 obtained in the region $x = O(1)$ and the zero value is governed by the equation

$$\frac{\partial s_0}{\partial Z} = \frac{\partial^2 s_0}{\partial Y^2} + \frac{\partial^2 s_0}{\partial Z^2}$$

with the boundary conditions

$$s_0 = \left(\frac{\partial^2 \phi_0}{\partial Y^2} + \frac{\partial^2 \phi_0}{\partial Z^2}\right),\ Z \to 0^+;\ s_0 \to 0,\ Z \to \infty.$$

The solution is

$$s_0 = \frac{1}{2\pi X}\int_{-\infty}^{\infty}\int_{-\infty}^{\infty} dY_1 dZ_1 \exp\{-[(Y-Y_1)^2+(Z-Z_1)^2]/4X\}\left(\frac{\partial^2 \phi_0}{\partial Y_1^2} + \frac{\partial^2 \phi_0}{\partial Z_1^2}\right).$$

In the next order, the following inhomogeneous linear problem is obtained:

$$-\frac{\partial \phi_0}{\partial T} \cdot \left(\begin{array}{c} e^x,\ x<0 \\ 0,\ x>0 \end{array}\right) + \frac{\partial \theta_1}{\partial x} = \frac{\partial^2 \theta_0}{\partial Y^2} + \frac{\partial^2 \theta_0}{\partial Z^2} + \frac{\partial^2 \theta_1}{\partial x^2}$$

$$-\left[\left(\frac{\partial \phi_0}{\partial Y}\right)^2 + \left(\frac{\partial \phi_0}{\partial Z}\right)^2\right]\delta(x) + \left[\left(\frac{\partial \phi_0}{\partial Y}\right)^2 + \left(\frac{\partial \phi_0}{\partial Z}\right)^2\right] \cdot \left(\begin{array}{c} e^x,\ x<0 \\ 0,\ x>0 \end{array}\right)$$

$$-\left(\frac{\partial^2 \phi_1}{\partial Y^2} + \frac{\partial^2 \phi_1}{\partial Z^2}\right) \cdot \left(\begin{array}{c} e^x,\ x<0 \\ 0,\ x>0 \end{array}\right) + \left[\frac{1}{2}s_1 + \frac{1}{2}\left(\frac{\partial \phi_0}{\partial Y}\right)^2 + \left(\frac{\partial \phi_0}{\partial Z}\right)^2\right]\delta(x);$$

$$-\frac{\partial\phi_0}{\partial T}\cdot\left(\begin{array}{c}2(x+1)e^x,\ x<0\\ 0,\ x>0\end{array}\right)+\frac{\partial s_1}{\partial x}=\frac{\partial^2}{\partial Y^2}(s_0-2\theta_0)+\frac{\partial^2}{\partial Z^2}(s_0-2\theta_0)$$

$$+\frac{\partial^2}{\partial x^2}(s_1-2\theta_1-2\theta_0)+\left[\left(\frac{\partial\phi_0}{\partial Y}\right)^2+\left(\frac{\partial\phi_0}{\partial Z}\right)^2\right]\cdot\left(\begin{array}{c}2(x+1)e^x,\ x<0\\ 0,\ x>0\end{array}\right)$$

$$-\left(\frac{\partial^2\phi_1}{\partial Y^2}+\frac{\partial^2\phi_1}{\partial Z^2}\right)\cdot\left(\begin{array}{c}2xe^x,\ x<0\\ 0,\ x>0\end{array}\right)-\left(\frac{\partial^2\phi_0}{\partial Y^2}+\frac{\partial^2\phi_0}{\partial Z^2}\right)\cdot\left(\begin{array}{c}2xe^x,\ x<0\\ 0,\ x>0\end{array}\right).$$

One can show that the solvability conditions of the obtained system of equation is

$$(\phi_0)_T+\frac{1}{2}(\nabla\phi_0)^2+\Delta\phi_0+4\Delta^2\phi_0=0. \tag{105}$$

Here

$$\nabla=\mathbf{e}_Y\frac{\partial}{\partial Y}+\mathbf{e}_Z\frac{\partial}{\partial Z},\ \Delta=\nabla^2.$$

Equation (105) is the famous *Kuramoto-Sivashinsky (KS) equation*, which is a paradigmatic example of the dynamical system which creates spatio-temporal chaotic regimes. It has been found that there exists a certain kind of energy cascade from *short* to *long* scales which determines the large-scale properties of the spatio-temporal chaos. More precisely, the large-scale behavior of the system is governed by the Burgers equation with a certain effective positive *turbulent viscosity*, while the instability manifests itself as an uncorrelated short-scale ("white") Gaussian noise. In other words, the large-scale behavior of the KS chaos can be modeled by the universal Kardar-Parisi-Zhang equation, and it is characterized by corresponding scaling properties (Bohr e.a. (1998)).

Oscillatory shortwave instability The case of the oscillatory shortwave instability is more difficult for the theoretical analysis. In the supercritical region, waves with $k_Y^2+k_Z^2\approx k_c^2$ moving in arbitrary direction along the front are spontanuosly generated. Near the instability threshold, $\beta=\beta_c+\delta^2$, each of these waves can be presented as a plane wave slowly modulated in time and space. In the particular case of one-dimensional modulations the envelope function $A(X,T)$, where $X=\delta x$ and $T=\delta^2 t$, is governed by the *complex Ginzburg-Landau equation* that can be transformed to the standard form,

$$A_t=A+(1+i\alpha)A_{xx}-(s+i\beta)|A|^2A. \tag{106}$$

Despite the apparent simplicity of equation (106), the dynamics governed by this equation is incredibly reach. The basic features of this equation have been summarized in a recent review paper by Aranson and Kramer (2002). It includes spontaneous destruction of waves periodicity, appearance of defects, different kinds of spatio-temporal chaos. Two-dimensional modulations of plane waves have been hardly investigated. One can expect development of topological defects and spiral-like waves. Waves propagating in different directions interact with each other in a nonlocal way (Knobloch and De Luca (1990), Matkowsky and Volpert (1992)). We conclude that the investigation of the nonlinear development of oscillatory shortwave instability is a developing field which is beyond the scope of the present course.

Bibliography

I.S. Aranson and L. Kramer. The world of the complex Ginzburg-Landau equation. *Rev. Mod. Phys.*, 74:99, 2002.

D.G. Aronson and H.F. Weinberger. Multidimensional nonlinear diffusion arising in population genetics. *Adv. Math.*, 30:33, 1978.

T. Bohr, M.H. Jensen, G. Paladin, and A. Vulpiani. *Dynamical Systems Approach to Turbulence.* Cambridge University Press, 1998.

R.A. Fisher. The wave of advance of advantageous genes. *Ann. Eugenics*, 7:355, 1937.

E. Knobloch and J. De Luca. Amplitude equations for travelling wave convection. *Nonlinearity*, 3:975, 1990.

A. Kolmogoroff, I. Petrovsky, and N. Piscounoff. Study of the diffusion equation with growth of the quantity of matter and its application to a biological problem. *Moscow University, Bull. Math.*, 1:1, 1937.

S.B. Margolis and B.J. Matkowsky. Flame propagation in channels: Secondary bifurcation to quasi-periodic pulsations. *SIAM J. Appl. Math*, 45:93, 1985.

B.J. Matkowsky and G.I. Sivashinsky. Propagation of a pulsating reaction front in solid fuel combustion. *SIAM J. Appl. Math*, 37:686, 1979.

B.J. Matkowsky and V.A. Volpert. Coupled nonlocal complex Ginzburg-Landau equations in gasless combustion. *Physica D*, 54:203, 1992.

J.D. Murray. *Mathematical Biology.* Springer, New York, 2002.

G.I. Sivashinsky. Nonlinear analysis of hydrodynamic instability in laminar flames - I. Derivation of basic equations. *Acta Astronautica*, 4:1177, 1977.

P.-F. Verhulst. Notice sur la loi que la population poursuit dans son accroissement. *Corr. Math. et Phys.*, 10:113, 1838.

Three Dimensional Film Dynamics

Michael Bestehorn

BTU Cottbus, Lehrstuhl für Theoretische Physik II, Erich-Weinert-Str 1, 03046 Cottbus, Germany

Abstract In this course, I wish to give an overview on methods applied in systems showing macroscopic, self-organized pattern formation (Haken, 2004; Pismen, 2006). The course is based on six lectures, organized as follows : 1 - Introduction. 2 - Thick films, undeformable surface – pure fluids. 3 - Thick films, undeformable surface – binary mixtures. 4 - Reduced description – order parameter equations. 5 - Thin films with a deformable surface. 6 - Spinodal dewetting.

Contents

1 Introduction

1.1 Basic equations

To be more specific we shall concentrate on liquid layers of Newtonian, viscous fluids which are described by the hydrodynamic basic equations for the field variables velocity $\vec{v}$, temperature T, density ρ, and pressure p. If in addition the fluid can be assumed to be incompressible, they read (Chorin and Marsden, 1993; Bestehorn, 2006)

$$\rho \left[\frac{\partial \vec{v}}{\partial t} + (\vec{v} \cdot \nabla)\, \vec{v} \right] = -\nabla p + \vec{f} + \eta\, \Delta \vec{v} \tag{1a}$$

$$\operatorname{div} \vec{v} = 0 \tag{1b}$$

$$\frac{\partial T}{\partial t} + (\vec{v} \cdot \nabla)\, T = \kappa\, \Delta T \tag{1c}$$

with the dynamic viscosity η, the thermal diffusivity κ and for the moment arbitrary external volume forces $\vec{f}$. Eqs. (1a) are the Navier-Stokes equations, (1b) is the continuity equation which reduces here to the incompressibility condition, and (1c) is the temperature (or heat) equation. The system (1) has to be completed by suitable boundary and initial conditions, as well as by some material laws.

To write down (1) in the present form, several assumptions and approximations have been made. Firstly, one assumes that viscosity and thermal diffusivity of the fluid are constant, where "constant" means constant with respect to space and time. Normally they are not but depend at least on temperature (in more sophisticated situations perhaps also on pressure and velocity) and therefore also on space and time. In these cases additional terms on the right hand sides of (1a) and (1c) would occur. Another assumption is made for the heat equation. Here one neglects heat production by friction, which can be important for turbulent flows.

1.2 Instabilities

As already mentioned, we shall consider a plane fluid layer. We assume that a motionless stationary state exists as a (stable or unstable) solution of (1). Its temperature and pressure distribution can then be computed from (1) by setting $\vec{v}$ and all time derivatives to zero:

$$\nabla p^0 = \vec{f} \tag{2a}$$

$$\Delta T^0 = 0 \tag{2b}$$

If the external force is provided by buoyancy, we may align the z-axis of the coordinate system along $\vec{f}$ which yields

$$\vec{f} = -g\,\rho(x,y,z)\,\hat{e}_z\ ,$$

where g is the gravitational acceleration. Eq. (2a) can only be solved if ρ does not depend on x and y. If one assumes that the density depends anyhow on temperature

$$\rho = \rho(T^0) \tag{3}$$

then T^0 can also depend only on z . Thus one finds from (2b)

$$T^0(z) = a + \beta z\ . \tag{4}$$

Taking a linear relation for (3)

$$\rho(T) = \rho_0\,[1 - \alpha(T - T_0)] \tag{5}$$

with the heat expansion coefficient $\alpha \equiv -\rho_0^{-1} d\rho/dT$ and ρ_0 as the density at the reference temperature T_0, one may integrate (2a) and find for the pressure of the motionless state

$$p^0(z) = -g\int \rho\,\mathrm{d}z = -g\rho_0\left(z - \frac{1}{2}\alpha\beta z^2\right) \tag{6}$$

where we put $a = T_0$.

A linear stability analysis, which we shall describe in more detail in the next lectures, shows that the motionless, non-equilibrium state (4) can get unstable if the temperature gradient β exceeds a certain critical value, depending on the fluid properties and the geometry of the layer. There are two different mechanisms, if the fluid layer is heated from below:

(1) *Buoyancy:* Hot fluid particles (volume elements) near the bottom are lighter than colder ones and want to rise. Colder particles near the top want to sink. If the stabilizing forces of thermal conduction and friction in the fluid are exceeded by the externally applied temperature gradient, a (regular, spatially periodic) fluid motion sets in.

(2) *Surface tension:* If the upper surface of the fluid is free, i.e. in contact with the ambient air, tangential surface tension normally increases with decreasing surface temperature, see fig. 1. If a fluid particle near the surface moves by fluctuations say to the right, then warmer fluid is pulled up from the bottom, increasing the surface temperature locally. Due to increasing laterally surface tension with respect to the neighbouring points, even more hot fluid is pumped up from the bottom and the fluid starts to move. This is called the Marangoni effect and works even without gravity,

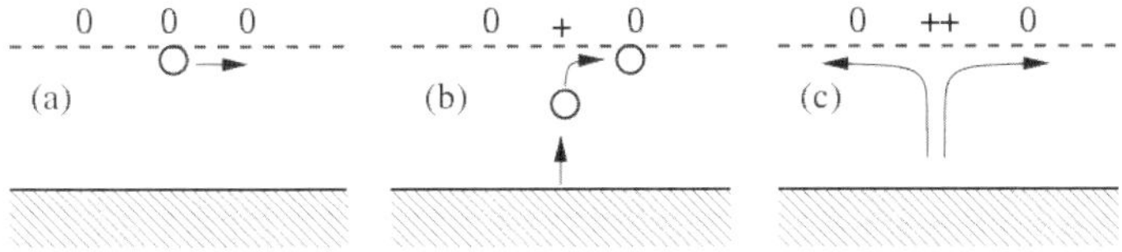

Figure 1. The Marangoni effect may destabilize a fluid layer at rest and may generate a (regular) fluid motion.

i.e. in space experiments. In both cases, the typical length of the structures which bifurcate from the motionless state are of the order of the layer depth. This instabilities are sometimes called *short scale instability.* The main emphasis of these lectures will lay on instabilities induced by surface or interface effects. In the situation described above, the surface can be assumed to be flat and undeformable. Of course this is only an approximation, but valid for not too thin fluid layers. If, on the other hand, the thickness of the fluid layer is less than a certain value which is in the order of 10^{-4} m for common silicone oils, another mechanism comes to the foreground. This mechanism is based on *Surface deformation.* If the surface is locally

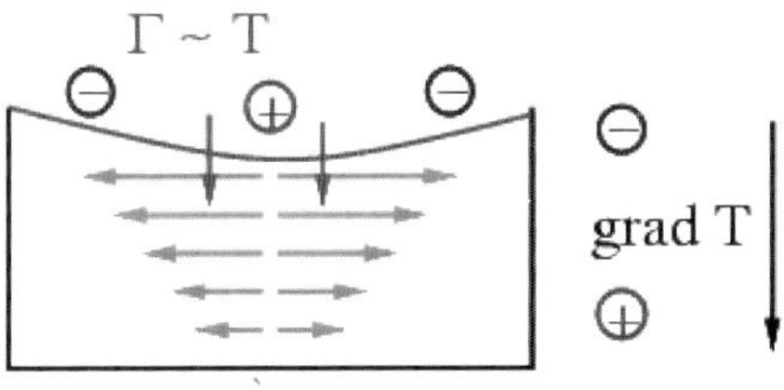

Figure 2. If the surface is deformable, a large scale instability may occur as a consequence of the Marangoni effect and mass conservation. The surface tension is denoted by Γ which often can be assumed as a linear function of temperature.

depressed by arbitrary fluctuations, the depressed part is heated up due to the vertical temperature gradient. A lateral surface tension gradient is formed which pulls the liquid outside the depressed region (see fig. 2). Since the continuity equation must hold, the surface gets even more depressed and an instability occurs. The same mechanism leads to the growth of elevated parts of the surface, under which fluid is pumped in from adjacent regions.

As linear stability analysis shows, the deformation mode belongs to the so-called *large-scale instability.* This means that the fastest growing modes have a wave length very large compared to the layer depth. It is the depth of the layer which distinguishes which instability occurs first if the temperature gradient is increased from the sub-critical region, as shown in fig. 3. In

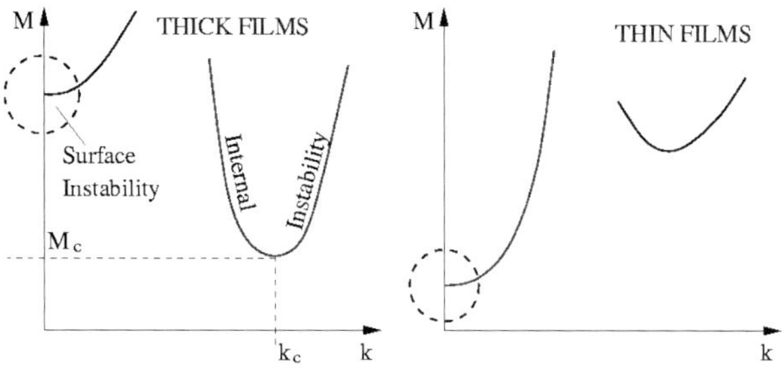

Figure 3. The two cases "thick film" and "thin film" used in the text are defined by the instability that comes first, if the temperature gradient is increased. The two instabilities differ in the wave length Λ (wave number $k = 2\pi/\Lambda$) of the growing structures.

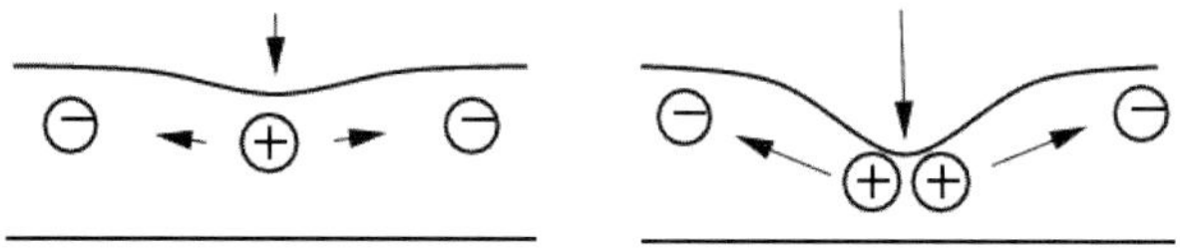

Figure 4. In ultra-thin films, van der Waals forces between free surface and solid substrate may destabilize a plane fluid layer even without an external temperature gradient (+/- denote relative values of the disjoining pressure).

ultra-thin films, this means a depth of the order 100 nm or less, other mechanisms are possible. Van der Waals forces between the free surface and the solid substrate then become important. They have a potential and can be expressed in the pressure by an extra term, often called *disjoining pressure*. If that pressure increases with decreasing layer depth, fluid is pressed out of depressed regions and pumped into elevated regions and an instability occurs, even for isothermal cases, see fig. 4.

The spatio-temporal behavior of thin films can be expressed by a single partial differential equation for the location of the surface $h(x, y, t)$. It is called *thin film equation* and can be derived systematically from (1) with additional boundary conditions, using the lubrication approximation (for detail see sections 4-6). It has the form

$$\partial_t h = \text{div}\,(Q(h)\,\text{grad}\,P(h))\,, \tag{7}$$

where $P(h)$ is the disjoining pressure at the free surface and $Q(h)$ a (positive definite) mobility. Both quantities depend on material parameters.

1.3 Pattern formation – Examples

What happens if the critical value for the temperature gradient is exceeded? Since the famous experiments of Henri Bénard in the beginning of the 20th century, one knows that hexagons as shown in fig. 5 occur if the surface is free and the layer is "thick". These kind of experiments were repeated many times under excellent conditions, for free and closed surfaces, with different fluids, even under micro gravity conditions (Koschmieder, 1993). We shall discuss these self-organized structures in the section 2 in more detail. Surprisingly, a secondary instability takes place for a larger

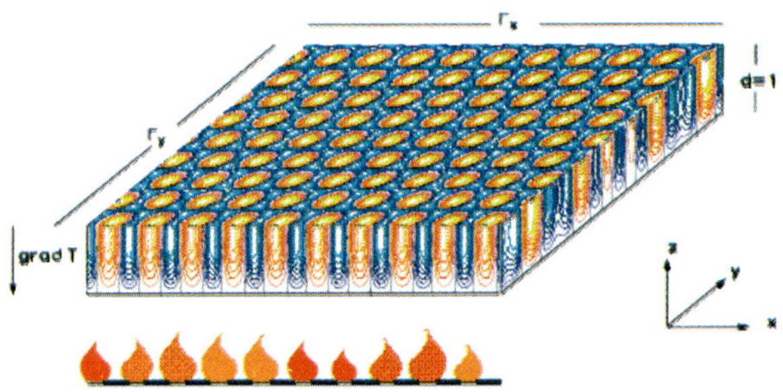

Figure 5. Hexagonal motion of a fluid heated from below, found by computer solution of eqs. (1). Shown are contour lines of the temperature field.

external temperature gradient, which was not known before 1995, almost 100 years after Bénard. This instability shows the occurrence of rather regular squares and was discovered by Eckert and Thess in Dresden, Germany and, in the mean time but independently, by Schatz and Swinney in Austin, Texas (Eckert et al., 1998). Lateron, direct numerical simulations of the hydrodynamic basic equations in 3D also confirmed this transition, see fig.6. (Bestehorn, 1996).

If the fluid is covered by a good thermal conductor (a sapphire plate, for instance), hexagons are not the typically found structure at onset but rather stripes or rolls are encountered. This can be understood in the frame

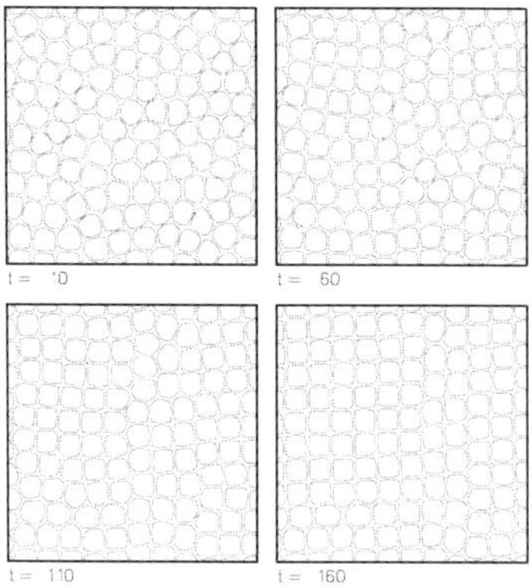

Figure 6. Regular squares as a secondary instability of hexagons. Numerical solution of the basic equations (1).

of reduced order parameter equations by simple symmetry arguments. We shall discuss this in detail in section 3. For small Prandtl numbers (the ratio between viscosity and thermal diffusivity of the fluid) more complicated and time dependent patterns are found in the form of spirals (fig. 7).

The rather fast growth of patterns with a typical horizontal length scale of the order of the depth of the fluid layer is typical for pattern formation in thick films. In the long time limit, these structures can be stationary or time dependent, depending on several control and fluid parameters (temperature gradient, material properties, etc.). On the other hand, the spatio-temporal behavior is completely different for thin and ultra-thin films. Here one finds after a rather short initial phase the formation of larger and larger structures, known as *coarsening*. Eventually, the dynamics converge to a stationary state that consists in a single elevation (drop) or suppression (hole) of the surface, as shown in fig 8, left panel. This development can be interrupted by rupture of the film. Rupture is obtained if the surface touches the substrate and the thickness reaches zero in certain domains. Then the surface cannot longer be expressed as a smooth function of the space variables and the description by a thin film equation like (7) breaks down. Rupture can be avoided by introducing a repelling disjoining pressure. In this situation, completely dry region cannot exist but the substrate is rather covered by a so-called (ultra thin) precursor film.

If in addition horizontal forces are applied, i.e. by simply inclining the fluid layer, interesting studies of falling films and front instabilities can be made in the frame of the thin film equation. A typical example is shown in

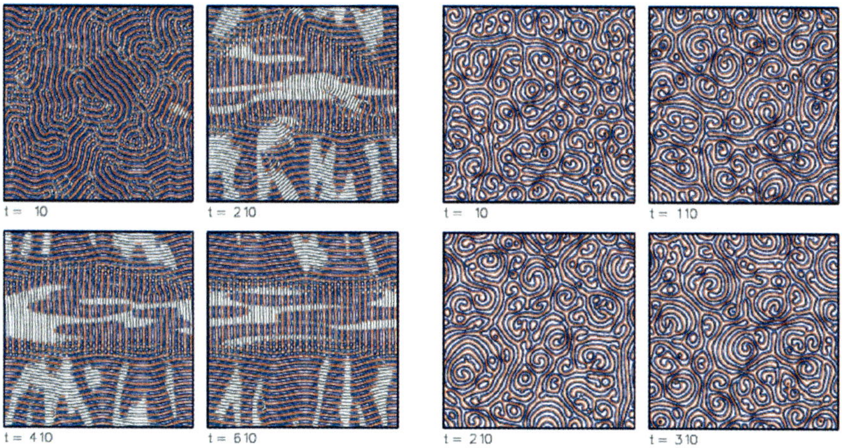

Figure 7. Rolls (left) for high Prandtl number (*Pr*) fluids, spirals (right) for low *Pr* are found if the surface is covered by a good thermal conductor.

fig. 8, right panel.

1.4 Types of instabilities

Different types of instabilities can be classified according to their linear behavior at onset. Several notions are used, we refer to those introduced by Cross and Hohenberg (1993). Consider a mode having the complex eigenvalue

$$\lambda(k^2) = i\omega(k^2) + \sigma(k^2) \tag{8}$$

with real valued frequency ω real valued growth rate σ. Due to rotation symmetry with respect to the horizontal coordinates, all values depend on the modulus of the wave vector of the unstable mode (assumed to be a plane wave in horizontal direction). We use the following notions:

Type III$_s$. "s" denotes stationary or monotonic and refers to the temporal behavior of the unstable mode close to onset. The type number specifies the spatial behavior of the modes. Type III means slowly varying or even constant in space ($k \approx 0$). The spatial structure beyond instability is then mainly dominated by the geometry and boundary conditions of the system

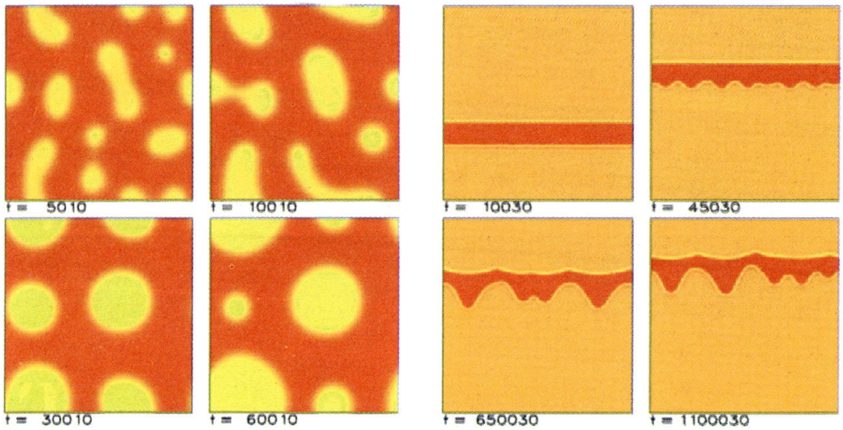

Figure 8. Left: Coarsening is the typical spatial behavior for a thin film. Finally, a stationary solution consisting of one single hole would survive. Right: If the layer is inclined, the motion of fronts and the development of front instabilities can be examined. Numerical solution of the thin film equation (7), red: elevation, yellow: depression.

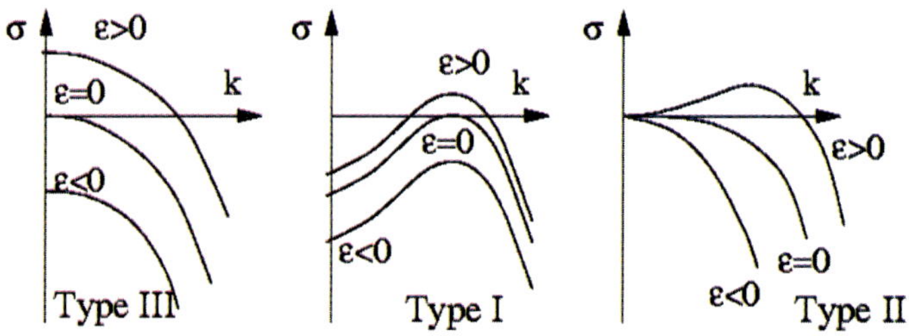

Figure 9. schematic drawing of the real part of the eigenvalue (7) as function of the wave vector for the three types of instabilities.

under consideration. For (8) this means[1]

$$\omega = 0 \qquad \text{and} \qquad \left.\frac{d\sigma}{dk}\right|_{k=0} = 0 \ ,$$

[1]In the following $d\sigma/dk = 0$ determines the (typically unique) *maxima* of σ, belonging to the fastest growing mode.

see fig. 9. A typical example for a III_s instability is the real Ginzburg-

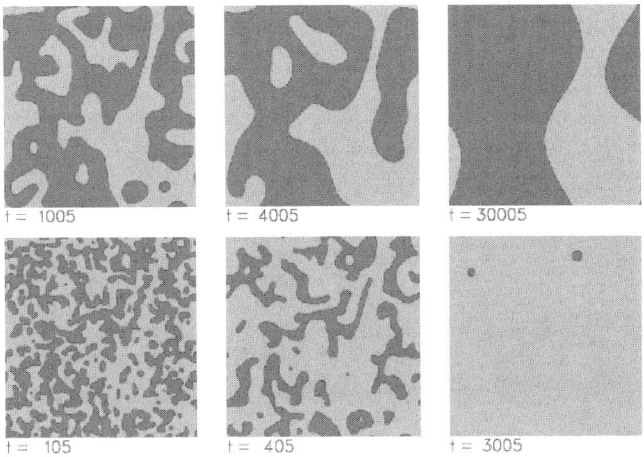

Figure 10. Numerical solution (time series) of the real Ginzburg-Landau equation which shows a III_s instability. The Ginzburg-Landau equation can be considered as a simple model for the magnetization of a ferro magnet. Then the two rows show the spatio-temporal evolution of the magnetization, top: without external field, bottom: with external field.

Landau equation. A computer solution showing clearly the spatially (and temporally) slowly varying behavior can be seen in fig. 10.

Type III_o. "o" stands for oscillatory and denotes a non-vanishing imaginary part of (8) at threshold. This type includes Hopf-instabilities which have the same slow spatial behavior than III_s. In (8) we have

$$\omega \neq 0 \qquad \text{and} \qquad \left.\frac{d\sigma}{dk}\right|_{k=0} = 0 \; .$$

For this kind of instabilities one needs at least two coupled diffusion equations. It is often encountered in reaction diffusion systems like for instance the "Brusselator".

Type I_s. The short scale pattern forming instabilities shown in figs. 5-7 with periodicity in space, $k_c \neq 0$ are of this type, see fig. 9, middle frame. Again one needs at least two coupled diffusion equations to obtain such an instability. For the eigenvalue,

$$\omega = 0 \qquad \text{and} \qquad \left.\frac{d\sigma}{dk}\right|_{k=k_c} = 0 \qquad \text{with} \qquad k_c \neq 0$$

holds. Sometimes this kind of patterns are called *Turing structures* or *Turing instabilities*, after the seminal work of Alan Turing, who predicted this patterns in skin, scales, or hair coating of certain animals, see fig. 11. For more details and pattern formation in biology see (Murray, 2002).

Figure 11. After a theory by A. Turing the painting on skin, scales, or coats of animals is organized by a chemical non-equilibrium reaction during the embryonal phase. Left: regular spots arranged in a hexagonal manner on the panther fish, right: stripes with defects on the lion fish.

Type $\mathbf{I}_o$ denotes oscillating Turing structures, sometimes also called *wave instabilities*. The eigenvalue λ then has the form

$$\omega \neq 0 \quad \text{and} \quad \left.\frac{d\sigma}{dk}\right|_{k=k_c} = 0 \quad \text{with} \quad k_c \neq 0 \; .$$

For this instability, the system has to be described by at least three coupled diffusion equations. We shall encounter this kind of instabilities again in section 3 on binary mixtures.

Type $\mathbf{II}_s$. This type is realized in the surface patterns of thin films, fig. 8. Here, λ depends on k as shown in the right frame of fig. 9. One has

$$\omega = 0 \quad \text{and} \quad \left.\frac{d\sigma}{dk}\right|_{k=k_c} = 0 \quad \text{with} \quad k_c \neq 0$$

and in addition

$$\sigma(k=0) = 0 \; .$$

From the last condition one sees that modes with $k = 0$, i.e. those which are homogeneous in space, are marginally stable, this means neither stable nor unstable. Then one may add a constant to the order parameter (the mode amplitude)

$$\xi' = \xi + \text{const} \ ,$$

where ξ' is still a solution of the linear part of the order parameter equation. This property has usually its origin in a symmetry of the basic problem. We shall discuss this instability type in the section on thin films. There, the symmetry is simply a global shift of the surface in vertical direction.

Type II$_0$. The same as II$_s$ but with an additional imaginary part $\omega \neq 0$. We mention that type only for the sake of completeness, there will be no further examples during the course. A possible application could be the instability of a binary mixture of a more complex fluid.

2 Thick films with undeformable surface – Pure fluids

In this section, we first state the basic equation in non-dimensional form, supplied by the boundary conditions for buoyancy-driven as well as for surface-driven convection. After examining the linear stability, we explain the numerical method and discuss several solutions. For a recent review on surface-driven instabilities see (Colinet et al., 2001).

2.1 The basic equations

Scaling to non-dimensional quantities Scaling of independent (x, y, z, t) and dependent $(\vec{v}, T)$ variables allows the reduction of the numbers of parameters. Scaling cannot be unique. We use that one which is common in literature

$$\vec{r} = \vec{\tilde{r}} \cdot d, \quad t = \tilde{t} \cdot (d^2/\kappa), \quad \vec{v} = \vec{\tilde{v}} \cdot (\kappa/d), \quad T = \tilde{T} \cdot \beta \cdot d \ , \tag{9}$$

with the constant depth d and the externally applied temperature gradient

$$\beta = \frac{T_1 - T_0}{d} \ .$$

The temperature of the substrate is T_0, that of the surface (either covered or free) of the *motionless* state T_1, both are uniform in space. Note that if the liquid his heated from below, $\beta < 0$.

Decomposition of the velocity field, basic equations For incompressible fluids, the velocity field has to be free of sources and sinks according to (1b). Such a field is sometimes called *solenoidal* and can be written as the sum of a *toroidal* and a *poloidal* part:

$$\vec{v}(\vec{r}, t) = \vec{v}_T(\vec{r}, t) + \vec{v}_P(\vec{r}, t) \ , \tag{10}$$

where each part can be represented by a single scalar function

$$\vec{v}_T = \nabla \times (\Phi \hat{e}_z) = \begin{pmatrix} \partial_y \Phi \\ -\partial_x \Phi \\ 0 \end{pmatrix}, \qquad \vec{v}_P = \nabla \times \nabla \times (\Psi \hat{e}_z) = \begin{pmatrix} \partial_z \partial_x \Psi \\ \partial_z \partial_y \Psi \\ -\Delta_2 \Psi \end{pmatrix} \tag{11}$$

with $\Delta_2 = \partial^2_{xx} + \partial^2_{yy}$. One can consider Φ as the stream function for the horizontal toroidal field (it has no vertical component), and Ψ as a kind of generalized potential for the poloidal part.

Taking the curl of the Navier-Stokes eq. (1a), the z-component reads in the scaling of (9)

$$\left\{\Delta - \frac{1}{Pr}\partial_t\right\} \Delta_2 \Phi(\vec{r},t) = -\frac{1}{Pr}\left[\nabla \times ((\vec{v}\cdot\nabla)\vec{v})\right]_z , \tag{12}$$

where we dropped all the tildes. Note that Φ (and later on Ψ) are also scaled to dimensionless quantities. For the sake of clarity, on the left hand side $\vec{v}$ was left as an abbreviation. It has also to be substituted by (10, 11).

An equation for Ψ is found by taking twice the curl of (1a):

$$\left\{\Delta - \frac{1}{Pr}\partial_t\right\} \Delta\Delta_2 \Psi(\vec{r},t) = -R\, \Delta_2 \Theta(\vec{r},t) - \frac{1}{Pr}\left[\nabla \times \nabla \times ((\vec{v}\cdot\nabla)\vec{v})\right]_z . \tag{13}$$

Two dimensionless numbers appear in (12) and (13). One is the material dependent *Prandtl number*

$$Pr = \frac{\nu}{\kappa} , \tag{14}$$

which measures the ratio of the diffusion times of heat and momentum , the other one is called *Rayleigh number* and turns out to be

$$R = -\frac{\beta g \alpha d^4}{\nu \kappa} . \tag{15}$$

To close the system, we need an equation for the temperature field. It is often of advantage to deal with small quantities. Therefore one introduces the deviation Θ from the thermally conducting state (4):

$$T(\vec{r},t) = T^0(z) + \Theta(\vec{r},t) = T_0 + \beta z + \Theta(\vec{r},t) . \tag{16}$$

Inserting this into the heat equation (1c) yields after scaling

$$\{\Delta - \partial_t\}\, \Theta(\vec{r},t) = -\Delta_2 \Psi(\vec{r},t) + (\vec{v}\cdot\nabla)\Theta(\vec{r},t) . \tag{17}$$

The system (12), (13), and (17) constitutes the basic equations for the three scalar fields Φ, Ψ, and Θ which describe convective motion and temperature of a plane fluid layer with flat and undeformable surface on a plane substrate.

Boundary conditions Next we need boundary conditions (b.c.). Since we wish to examine laterally extended systems, we can use periodic boundary conditions for x, y which a large periodicity length (aspect ratio) compared to d. In vertical direction, the situation deserves more attention.

A. Bottom plate

For viscous fluids, all velocity components have to vanish at a solid wall (bottom plate):

$$\vec{v}\big|_{z=0} = 0 \; . \tag{18}$$

These are the so-called no-slip conditions. In terms of Φ and Ψ they read

$$\Psi\big|_{z=0} = \Phi\big|_{z=0} = 0 \; . \tag{19}$$

(In fact a constant would also do it, but this can be put to zero). For vanishing $\vec{v}_P$ one has to require in addition

$$\partial_z \Psi\big|_{z=0} = 0 \; . \tag{20}$$

The temperature on the bottom is given by T_0, so the deviations have to vanish:

$$\Theta\big|_{z=0} = 0 \; . \tag{21}$$

B. Surface of the fluid

On a free, flat surface with normal $\hat{e}_z$, only the vertical velocity component has to vanish. In addition, the xz and yz components of the viscous stress tensor have to be balanced by surface stress forces:

$$v_z|_{z=1} = 0 \tag{22a}$$

$$\eta\, \partial_z v_x|_{z=1} = \partial_x \Gamma|_{z=1} \tag{22b}$$

$$\eta\, \partial_z v_y|_{z=1} = \partial_y \Gamma|_{z=1} \; , \tag{22c}$$

with the surface tension Γ. Differentiating (22b) with respect to x, (22c) with respect to y and adding both equations yields with (1b)

$$\eta\, \partial^2_{zz} v_z|_{z=1} = -\Delta_2 \Gamma|_{z=1} \; . \tag{23}$$

Spatial surface tension inhomogeneities may be caused by a temperature gradient. To include the Marangoni effect one assumes that the surface tension is a linear function of temperature:

$$\Gamma(T) \approx \Gamma(T_0) + \gamma_1 \cdot (T - T_0) \; . \tag{24}$$

Inserting this and the velocity decomposition (10) into (23) gives after scaling the boundary conditions for Ψ:

$$\partial^2_{zz}\Psi|_{z=1} = Ma \cdot \Theta|_{z=1} \tag{25}$$

with the non-dimensional *Marangoni number* defined as

$$Ma = \frac{\gamma_1 \beta d^2}{\kappa\eta} = \frac{\gamma_1 \beta d^2}{\rho\kappa\nu} \ . \tag{26}$$

For Φ one finds from the z-component of the curl of (22)

$$\partial_z \Phi|_{z=1} = 0 \ . \tag{27}$$

The last boundary condition we need is that for the temperature. If the surface is not a good thermal conductor one can derive it by solving the heat equation in the gas layer above the fluid. If motion is neglected in this layer, one finds

$$\partial_z \Theta|_{z=1} = -Bi \cdot \Theta|_{z=1} \tag{28}$$

with the non-dimensional (positive) *Biot number* Bi which describes the ratio of thermal conduction of the boundary (here the gas or atmosphere) to that of the fluid. For perfectly conducting boundaries $Bi \to \infty$, for bad conductors, $Bi << 1$.

2.2 Linear stability analysis

Next we wish to compute the critical temperature gradient for which the motionless state looses stability and convection sets in. Therefore we linearize (12,13,17) around its trivial solution and obtain the linear PDE system

$$\left\{\Delta - \frac{1}{Pr}\partial_t\right\}\Delta_2\Phi(\vec{r},t) = 0 \tag{29a}$$

$$\left\{\Delta - \frac{1}{Pr}\partial_t\right\}\Delta\Delta_2\Psi(\vec{r},t) = -R\,\Delta_2\Theta(\vec{r},t) \tag{29b}$$

$$\{\Delta - \partial_t\}\,\Theta(\vec{r},t) = -\Delta_2\Psi(\vec{r},t) \ . \tag{29c}$$

Clearly, the first equation separates and can be solved independently from the other two. It is a diffusion equation for the scalar field $\Delta_2\Phi$ and has in the long time limit only the trivial solution $\Phi = 0$, at least with the boundary conditions discussed in the last paragraph. But even in the weak non-linear region this field can become important if the "diffusion constant"

Pr is small enough. For low Prandtl number fluids (or gases) this field is the origin of the so-called spiral turbulence we already presented in fig. 7 and to which we shall come back later.

Therefore it is sufficient to examine the linear stability using only (29b) and (29c). We do the separation (periodic b.c. or infinitely extended layer in the horizontal direction)

$$\Psi(\vec{r},t) = \psi(z)\, \mathrm{e}^{i\vec{k}\vec{x}}\, \mathrm{e}^{\lambda t}, \qquad \Theta(\vec{r},t) = \theta(z)\, \mathrm{e}^{i\vec{k}\vec{x}}\, \mathrm{e}^{\lambda t} \tag{30}$$

with the plane vectors $\vec{k} = (k_x, k_y)$, $\vec{x} = (x, y)$ and obtain from (29b,29c) the eigenvalue problem

$$\lambda \frac{1}{Pr} (d_{zz}^2 - k^2)\psi(z) = (d_{zz}^2 - k^2)^2 \psi(z) + R\,\theta(z) \tag{31a}$$

$$\lambda\,\theta(z) = (d_{zz}^2 - k^2)\,\theta(z) - k^2\psi(z)\;. \tag{31b}$$

The system (31) with boundary conditions from the preceding paragraph can only be solved numerically (for a treatment with no-slip conditions on both sides see (Chandrasekhar, 1981). It yields the eigenvalues

$$\lambda = \lambda_\ell(k^2, Pr, R, Ma)\;, \tag{32}$$

where the index ℓ counts the (infinitely many) eigenvalues belonging to different eigenfunctions. The critical control parameter (R or Ma) as a function of the wave vector k is found by solving the equation

$$\lambda_\ell(k^2, Pr, R_c, Ma_c) = 0 \tag{33}$$

for R_c or Ma_c. Fig. 12 (left) shows R_c as a function of k for the case of a solid (glass) plate on top of the liquid (no Ma). Fig. 12 (right) shows the critical lines in the $Ma_c - R_c-$ plane for several Biot numbers. It turns out that the instability in both cases (free and closed surface of the fluid) is of type I_s.

For further purpose we define the *reduced control parameter*

$$\varepsilon = \frac{Ma - Ma_c}{Ma_c} = \frac{R - R_c}{R_c}\;. \tag{34}$$

2.3 Numerics

Before presenting some numerical results, we briefly explain the procedure applied. It is able to solve the basic equations in 3D for rather large aspect ratios on small computers (PC or work station). For details and program packages see (Press et al., 1992; Hirsch, 1988; Fletcher, 1991).

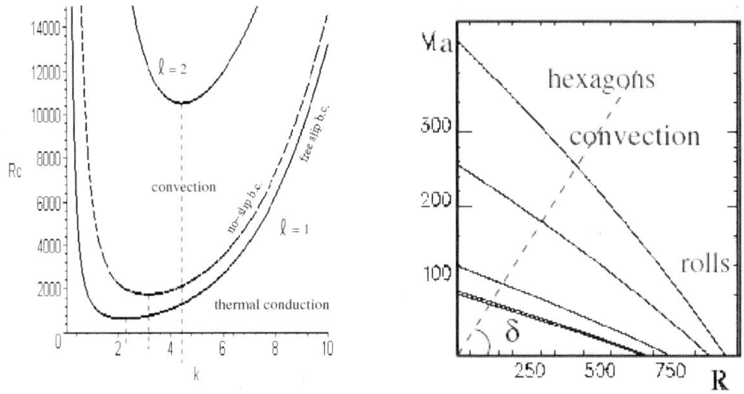

Figure 12. Left: critical lines, where a pattern forming instability sets in if the surface is covered by a good thermal conductor. The mode without nodes ($\ell = 1$) gets unstable first. The solid line is for free-slip b.c. where the system (31) can be solved analytically by trigonometric functions. Right: plane of control parameters for Marangoni convection. Shown are the critical lines for different Biot number $Bi = 10, 5, 1, 0.1, 0$ (top to bottom). If the external temperature gradient is changed, the system moves on the "physical line" (dashed). Right of the critical line, convection sets in. The slope $\tan\delta$ of the physical line is inverse to the square of the depth of the fluid.

The method The method we describe here was first published in Bestehorn (1993). We treat eq.(13,17) for the case of infinite Pr ($\phi = 0$) in three spatial dimensions under periodic b.c. in horizontal directions. To perform the time integration, we use a semi-implicit one step forward scheme, where the linear parts of (13,14) are treated implicitly, the non-linear ones explicitly:

$$\begin{aligned} \Delta^2\Psi(\vec{r}, t+\delta t) + R\Theta(\vec{r}, t+\delta t) &= 0 \qquad (35) \\ \left[\frac{1}{\delta t} - \Delta\right]\Theta(\vec{r}, t+\delta t) - \Delta_2\Psi(\vec{r}, t+\delta t) &= \frac{\Theta(\vec{r}, t)}{\delta t} - \vec{v}(\vec{r}, t)\cdot\nabla\Theta(\vec{r}, t) \end{aligned}$$

with δt being the time step. To invert the right hand side of (35), we apply a fast Fourier transform (FFT) in the horizontal coordinates $\vec{x}$. To calculate the derivatives with respect to the vertical direction z, we use a finite difference method. Eqs.(35) take the discretized form in each layer i:

$$\begin{aligned}\left[\delta_{ij}k^4 - 2k^2 D_{ij}^{(2)} + D_{ij}^{(4)}\right]\Psi_j(\vec{k}, t+\delta t) + R\,\Theta_i(\vec{k}, t+\delta t) &= 0 \\ \left[\delta_{ij}\left(\frac{1}{\delta t} + k^2\right) - D_{ij}^{(2)}\right]\Theta_j(\vec{k}, t+\delta t) + k^2\Psi_i(\vec{k}, t+\delta t) &= F_i(\vec{k}, t)\ ,\end{aligned} \qquad (36)$$

where one has to sum up over j on the right hand side and the indices i, j denote the N mesh points in z-direction, δ_{ij} the Kronecker symbol. $D_{ij}^{(n)}$ stands for the differential matrix of the nth derivative. The b.c. in vertical direction are incorporated in $D_{ij}^{(n)}$, $D_{ij}^{(4)}$ depends on the Marangoni number, $D_{ij}^{(2)}$ in the second equation on the Biot number. We used central differences in the order δz. F_i denotes the Fourier transform of the left hand side of (35) in layer i. If we combine Θ_i and Ψ_i to a $2N$-component vector $\vec{U}$ of the form

$$\vec{U} = (\Theta_1, \Psi_1, ...\Theta_N, \Psi_N)$$

the task we are left with is the inversion of a band matrix $\underline{L}$ at each time step:

$$\underline{L}\vec{U}(\vec{k}, t+\delta t) = \vec{G}(\vec{k}, t) \qquad (37)$$

that can be performed effectively using standard routines. Here, $\vec{G}$ is introduced as

$$\vec{G} = (F_1, 0, F_2, 0.......F_N, 0)$$

An inverse FFT of $\Theta_i(\vec{k}, t+\delta t)$ and $\Psi_i(\vec{k}, t+\delta t)$ to real space completes one iteration step. To avoid convolution sums in Fourier space, the non-linearities are calculated in real space using again a finite difference method with grid spacing $\Delta x, \Delta z$ for their derivatives. Switching between real and Fourier space is also known as "pseudo-spectral method".

Stability We wish to show the relation between the size of $\delta t, \Delta x, \Delta z$ and numerical stability. To this end we have to linearize (35) around a steady stable solution and examine the temporal behavior of an arbitrary disturbance. For the sake of simplicity we restrict the calculation here to the two-dimensional case and to free/free b.c. in vertical direction. Then a stationary solution of (13,17) consists of parallel rolls and reads up to the order ε:

$$\begin{aligned}\Theta_0(x, z) &= \frac{4}{\pi}\sqrt{\frac{\varepsilon}{3}}\sin\pi z\cos k_c x + \frac{\varepsilon}{\pi}\sin 2\pi z \\ \Psi_0(x, z) &= \frac{4}{\pi}\sqrt{3\varepsilon}\sin\pi z\cos k_c x\end{aligned} \qquad (38)$$

with $k_c = \pi/\sqrt{2}$. Inserting (38) into (35) and linearizing with respect to small disturbances $\vec{u}(\vec{k}, t)$ added to (38), (37) yields a linear iteration rule for $\vec{u}$ in Fourier space:

$$\underline{L}\vec{u}(\vec{k}, t + \delta t) = \underline{N}\vec{u}(\vec{k}, t) \tag{39}$$

with the matrix elements of $\underline{\mathrm{N}}$ given as

$$N_{ij} = \left. \frac{\delta G_i}{\delta U_j} \right|_{\Theta=\Theta_0, \Psi=\Psi_0} .$$

The condition for numerical stability is now that the spectral radius of the *amplification matrix* $\underline{Q}$

$$\underline{Q} = \underline{L}^{-1}\underline{N} \tag{40}$$

is less than one. We evaluated the spectrum of (40) numerically for several

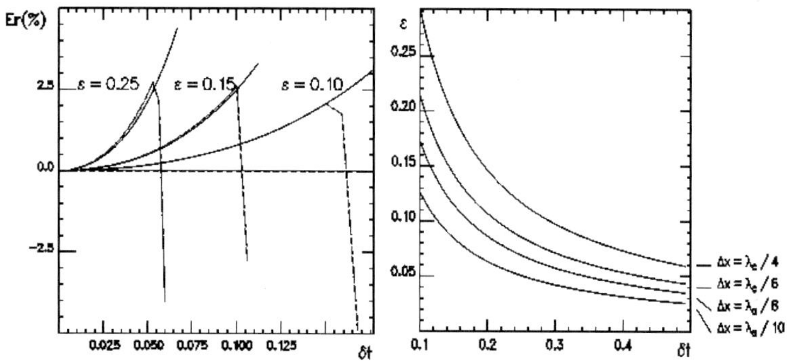

Figure 13. Left: numerical error in per cent of the two eigenmodes (solid and dashed) that belong to the two largest eigenvalues as a function of time step for several ε. Right: the numerical code for the 3D integration of Navier Stokes and temperature equation is stable under the hyperboles that depend on δx but that are more or less independent on δz.

step sizes $\Delta z, \Delta x$, and δt (fig.13, right frame). Up to rather large values of ε, we found the restriction

$$\delta t < \frac{C(\Delta x)}{\varepsilon}$$

where C increases monotonically with Δx but is more or less independent on Δz. Some values for C and Δx are $C = 0.030$, $\Delta x = \lambda_c/4$, $C = 0.021$, $\Delta x = \lambda_c/6$, $C = 0.017$, $\Delta x = \lambda_c/8$, and $C = 0.013$, $\Delta x = \lambda_c/10$ with $\lambda_c = 2\pi/k_c$. Finally, we note that the restriction on time step to ensure numerical stability is much weaker then that for a fully explicit scheme. This justifies the larger numerical effort of the semi implicit method.

Accuracy The single step forward method yields an accuracy of order δt in time. The remaining relative truncation error is of order $(\delta t)^2$ and is determined near the above calculated fixed point by:

$$E_{\rm err}(\delta t) = 2 \max_i \left[\Lambda_i^2(\delta t/2) - \Lambda_i(\delta t)\right] \tag{41}$$

where $\Lambda_i(\delta t)$ denote the eigenvalues of $\underline{Q}$ for given δt. Fig. 13, left frame shows values for the truncation error in per cents for several values of ε as function of the time step. Due to the increase of the spectral radius of $\underline{Q}$ when the product $\varepsilon\delta t$ approaches the region where the numerical method gets unstable, the truncation error changes sign and increases strongly in magnitude. From fig. 13, the relation

$$E_{\rm err}(\delta t) \approx \text{const} \cdot (\varepsilon\delta t)^2$$

with const ≈ 75 for small $\varepsilon\delta t$ can be extracted.

2.4 Results

Closed upper surface If the surface is closed by a good conductor, rolls are the preferred structure close to onset, as already shown in fig. 7. Depending on Pr, these rolls are either parallel (with some possible defects and grain boundaries) or have the tendency to form spirals. The bending and formation of spirals is caused by the so-called *mean flow*, this is the horizontal large scale motion of the fluid which is represented by the stream function Φ, compare fig. 14. For large Pr and close to onset the stream function can be neglected. We complete this study by presenting a time dependent, weakly turbulent series for small Pr relatively far above threshold in fig.15.

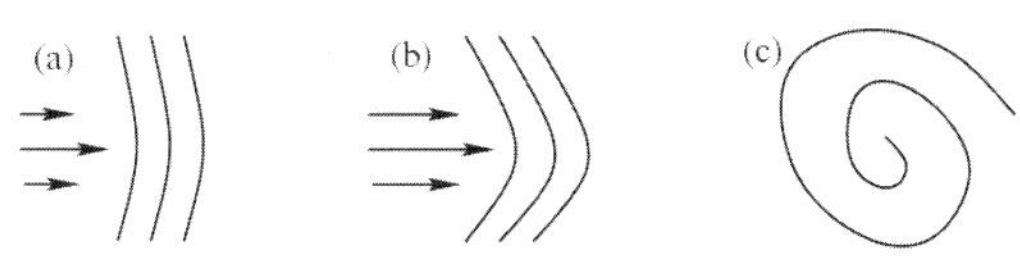

Figure 14. Slightly bended rolls produce a mean flow (a) which further increases the deformation of the rolls (b). This kind of instability may lead to a break up of the rolls and the formation of spirals (c).

For bad conductors as vertical boundaries, squares are also possible, we refer to the literature (Busse and Riahi, 1980).

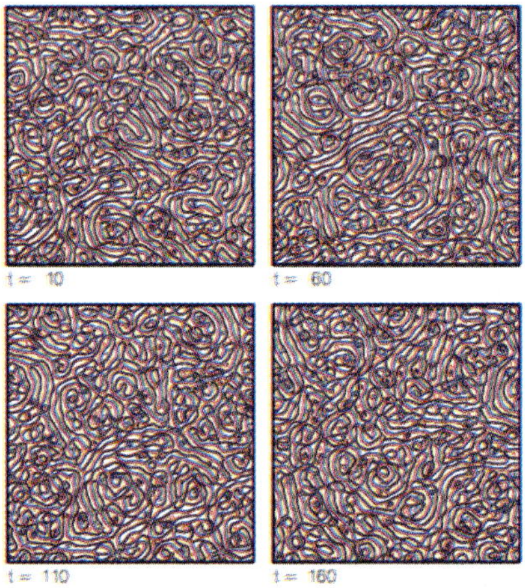

Figure 15. Spirals become again unstable and a turbulent structure is found for small *Pr* and larger Rayleigh number $Pr = 0.2$, $R = 2R_c$. In contrast to fully developed turbulence, a typical length scale is still present.

Free upper surface If the Marangoni effect is the main driving mechanism, hexagons are obtained at threshold as already seen in fig. 5. Defects my travel through the structure and survive for a long time, even in high *Pr* fluids, compare fig. 16 left. A secondary instability takes place to squares, fig. 6. The threshold of this instability depends on *Pr*, as schematically shown in fig. 17. For even larger values of *Ma*, time dependent structures emerge as shown in fig. 16, right. For very small *Pr*, hexagons with mirror symmetry are stable, where the fluid rises on the walls and sinks down in the center. In contrast to the former ones (called ℓ-hexagons, from liquid), they are sometimes named g-hexagons (from "gas", because they are also found in gases under certain non-Boussinesq conditions). The notions "up" and "down" hexagons are also in use. At a certain critical $Pr \approx 0.27$, hexagons are not found at all but rolls turn out to be stable again.

3 Thick films with undeformable surface – binary mixtures

The instabilities discussed so far were non-oscillating or monotonic, i.e. of Type I_s. Now we wish to extend the system into such a way that the

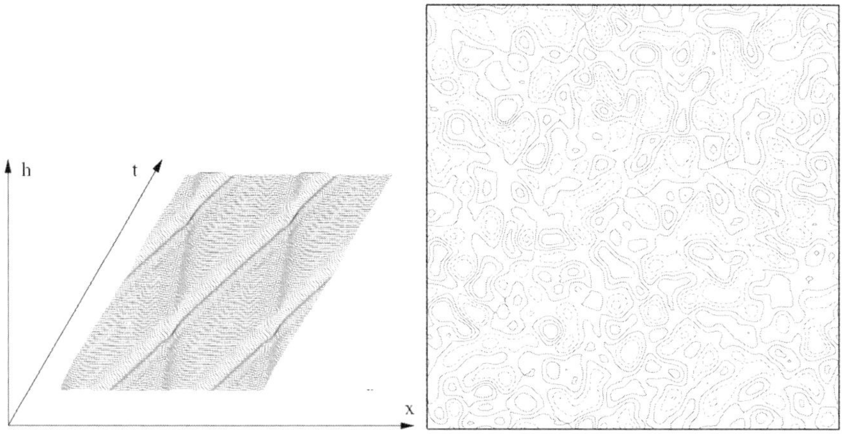

Figure 27. Numerical solutions of the shallow water equations, left in one dimension, right in two. Dashed contour lines mark troughs, solid ones correspond to peaks of the sea.

5.2 Reduced two-dimensional description – viscous fluids

In the last paragraph, viscosity was neglected. Now we turn to the other limit, where viscosity is large (or better: the Reynolds number is small). This is the case in thin films, where the fluid motion is usually very slow. Then one can neglect the inertial part of the Navier-Stokes equations and obtains the *Stokes equation*

$$\eta\,\Delta\vec{v} = \nabla P\ . \tag{98}$$

Again we assume the fluid to be incompressible. Further we restrict us to the case that external forces have a potential. They can be included into the "generalized" pressure

$$P = p + U\ .$$

The lubrication approximation For sake of simplicity we shall restrict the derivation of this section on one spatial dimension, say x. The final result will be given also in 2D (for more details see (Oron et al., 1997)). As above, the aim is to find an equation for the surface location $h(x,t)$ of the liquid. Such an equation is already provided by the kinematic boundary condition

$$\partial_t h = v_z|_h - v_x|_h\,\partial_x h\ . \tag{99}$$

To compute v_z at the surface $z = h$, one integrates the continuity equation with respect to z and finds with the b.c. $v_z(z=0) = 0$

$$\int_0^{h(x,t)} \partial_x v_x \, \mathrm{d}z \; + \; v_z|_h \; = \; 0 \; .$$

Extracting the derivative from the integral yields

$$v_z|_h \; = \; -\partial_x \int_0^{h(x,t)} v_x \, \mathrm{d}z \; + \; v_x|_h \, \partial_x h \; ,$$

and inserted into (99) the desired equation for $h(x,t)$

$$\partial_t h \; = \; -\partial_x \int_0^{h(x,t)} v_x \, \mathrm{d}z \tag{100}$$

or in two horizontal dimensions

$$\partial_t h \; = \; -\nabla_2 \cdot \int_0^{h(x,y,t)} \vec{v}_H \, \mathrm{d}z \tag{101}$$

where $\vec{v}_H = (v_x, v_y)$ denotes the horizontal velocity components.

To close the equation, it is necessary to compute v_x (or $\vec{v}_H$) as a function of h. This can be done solving the Stokes equation. It reads using the scaling (91)

$$\left(\delta^2 \partial^2_{\tilde{x}\tilde{x}} + \partial^2_{\tilde{z}\tilde{z}}\right) \tilde{v}_x = \partial_{\tilde{x}} \tilde{P} \tag{102a}$$

$$\delta^2 \left(\delta^2 \partial^2_{\tilde{x}\tilde{x}} + \partial^2_{\tilde{z}\tilde{z}}\right) \tilde{v}_z = \partial_{\tilde{z}} \tilde{P} \; . \tag{102b}$$

with dimensionless velocity and pressure

$$v_x = \tilde{v}_x \cdot \frac{\ell}{\tau}, \qquad v_z = \tilde{v}_z \cdot \frac{d}{\tau}, \qquad P = \tilde{P} \cdot \frac{\eta}{\delta^2 \tau}$$

(we use now d instead of h_0 for the uniform depth of the layer in rest). In the limit $\delta \to 0$ it follows from (102b)

$$\partial_{\tilde{z}} \tilde{P} = 0 \qquad \text{or} \qquad \tilde{P} = \tilde{P}(\tilde{x}) \; .$$

Thus one can integrate (102a) twice over $\tilde{z}$ and finds with the no-slip condition $\tilde{v}_x(0) = 0$

$$\tilde{v}_x(\tilde{x}, \tilde{z}) = f(\tilde{x}) \cdot \tilde{z} + \frac{1}{2}(\partial_{\tilde{x}} \tilde{P}(\tilde{x})) \cdot \tilde{z}^2 \tag{103}$$

with a function $f(\tilde{x})$ which can be determined by the b.c.. To this end we consider an inhomogeneous surface tension (caused e. g. by a temperature gradient) as in (22b). Inserting (103) there yields

$$f(\tilde{x}) = \partial_{\tilde{x}}\tilde{\Gamma} - (\partial_{\tilde{x}}\tilde{P}) \cdot \tilde{h}$$

with the non-dimensional surface tension

$$\tilde{\Gamma} = \Gamma \, \frac{\tau d}{\eta \ell^2} \ .$$

Inserting everything into (100) and integrating by $\tilde{z}$ finally yields (all tildes omitted)

$$\partial_t h \ = \ -\partial_x \left[-\frac{h^3}{3} \partial_x P + \frac{h^2}{2} \partial_x \Gamma \right] \tag{104}$$

or in two dimensions

$$\partial_t h \ = \ -\nabla_2 \cdot \left[-\frac{h^3}{3} \nabla_2 P + \frac{h^2}{2} \nabla_2 \Gamma \right] \ . \tag{105}$$

This is the basic equation for the evolution of the surface of a thin film in the lubrication approximation. It has the form already presented in (7) and is sometimes called *thin film equation.*

Laplace pressure and gravity To discuss eq (105) further, one has to express both pressure and surface tension as functions of the depth

$$P = P(h) \ , \qquad \Gamma = \Gamma(h)$$

what we shall do now.

To include the influence of gravity, one adds the gravity potential $\rho g z$ to the pressure and evaluates it at the surface $z = h$. In the scaling of (105) the total pressure then reads

$$P = p_0 + G(h-1) \tag{106}$$

with the *gravitation number*

$$G = \frac{d^3 g \tau}{\ell^2 \nu} \ .$$

If the films are very thin, we expect to have a surface structure with a length scale in the range or even well below the capillary length

$$a = \sqrt{\frac{\Gamma}{g\rho}} \ . \tag{107}$$

Then one has to take into account the additional pressure which origins from the curvature of the surface. Therefore we add the *Laplace pressure* to P which reads for weakly curved surfaces

$$-q\,\Delta_2\,h, \qquad q > 0\;. \tag{108}$$

The constant q is linked to the so-called capillary number C and follows from scaling as

$$q = \Gamma\,\frac{\tau d^3}{\ell^4 \eta} = C^{-1}\,\delta^2\;.$$

Thus we have to substitute for $P(h)$ in (105) the expression

$$P = p_0 + G(h-1) - q\,\Delta_2\,h\;. \tag{109}$$

Assuming for the time being that the surface tension Γ is constant along the surface, (105) then reads

$$\partial_t h \;=\; \nabla_2 \cdot \left[\frac{h^3}{3}\nabla_2\left(Gh - q\,\Delta_2\,h\right)\right]\;. \tag{110}$$

The stability of a flat surface with thickness $h = 1$ is examined be introducing small deviations

$$h(x,y,t) = 1 + \eta(x,y,t)$$

and linearizing with respect to η:

$$\partial_t \eta = \frac{1}{3}\left(G\,\Delta_2\,\eta - q\,\Delta_2^2\,\eta\right)\;. \tag{111}$$

Assuming plane waves with exponential behavior in time

$$\eta \sim \mathrm{e}^{\lambda t + ikx}$$

one finds the for the growth rate the dispersion relation

$$\lambda = -\frac{1}{3}\left(Gk^2 + qk^4\right)\;.$$

Obviously, λ is always real and for positive G and q and finite k always less than zero. Thus the flat film is always stable, both gravity and Laplace pressure act stabilizing.

However, this can be changed if the system is put "upside down", leading to negative G. In the experiments this can be realized by fixing a thin film

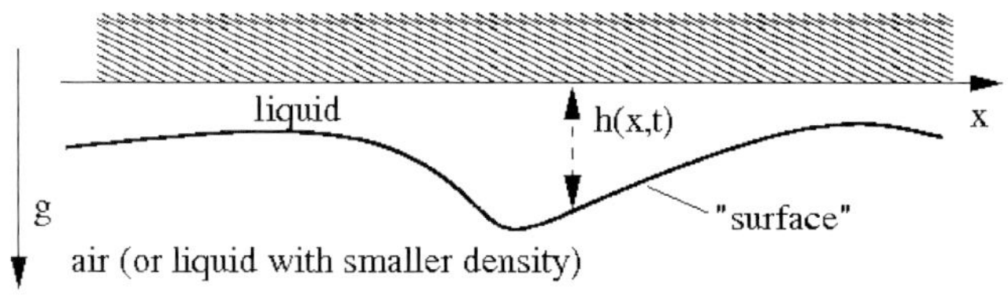

Figure 28. Sketch of a thin liquid film fixed under a solid plane. The plane surface is unstable due to gravity forces. Fingers are formed and finally the film ruptures.

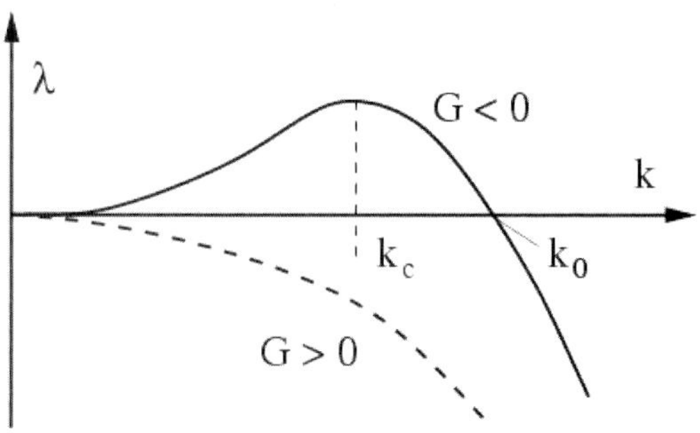

Figure 29. Growth rates of periodic disturbances of the plane surface with wave number k. The situation shown in fig. 28 corresponds to the solid line $G < 0$. Waves having a wave number $k < k_0$ grow exponentially, the mode with $k = k_c$ has the largest growth rate (most dangerous mode). For $G > 0$, the flat film is always stable.

on the underside of a horizontal plane, fig. 28. Gravity is now destabilizing for all wave vectors which have a length less than the cut-off

$$k_0 = \sqrt{-\frac{G}{q}} ,$$

a value that can be expressed by the already in (107) defined capillary length a if one scales back to dimensional values:

$$k_0 = \sqrt{\frac{g\rho}{\gamma}} = \frac{1}{a} .$$

The numerical solution of the full equation (110) is not very instructive. In the first phase periodic instabilities with $k \approx k_c$ occur and grow until h

takes negative (!) values. Physically this corresponds to a rupture of the film beyond which the thin film equation cannot be used.

The disjoining pressure and ultra-thin films Another instability mechanism is encountered in very thin (ultra-thin) films where the thickness is some 100 nm or even less. Then, van der Waals forces between free surface and solid substrate cannot longer be neglected (Israelachvili, 1992).

The van der Waals force depends on the distance of surface and support which is the depth h. One can show that it is proportional to h^{-4}. It can be taken into account by adding to the generalized pressure (109) the interaction potential

$$\Phi(h) = \frac{A_H}{h^3} ,$$

where A_H is the *Hamaker constant.* The function Φ is called *disjoining pressure.* If we neglect gravitation for the moment (it plays no role in ultra-thin films) the expression for the total pressure reads

$$P = p_0 + \frac{A_H}{h^3} - q\,\Delta_2\,h\ . \tag{112}$$

If $A_H > 0$, the pressure increases with decreasing depth and an instability may occur as already discussed in sect 1.2, see fig. 4. Mathematically this corresponds to the condition (prime means derivative with respect to h)

$$P' = \Phi' < 0\ ,$$

which is the case if $A_H > 0$. The flat film is for all initial depths unstable if the Hamaker constant is positive. As in the case of the Rayleigh-Taylor instability the dynamics start with the formation of periodic surface structures followed by rupture.

On the other hand, there can also exist a repelling force between surface and substrate which is modelled taking $A_H < 0$ and stabilizes the flat surface. Normally, attractive and repelling forces have different ranges. Usually, the repelling force is short range, the attractive long range. Then, the initially "thick" film can be unstable due to attraction but rupture can be avoided by repulsion. In this way completely dry regions cannot exist but the substrate remains always covered by an extremely thin film (some nm), called *precursor film.*

The complete expression for such a attractive/repulsive disjoining pressure may read

$$\Phi(h) = \frac{A_n}{h^n} - \frac{A_m}{h^m} , \qquad m > n$$

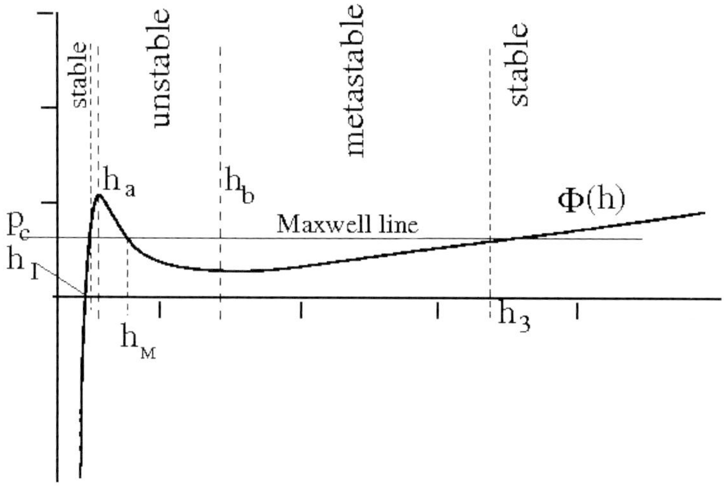

Figure 30. The disjoining pressure including gravitation, $A_3 = 3$, $A_9 = 1$, $G = 0.1$. The region of unstable films is bounded by h_a and h_b. The critical pressure (depth) P_c (h_M) where drops turn into holes is determined by a Maxwell construction (see section 6).

with the two positive Hamaker constants A_n and A_m. Different models were discussed in literature in detail. We shall restrict us here to the "Lennard-Jones potential" which results for $n = 3$ and $m = 9$. Examining the derivative of Φ shows that layers which have a depth bigger than

$$h_a = \left(\frac{3A_9}{A_3}\right)^{1/6}$$

are always unstable. Especially arbitrary thick films are unstable in this model. This is of course not realistic and has its reason in the neglect of the gravitational force in (112). If this is also included, one finds instead of (112)

$$P = p_0 + \frac{A_3}{h^3} - \frac{A_9}{h^9} + G(h-1) - q\,\Delta_2\,h\,. \tag{113}$$

The lower and upper stability bounds h_a, h_b for the depth are then solutions of the polynomial

$$Gh^{10} - 3A_3h^6 + 9A_9 = 0$$

or can be taken graphically from fig. 30.

6 Spinodal dewetting

If a thin liquid film is exposed to a non- or partially wetting substrate, a small perturbation is sufficient to destabilize the flat surface. The fluid then bubbles and many small drops are formed. This phenomenon can be seen for instance if rain falls on a waxed cloth or on a well polished car roof. Such a process is called *spinodal dewetting* and can be described in the frame of eq. (105) if the disjoining pressure has a negative slope for a certain region of the initial depth h (Bestehorn and Neuffer, 2001).

6.1 Ultra-thin isothermal films

For the following study we take a attracting/repelling disjoining pressure discussed in the last section (eq. (113))

$$\Phi(h) = \frac{A_3}{h^3} - \frac{A_9}{h^9} + Gh \ .$$

This inserted into (105) yields the thin film equation (Γ =const):

$$\partial_t h \ = \ -\nabla_2 \cdot \left[-\frac{h^3}{3} \nabla_2 \left(\Phi(h) - \Delta_2 h \right) \right] \ . \tag{114}$$

Note that we put $q = 1$. This is possible by a suitable choice of τ and δ. We shall come back to this issue in more detail below.

Normal form Instead of further examining (114) we first wish to convert it to a more convenient form, showing the canonical form of a Type II$_s$ instability. It will have also the advantage to be more easy transformable into a numerical scheme.

To this end we introduce a variable that describes deviations from the average reference (initial) depth h_0 (a conserved quantity):

$$\eta(x, y, t) = h(x, y, t) - h_0 \ .$$

Inserting this into (114) one can separate on the right hand side linear and nonlinear expressions in η:

$$\partial_t \eta = \frac{1}{3} h_0^3 D(h_0) \Delta_2 \eta - \frac{1}{3} h_0^3 \Delta_2^2 \eta - \nabla_2 \cdot \vec{j}_{NL}(\eta) \tag{115}$$

with the nonlinear current

$$\vec{j}_{NL}(\eta) = -\frac{1}{3} \left(q_2 \nabla_2 \eta - q_1 \Delta_2 \nabla_2 \eta \right)$$

and the abbreviations

$$D(h) = d_h \Phi, \quad q_1 = 3h_0^2\eta + 3h_0\eta^2 + \eta^3, \quad q_2 = (h_0+\eta)^3 D(h_0+\eta) - h_0^3 D(h_0)\,.$$

Both functions q_1 and q_2 vanish as $\sim \eta$. From (115) it is evident how the linearized problem near h_0 looks. One has just to drop the nonlinear expression $\vec{j}_{NL}$ and obtains a linear equation of the form (111). The same ansatz used there yields the dispersion relation

$$\lambda = \frac{1}{3} h_0^3 \left(-D(h_0)k^2 - k^4\right)\,. \tag{116}$$

An instability occurs if the "diffusion coefficient" D is less than zero, the dispersion relation then has the form of the solid line in fig. 29 and is exactly that of a type II_s instability (compare. fig. 9, right frame). This of course coincides with the reasoning of section 5.2. An instability occurs, where the slope of the pressure is negative. Numerically one finds from

$$D(h_0) = 0$$

the two limits (for the special choice of $A_3 = 3,\ A_9 = 1,\ G = 0.1$)

$$h_0^{(1)} = h_a = 1.002, \qquad h_0^{(2)} = h_b = 3.08\,.$$

Numerical solutions In fig. 31, we present solutions of the fully nonlinear eq. (115) for the parameters of fig. 30 and several initial depths h_0. As initial condition a random distribution around the average depth h_0 was chosen.

In the early stage of the evolution, structures having a length scale of the critical wave length $\Lambda = 2\pi/k_c$, occur, where k_c is the wave number of the fastest growing mode

$$k_c = \sqrt{-\frac{D}{2}}\,.$$

This can be called "linear phase" since the amplitudes are still small and nonlinearities play no important role. The structure grows on the typical time scale

$$\tau = \lambda^{-1}(k_c) = \frac{12}{h_0^3 D^2} = \frac{12}{h_0^3}\left(\Phi'(h_0)\right)^{-2}\,,$$

which is inverse to the square of the slope of the disjoining pressure. This is the reason why pattern formation in thicker films takes much longer (right column in fig. 31). As a consequence, the small scale (linear phase) structures are overlayed by holes created by certain seeds (see also next

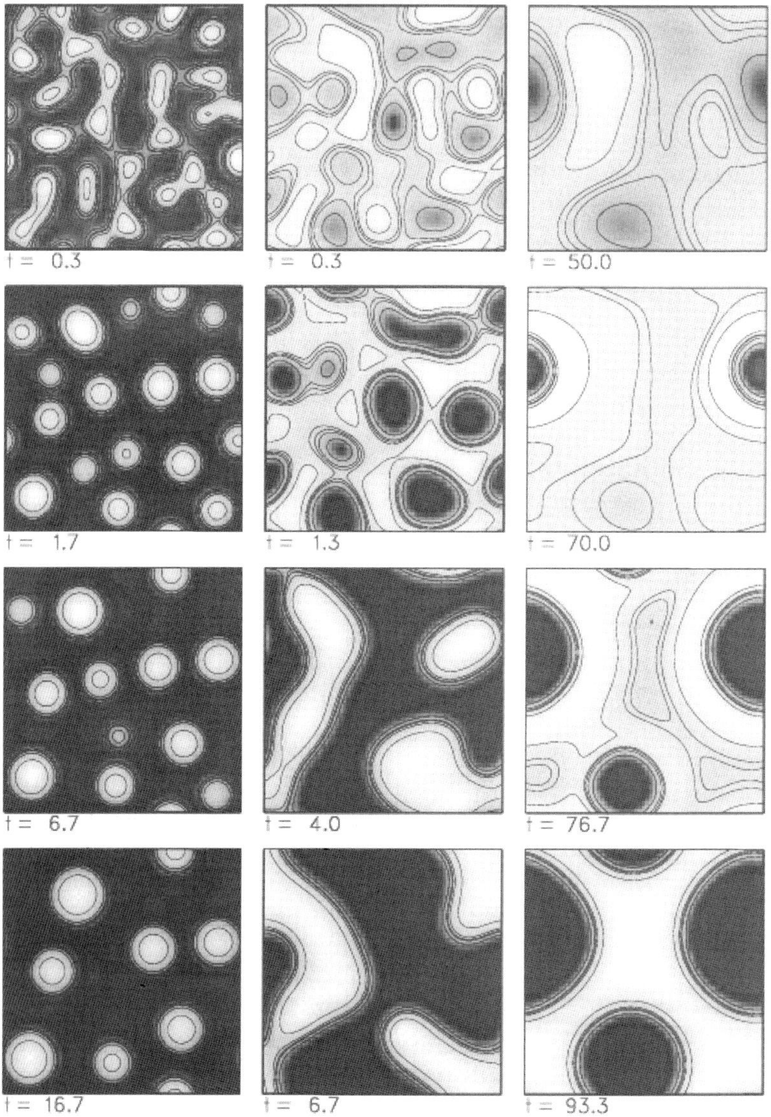

Figure 31. Time series found by numerical integration of (115) for $h_0 = 1.2$ (left column), 1.862 (middle), and 2.8 (right). Light areas correspond to elevated regions of the surface.

paragraph). After the linear phase, the position of h_0 with respect to the Maxwell point shown in fig. 30 is decisive. If $h_0 > h_M$, holes are formed, for $h_0 < h_M$, one finds drops. If $h_0 \approx h_M$, maze-like patterns are obtained in form of bended, rather irregular stripes, see fig. 31, middle column. In a last, strongly nonlinear, phase, so-called *coarsening* is observed. This is a slow increase of the length scale of the structures (holes, drops, or mazes), accompanied by fusion of smaller objects to larger ones. The final structure (long time limit) is often a single entity, one big drop or hole, which is finally time stable. The whole spatio-temporal evolution is transient and can be formulated as a gradient dynamics. The potential plays the role of a generalized free energy reaching its minimum in the steady end state.

The critical pressure as well as h_M can be found by a Maxwell construction in fig. 30. One has

$$P_c \cdot (h_3 - h_1) = \int_{h_1}^{h_3} \Phi(h)\, \mathrm{d}h \ ,$$

with h_1 (h_3) as left (right) intersection of P with the Maxwell line:

$$\Phi(h_1) = \Phi(h_3) = P_c \ .$$

From these three conditions one can determine P_c and the three intersection points by numerical iteration:

$$P_c = 0.647, \quad h_1 = 0.850, \quad h_M = 1.862, \quad h_3 = 6.56 \ .$$

Metastable region and nucleation

The flat film is unstable with respect to infinitesimal disturbances if h_0 is in the region between h_a and h_b. On the other hand, two metastable domains exist, where the flat film is stable, although the free energy could be lowered by pattern formation. Then, a finite disturbance is necessary, which can be caused by seeds coming for instance from impurities. Such a process is called *nucleation* and can be seen in the right column of fig. 31. There, the seeds were provided by the random dot initial conditions and two holes are formed. Both processes (nucleation and wetting) concur in this region and it is a question of time scales which one emerges first. In experiments, the formation of holes by nucleation is seen quite often. The reason is that the metastable hole region is much larger compared to that of drops, see fig. 30.

Physical values

To get an idea of the spatial and temporal scales of the experiments, one has to rescale to dimensional variables. For the Hamaker constant A_3 this means

$$\tilde{A}_3 = \frac{\delta^2 \tau}{\eta d^3} A_3 \ . \tag{117}$$

where the variables bearing a tilde are non-dimensional (in our simulations e.g. $\tilde{A}_3 = 3$). Since we chose $q = 1$, one can determine δ and τ from

$$q = \frac{\tau d^3}{\ell^4 \eta}\gamma = \frac{\delta^4 \tau}{\eta d}\gamma = 1$$

together with (117):

$$\delta = \frac{1}{d}\left(\frac{A_3}{\gamma \tilde{A}_3}\right)^{1/2} , \qquad \tau = \left(\frac{\tilde{A}_3}{A_3}\right)^2 \eta \gamma d^5 \ .$$

Knowing δ, the horizontal scale ℓ can be specified:

$$\ell = d/\delta = d^2 \left(\frac{\gamma \tilde{A}_3}{A_3}\right)^{1/2} \ .$$

To compute the scaling, we need values for the Hamaker constant A_3 and for the depth d. From literature we take

$$A_3 = \frac{0.5}{6\pi} \cdot 10^{-20} J \ .$$

For the depth of the precursor film h_a this gives about 5.5 nm. Then the depths (water as a working substance) of the runs of fig. 31 correspond to

$$d \approx 7 \text{ nm} \quad \text{(left)}, \qquad d \approx 10 \text{ nm} \quad \text{(middle)}, \qquad d \approx 16 \text{ nm} \quad \text{(right)} \ .$$

For the time scales

$$\tau \approx 0.15 \text{ s} \quad \text{(left)}, \qquad \tau \approx 1 \text{ s} \quad \text{(middle)}, \qquad \tau \approx 10 \text{ s} \quad \text{(right)} \ .$$

This means that the evolution shown in the left column takes about two seconds, the middle one seven seconds and on the right hand side 15 minutes. Finally we mention the horizontal scales:

$$\ell \approx 1.4 \ \mu\text{m} \quad \text{(left)}, \qquad \ell \approx 3.2 \ \mu\text{m} \quad \text{(middle)}, \qquad \ell \approx 7.5 \ \mu\text{m} \quad \text{(right)} \ ,$$

corresponding to the horizontal dimensions of the shown areas (64x64 mesh points with distance $\Delta x = 0.5$):

$$\Gamma \approx 45 \ \mu\text{m} \quad \text{(left)}, \qquad \Gamma \approx 100 \ \mu\text{m} \quad \text{(middle)}, \qquad \Gamma \approx 240 \ \mu\text{m} \quad \text{(right)} \ .$$

6.2 Externally heated thin films

In the last section we saw that for extremely thin films of thickness in the range of some 100 nm the effective molecular interaction between the free surface and the solid support, phenomenologically described by the disjoining pressure, provides an instability mechanism. Another one is present if the fluid is heated from below. Then the Marangoni effect can be responsible for a deformation instability, as already mentioned in section 1, see fig. 2 (Nepomnyashchy et al., 2002).

The purpose of this section is to examine a fluid film in three dimensions under a vertical temperature gradient including a stabilizing effect which is obtained by introducing a disjoining pressure which becomes effective for very small film thickness. Instead of rupture and completely dry domains the solid support is always covered by a thin precursor film of thickness of some 10 nm. This allows to study the long-time evolution of a film unstable due to the Marangoni effect.

In addition we shall analyze the spatio-temporal evolution of the film surface when a horizontal force is applied externally. This situation is found when the fluid layer is inclined with respect to the vertical. We discuss the formation of periodic structures perpendicular to the slope as well as the instability of fronts moving downwards the inclined plane. More details can be found in (Bestehorn et al., 2003).

Thin film equation and parameters Tangential surface stresses come into play if the surface tension Γ is assumed to be proportional to the surface temperature T_s (compare (24)). To determine the surface temperature, one has to solve the heat equation (1c), which reads in lowest order of the lubrication approximation simply

$$\partial^2_{zz}T(z) = 0 \ .$$

This is solved by a linear profile subject to the boundary conditions

$$T(0) = T_0, \qquad \partial_z T(z)|_h = -Bi \cdot T(h)$$

with the Biot number Bi. The solution reads

$$T(z) = T_0 - \frac{Bi \cdot T_0}{1 + Bi \cdot h} z \ . \tag{118}$$

To connect with section 2 we introduce the temperature gradient of the conducting, here undeformed ($h = 1$) state according to ($T_1 = T(1)$)

$$\beta = \frac{T_1 - T_0}{d} = -\frac{Bi \cdot T_0}{d \cdot (1 + Bi)}$$

and can express (118) at the surface as

$$T(h) = T_s = T_0 + \frac{\beta \cdot d \cdot (1 + Bi)}{1 + Bi \cdot h} h \ . \tag{119}$$

Inserting this in (24) and then in (105) results in the thin film equation (including surface tension and gravity) (Burelbach et al., 1988)

$$\partial_t h = \nabla_2 \cdot \left\{ \left[h^3 (B_o - \frac{\mathrm{d}\Pi}{\mathrm{d}h} - \Delta_2) - \frac{3}{2} h^2 \frac{C \cdot Ma \cdot (Bi + 1)}{(1 + Bi \cdot h)^2} \right] \nabla_2 h \right\} \tag{120}$$

with the additional numbers

$$C = \frac{\eta \kappa}{\Gamma d}, \quad B_o = \frac{g \rho d^2}{\Gamma}$$

as Crispation number (surface tension) and static Bond number (gravity), respectively. Note that (120) is in a different scaling than (105). Now, h and x, y are in units of d, time scales with $3\rho\nu d/\Gamma$.

In Eq. (120) the effective molecular interaction between substrate and film surface is taken into account by the disjoining pressure $\Pi(h)$. It will be discussed in detail below. A linear stability analysis of the flat film without disjoining pressure ($\Pi = 0$) shows that it gets unstable if Ma exceeds a certain critical value

$$Ma_c = \frac{2}{3} \frac{B_o}{C} (1 + Bi) = \frac{2}{3} G \cdot (1 + Bi) \tag{121}$$

where $G = B_o/C$ is the *Galileo number.*

If Ma exceeds Ma_c, a type II_s instability occurs, the dispersion relation has the same form than that of fig. 29. On the linear (short-time) regime one expects patterns with the typical length scale

$$\ell = 2\pi / \sqrt{\varepsilon B_o} \tag{122}$$

where ε is defined as above as the reduced distance from threshold

$$\varepsilon = (Ma - Ma_c)/Ma_c \ . \tag{123}$$

Previous work by Oron (2000) shows that above the onset of the instability holes are formed and film rupture occurs after a finite evolution time.

The disjoining pressure Now we extend the evolution equation by a disjoining pressure that is of repelling character. As above, it stabilizes a very thin film and avoids film rupture that otherwise would restrict the simulations to investigations of the short-time behavior only. With this extended model we are able to study pattern formation in three dimensions in the long-time limit.

Again we take a disjoining pressure of the van der Waals form

$$\Pi(h) = \frac{A}{h^3} \tag{124}$$

where $A > 0$. In our scaling, one finds the relation

$$A = \frac{A_h}{6\pi\gamma_0 d^2} \tag{125}$$

where A_h is a (dimensional) Hamaker constant.

With (124), the range of unstable flat films is also bounded from below, i.e. flat films with the height $h = h_0$ are unstable if

$$h_c^b \leq h_0 \leq h_c^t \,. \tag{126}$$

The two values h_c^b and h_c^t are named *spinodals* and the same as h_a and h_b of section 6.1. For thick films in the millimeter range, one finds the scaling $h_c^b \propto d^{-1/3}$. The values of h_c^b are very small in the range of $10^{-7}...10^{-8}$ m (see Table 1). Approximately flat parts of the film profile which have a thickness in the stable region below h_c^b can be considered as the precursor film.

We note that the inclusion of the disjoining pressure changes the value for Ma_c to

$$Ma_c = \frac{2}{3}\frac{B_o}{C} + 2\frac{A}{C} \,. \tag{127}$$

Here we approximated $(1 + Bi)$ by 1, which is good for small Biot number, i.e. for a thermally almost insulating upper boundary. We shall use this approximation for the rest of this section.

Fluid parameters To compare with the experiments we list in Table 1 the values of the non-dimensional parameters introduced above as well as those of some important properties of the fluid film. To demonstrate also the dependence on the film thickness, we choose the four cases water and silicone oil, each with layer thicknesses of 1 mm and 1 μ. For the Hamaker constant we took the value $A_h = 10^{-20} J$ from the literature. The typical length scale $\ell \cdot d$ is computed from (122) for $\varepsilon = 1$. It is remarkable that for thin films the critical Marangoni number as well as the applied temperature gradients seem to be very small. This means in real experiments one would usually exceed the critical point by a factor of some thousands. However, in a film of 1 mm depth, the value of M_c in silicone oil is already much larger than that for small scale convection (about 80..100). Consequently to obtain a pure surface instability without small convective cells (hexagons) the liquid depth should be below 1 mm. We note that in the case of a fluid depth where both instabilities may occur, pattern formation is expected to be much more involved and the small scale structure cannot longer be eliminated using the lubrication approximation. In this case the full system including Navier-Stokes as well as heat equation has to be considered. Experiments in this codimension-two region have been performed by Van Hook et al. (1997).

Table 1. Values of the parameters for water and silicone oil for two different depth of the film.

	water $60^\circ\,C$ $\rho = 983Kg/m^3$, $\nu = 4.7\cdot10^{-7}m^2/s$, $\gamma_0 = 6.6\cdot10^{-2}N/m$, $\gamma = 1.74\cdot10^{-4}N/mK$ $\kappa = 1.6\cdot10^{-7}m^2/s$		silicone oil $50cS$, $25^\circ\,C$ $\rho = 960Kg/m^3$, $\nu = 5\cdot10^{-5}m^2/s$, $\gamma_0 = 2.08\cdot10^{-2}N/m$, $\gamma = 6.8\cdot10^{-5}N/mK$ $\kappa = 10^{-7}m^2/s$	
	$d = 10^{-3}m$	$d = 10^{-6}m$	$d = 10^{-3}m$	$d = 10^{-6}m$
ΔT_c	$37^\circ\,C$	$4.3\cdot10^{-5}\,^\circ C$	$92^\circ\,C$	$1.07\cdot10^{-4}\,^\circ C$
M_c	86970	$1.01\cdot10^{-4}$	1308	$1.53\cdot10^{-6}$
B_o	0.146	$1.46\cdot10^{-7}$	0.453	$4.53\cdot10^{-7}$
C	$1.12\cdot10^{-6}$	$1.12\cdot10^{-3}$	$2.31\cdot10^{-4}$	0.23
A	$8.0\cdot10^{-15}$	$8.0\cdot10^{-9}$	$2.6\cdot10^{-14}$	$2.6\cdot10^{-8}$
A/B_o	$5.5\cdot10^{-14}$	$5.5\cdot10^{-2}$	$5.6\cdot10^{-14}$	$5.6\cdot10^{-2}$
$\ell\cdot d$	$1.6\cdot10^{-2}m$	$1.64\cdot10^{-2}m$	$9.3\cdot10^{-3}m$	$9.3\cdot10^{-3}m$
$h_c^b\cdot d$	$5.5\cdot10^{-8}m$	$7.1\cdot10^{-7}m$	$5.5\cdot10^{-8}m$	$7.3\cdot10^{-7}m$

Holes or drops? For the experiment as well as for the numerical solutions of the next paragraph it is important to know whether drops or holes are formed in the unstable range of the flat film $\varepsilon > 0$. To elaborate this further, we first introduce the potential

$$V[h] = \int \left\{ \frac{1}{2}h^2 + \frac{\alpha}{6}\frac{1}{h^2} - (1+\varepsilon)(1+\alpha)h(\ln h - 1) + \frac{1}{2B_0}(\nabla_2 h)^2 \right\} dxdy \tag{128}$$

where the reduced Hamaker constant α

$$\alpha = \frac{3A}{B_o}$$

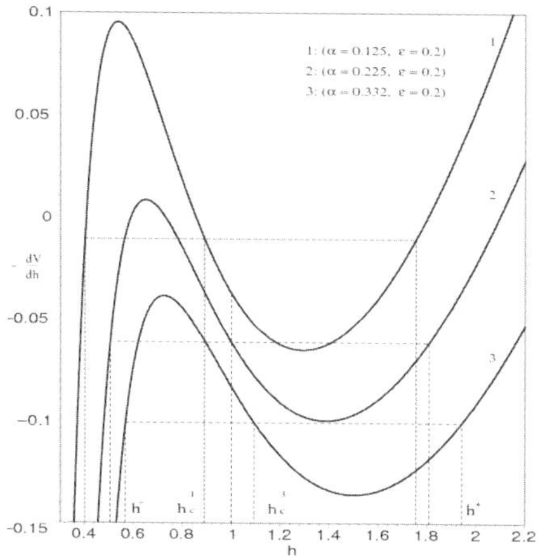

Figure 32. The disjoining pressure $\delta V(h)/\delta h$ according to (128) for uniform h and different values of α and $\varepsilon = 0.2$. The critical film depth h_c is obtained by a Maxwell-equal-area construction for each potential. If $h_c < 1$ (thick film) holes are energetically preferred, otherwise drops will be formed in the nonlinear stage. Between the two extrema (the spinodals) the film is absolutely unstable, between h^- and h^+ (the binodals) it is bistable.

enters as the only parameter that depends on the material. With the help of (128) the disjoining pressure reads

$$P(h) = \frac{\delta V(h)}{\delta h}$$

and the thin film equation (120) can be written in the form

$$\frac{1}{B_0}\partial_t h = \nabla_2 \left[h^3 \nabla_2 \left(\frac{\delta V(h)}{\delta h} \right) \right] , \tag{129}$$

where δ denotes the functional derivative. It can be easily shown that V is monotonically decreasing if h develops as (129). Then, V has the properties of a Lyapunov functional or a free energy and a Maxwell construction in the P-h-diagram can be used to determine the most stable homogeneous state, compare fig. 32.

If we fix the depth of the flat film with $h_0 = 1$ the shape of the potential can depend only on the values of ε and α. Fig. 32 shows plots of the pressure $\mathrm{d}_h V(h)$ for several values of these parameters. If $h_0 = 1$ lies on the left-hand-side of the critical film depth $h = h_c$ (Maxwell point), drops are

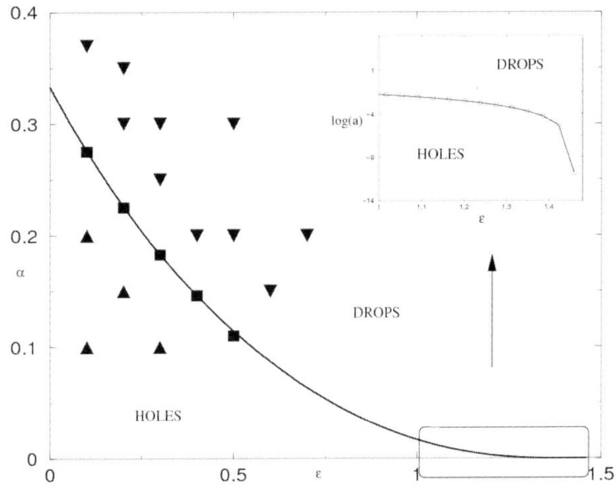

Figure 33. Phase diagram in the parameter plane. The codimension-one line results from the condition $h_c = h_0$ (cf. Fig. 32 and text). Solving the fully time dependent eq. (130) numerically, one obtains mazes (squares), holes (up-triangles) and drops (down-triangles) in excellent agreement with the theory. The inset shows that there is a finite $\varepsilon \approx 1.47$ above which drops mathematically exist even at $\alpha = 0$.

energetically preferred, in the other case holes are expected. The condition that $h_0 = 1$ coincides with the Maxwell point allows to compute α as a function of ε, as done in Fig. 33. From that figure it is clear that for small values of the reduced Hamaker constant holes should be observed. But it is interesting that for $\alpha > 1/3$ drops are the only possible structures. Since $\alpha \propto d^{-4}$ this means that for very thin films always drops evolve, whereas for thicker ones the pattern at onset consists of holes but turns to drops for larger temperature gradients far from threshold. From these calculations one would expect drops even in rather thick films, but further from threshold as a kind of secondary instability.

6.3 Time dependent numerical solutions

Normal Form

As in section 6.1, we first transform eq. (120) into a kind of normal form. Introducing the reduced control parameter ε and after rescaling of space and time variables, it reads

$$\partial_t u = -\varepsilon\Delta u - \Delta^2 u + \nabla \cdot [f(u,\Delta)\nabla u]\,. \tag{130}$$

Here,

$$u(x,y,t) = h(x,y,t) - 1$$

is the normalized, shifted height with vanishing mean value. Space and time are scaled again to (the primes are omitted in (130))

$$x = x'(B_o + 3A)^{-1/2}, \qquad t = t'(B_o + 3A)^{-2}\,.$$

In (130), $f(u,\Delta)$ stands for the operator function

$$\begin{aligned} f(u,\Delta) = \frac{1}{1+\alpha}\left[(1+u)^3 - 1 - \frac{\alpha u}{1+u}\right] \\ - (1+\varepsilon)\left[(1+u)^2 - 1\right] - \left[(1+u)^3 - 1\right]\Delta \end{aligned} \tag{131}$$

which vanishes with u. The linearly fastest growing mode has the wave vector $k_m = \sqrt{\varepsilon/2}$ and grows with the typical rate $\tau = 4/\varepsilon^2$. The nonlinear part (as well as the linear one) has the form of the divergence of a flux and clearly conserves the mean value of u to zero.

Results: the horizontal layer

To show the temporal evolution of the film we present runs for several parameter values in the $\varepsilon - \alpha$ plane. For all runs that follow random dot initial conditions have been used with vanishing mean value. We take a large aspect ratio of 520, in units of eq. (130), what corresponds to about 0.5 m for the silicon oil of Table 1. This of course is a very large value, expressing the huge scale of the structure, which is due to the small supercriticality of $\varepsilon = 0.1$. Starting with the relatively large value of $\alpha = 0.35$ (rather thin film) drops are expected even at onset. This can be seen in the evolution of fig. 34 left, where larger and larger drops are found with increasing time. We note that the evolution times are extremely long, also a consequence of the small supercriticality.

Next we use the smaller value $\alpha = 0.05$, corresponding to a thicker film. Holes are formed now quite early, see fig. 34 right. As can be also seen a small depression is formed along the border of each hole. The long-time behavior can be compared with that of drops. One observes a coarsening of the holes as for the drops. Eventually all holes (drops) merge into a single big hole (drop). However, this process may take a very long time, depending on the several fluid parameters and the geometry of the layer.

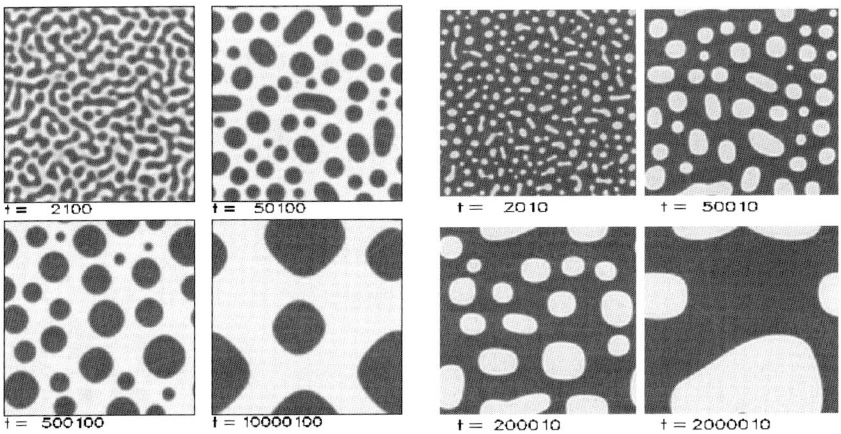

Figure 34. Left: time series found by numerical integration of (130) in the drop regime at $\varepsilon = 0.1$ and $\alpha = 0.35$. The numerical resolution is 256×256 mesh points, the aspect ratio (length to depth) 520. Dark regions correspond to an elevated surface. Periodic boundary conditions are assumed in the lateral directions. Right: time series in the hole regime at $\varepsilon = 0.1$ and $\alpha = 0.05$.

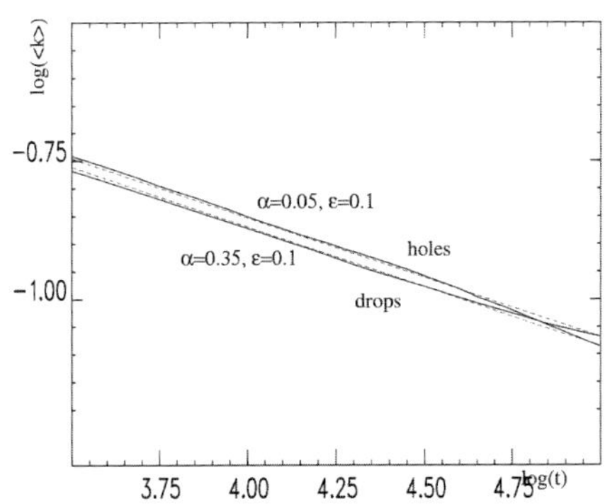

Figure 35. The solid lines show the typical scaling of the mean modulus of the wave vector in the temporal evolution for parameters as in fig. 34. The scaling exponent for both series is the same, $\beta \approx 0.21$ as indicated by the dashed lines. For this computation, we used twice the system size of fig. 34 and a numerical resolution of 512×512 points.

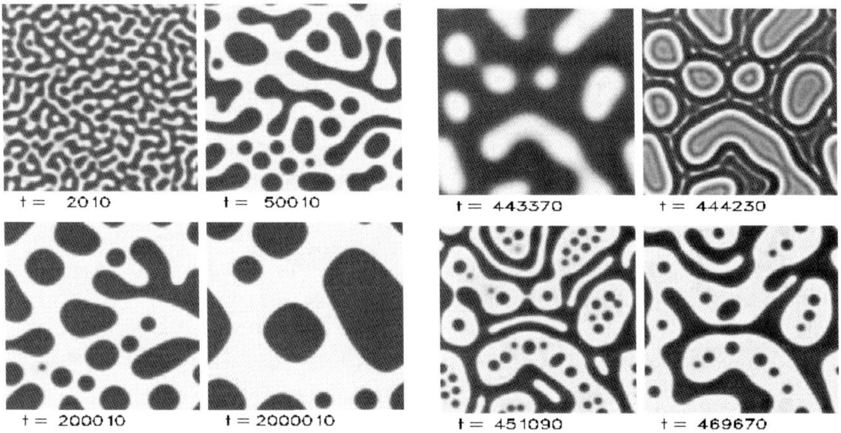

Figure 36. Left: Time series in the maze regime at the codimension-one line of fig.33 ($\varepsilon = 0.1$, $\alpha = 0.28$). Right: Small drops inside big holes are formed if the heating is suddenly increased and ε crosses the phase line of fig. 33. The series was started with $\varepsilon = 0.01$ and $\alpha = 0.3$, After $t = 443370$ we increased ε to $\varepsilon = 0.15$.

To examine the temporal behavior further, we compute the mean modulus of the wave vector in Fourier space according to

$$< k >= \frac{\int d^2\vec{k}\, |u_k|^2 |\vec{k}|}{\int d^2\vec{k}\, |u_k|^2}.$$

From fig. 35, a scaling law of the form

$$< k >= ct^{-\beta} \tag{132}$$

can be clearly extracted. It is remarkable that the exponent for both series, i.e. drops and holes, is almost the same. We found it to be $\beta \approx 0.21$.

Next we present a parameter pair directly on the critical line of fig. 33. As expected, the decision between drops and holes is not clear and a kind of maze structure remains for long times, see fig. 36 left.

In section 6.2 we discussed the possibility of a secondary instability of holes with respect to drops if ε is increased and the codimension-one line of fig. 32 is crossed. To examine this situation further we compute a temporal evolution of a randomly distributed initial pattern with $\varepsilon = 0.01$ and $\alpha = 0.3$ in the hole region. fig. 36 right panel, first frame, shows the formation of holes. After switching ε to a larger value in the drop region small drops are formed rather quickly inside the holes. In this way, drops with a superstructure left from the holes emerge.

6.4 The inclined layer

In this section, we study the influence of a constant external force in a certain horizontal direction. This can be due either to an inclination of the layer by an angle φ, or to an additional horizontal temperature gradient. Here, we shall concentrate on the first case. A constant body force gives an additional term of the form

$$B_o \sin\varphi \left(\vec{n} \cdot \nabla \left(h^3\right)\right)$$

on the right hand side of (120). Here, $\vec{n}$ is the direction of inclination. For the simulation, periodic boundary conditions are kept in all directions.

The inclination even by a very small angle φ completely changes the pattern morphology in the long time limit. This is shown in fig. 37 left, where all parameters are equal to that of fig.34 left, but the plane was inclined by an angle of order $0.1°$ (for a silicone film of thickness 1 mm). At the beginning both evolutions seem to be similar but after $t \approx 30.000$ in the inclined film coarsening is retarded orthogonally to the inclination and finally a certain wavelength is stabilized. The stripes orientate more and more along the direction of inclination, forming a structure of more or less equally spaced pipes or rivulets where the fluid flows down inside.

Finally we study the influence of a constant force perpendicular to a channel of depressed liquid. To this end we start the numerical integration with the initial condition

$$u(x,y,t=0) = \begin{cases} -0.77 & \text{for} \quad 0 \le y \le L/8 \\ 0.11 & \text{for} \quad L/8 < y < L \end{cases}$$

where L is the total length of the layer. Both heights, the elevated as well as the depressed one lay outside the spinodal region, i.e. the flat film in these parts is stable. To avoid the unphysical homogeneity in x-direction of the initial condition we add small fluctuations of about 1 per cent. The layer in this numerical experiment is inclined by about $1°$ (if the fluid is a silicone oil film with depth 1 mm). Then the channel moves with the average velocity of ≈ 1 mm/sec. After about $t = 10.000$ (corresponding to about 300 sec) a phase instability of the back front of the channel (or the leading front of the elevated part) can be clearly observed in fig. 37, right panel. Later on, the opposite front gets unstable with a smaller wavelength. At this stage, both fronts are well separated and can be considered as being independent from each other. This changes if the front instabilities evolve further. The channel

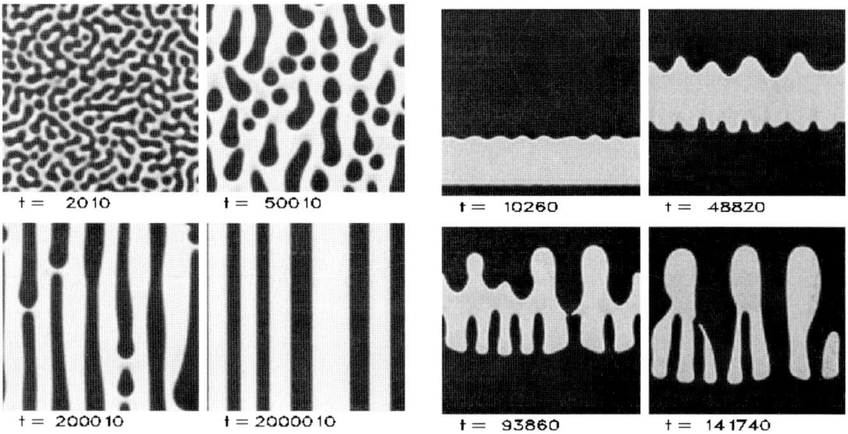

Figure 37. Left: Pattern formation in the inclined film. Parameters are those of fig. 34 left, but the fluid is very slightly inclined in vertical direction. After the linear stage, the pattern gets anisotropic and finally a periodic structure of parallel fluid pipes with a certain wavelength is stabilized. Right: Evolution of a channel on an inclined plane. Parameters are those of fig. 34 right, in the hole regime. Both edges of the channel get unstable on different time scales. The trailing edge is disturbed faster with a larger wavelength. Finally the two edges meet on several points and the channel separates.

gets more and more restricted by the respective advancing and receding fingers and finally breaks in several isolated holes. The final situation after a rather long evolution resembles that of fig. 37, left panel.

Bibliography

M. Bestehorn, *Phys. Rev.* E48:3622, 1993.

M. Bestehorn, *Phys. Rev. Lett.* 76:46-49, 1996.

M. Bestehorn, *Hydrodynamik und Strukturbildung*, Springer Press Berlin, 2006.

M.Bestehorn, P.Colinet (2000), *Physica* D145:84

M. Bestehorn, R. Friedrich, *Phys. Rev.* E59:2642, 1999.

M. Bestehorn, K. Neuffer, *Phys. Rev. Lett.* 87:046101, 2001.

M. Bestehorn, A. Pototsky, U. Thiele, *Eur. Phys. J.* B33:457, 2003.

J. Burelbach, S. Bankoff, S. Davis, *J. Fluid Mech.* 195:463, 1988.

F.H. Busse, N. Riahi, *J. Fluid Mech.* 96:243-256, 1980.

S. Chandrasekhar, *Hydrodynamic and Hydromagnetic Stability*, Dover Press New York, 1981.

A. J. Chorin, J. E. Marsden, *A Mathematical Introduction to Fluid Mechanics*, Springer Press Berlin, 1993.

P. Colinet, J. P. Legros, M. G. Velarde, *Nonlinear Dynamics of Surface-Tension-Driven Instabilities*, Wiley-VCH, 2001.

M. C. Cross, P. C. Hohenberg, *Rev. Mod. Phys.* 65:851, 1993.

K. Eckert, M. Bestehorn, A. Thess, *J. Fluid Mech.* 356:155, 1998.

T. E. Faber, *Fluid Dynamics for Physicists*, Cambridge University Press, 1995.

C. A. J. Fletcher, *Computational Techniques for Fluid Dynamics*, Vol.1,2, Springer Press Berlin, 1991.

H. Haken, *Synergetics: Introduction and Advanced Topics.* Springer Press Berlin, 2004.

C. Hirsch, *Numerical Computation of Internal and External Flows*, Vol.1,2 Wiley, 1988.

J. N. Israelachvili, *Intermolecular and Surface Forces*, Academic Press, 1992.

E. L. Koschmieder (1993), *Bénard Cells and Taylor Vortices*, Cambridge University Press, 1993.

M. P. K. Kundu, *Fluid Mechanics*, Academic Press, 2004.

P. Manneville, *Dissipative Structures and Weak Turbulence*, Academic Press, 1990.

J. D. Murray, *Mathematical Biology I,II*, Springer Press Berlin, 2002.

A. A. Nepomnyashchy, M. G. Velarde, P. Colinet, *Interfacial phenomena and convection*, Chapman & Hall/CRC, Boca Raton, 2002.

A. Oron, *Phys. Fluids* 12:1633, 2000.

A. Oron, S. H. Davis, S. G. Bankoff, *Rev. Mod. Phys.* 69:931, 1997

L. M. Pismen, *Patterns and Interfaces in Dissipative Dynamics*, Springer Press Berlin, 2006.

W. H. Press, B. P. Flannery, S. A. Teukolsky, W. T. Vetterling, *Numerical Recipes*, Cambridge University Press, 1992.

S. J. Van Hook, M. F. Schatz, J. B. Swift, W. D. McCormick, H. L. Swinney, *J. Fluid Mech.* 345:45, 1997.

Thin Film and Droplet Patterns Shaped by Surface Forces

L.M. Pismen

Department of Chemical Engineering and
Minerva Center for Nonlinear Physics of Complex Systems,
Technion – Israel Institute of Technology, 32000 Haifa, Israel

Abstract We consider structure and dynamics of thin films and droplets dominated by interaction with the substrate. Droplet velocities are computed analytically in lubrication approximation using an integral solvability condition. This allows us to describe slow motion and rearrangement of droplet patterns due to interactions mediated by the substrate or precursor layer. A droplet may exhibit self-propelled motion by modifying wetting properties of the substrate, resulting in asymmetry in the contact angles. Long-time dynamics in particular systems may lead, depending on relative time scales of motion and substrate modification, to coarsening, pattern formation or persistent wandering of droplets. Similar techniques are applied to describe coexistence of film domains of different thickness and instabilities of domain boundaries in an evaporating liquid film.

Contents

1 Basic Equations

1.1 Evolution Equation

Thin liquid films are distinguished by the dominant role of interactions with the substrate and of *disjoining pressure* (Derjaguin *et al*, 1987) generated by these interactions. They are commonly described in hydrodynamic lubrication approximation (Oron *et al*, 1997), which corresponds to Derjaguin's approximation of disjoining pressure and implies that the film has a large aspect ratio, so that gradients in "horizontal" (i.e. parallel to the substrate) direction are weaker than those in the "vertical" (normal to the substrate) direction.

The evolution equation of a *non-volatile* film is written in the form

$$\partial_t h = -\nabla \cdot \boldsymbol{j}, \qquad \boldsymbol{j} = -q(h)\nabla P, \tag{1}$$

where h is the layer thickness, $\boldsymbol{j}$ is the volumetric flux, $q(h)$ is the mobility coefficient, and P is pressure in the film expressed as

$$P = -\gamma_0 \nabla^2 h + \Pi(h) + V(h, \boldsymbol{x}). \tag{2}$$

The three terms in this expression describe, in this order, effects of interfacial curvature (γ_0 being the standard surface tension of the bulk fluid), disjoining pressure $\Pi(h)$ (which can be defined as the derivative of the interfacial energy with respect to the layer thickness, $\Pi = \mathrm{d}\gamma/\mathrm{d}h$), and potential $V(\boldsymbol{x})$ of external forces (e.g. gravity).

This form is universal for both sharp interface theories with intermolecular forces and diffuse interface theories with nominally defined h (Pismen, 2002); all that varies is the form of the dependence of the disjoining pressure on thickness $\Pi(h)$ and the effective mobility $q(h)$. In the following, we shall concentrate on evolution driven by surface forces and set $V(\boldsymbol{x}) = 0$; the influence of gravity is considered elsewhere in this volume (Bestehorn, this volume).

1.2 Disjoining Pressure

The dependence of disjoining pressure on the layer thickness follows from an appropriate fluid-substrate interaction model. A weaker dependence on the horizontal coordinate can be added if the substrate is inhomogeneous; its properties may be allowed to vary on a long scale to keep the lubrication approximation applicable. For either totally or partially wetting fluid the curve $\Pi(h)$ is monotonic. In the first case, attractive interaction between the substrate and fluid molecules is stronger than attractive fluid-fluid interaction; in the second, the opposite is true; in both cases the disjoining pressure grows indefinitely by its absolute value as the layer becomes thinner, until a cutoff thickness of the order of molecular size is reached. We shall further call these cases, respectively, "wetting" and "nonwetting", extending the latter term to all cases when the contact angle is finite. All curves $\Pi(h)$ approach zero at $h \to \infty$, as disjoining pressure vanishes in thick layers (practically, beyond a nanometer range) where the interfacial energy approaches its standard bulk value $\gamma_0 = \gamma(\infty)$.

A purely "nonwetting" case may be, however, unreachable theoretically, since fluid-fluid interaction energy is proportional to squared density and is much weaker in the gas than in the liquid phase. Therefore the curve $\Pi(h)$ for a nonwetting liquid should plunge down at $h \to 0$ as in Figure 1 (left). The layer with the thickness h_m corresponding to the intersection point with the abscissa is in equilibrium with a flat thick layer. If this value is larger than the molecular size d, it signals the formation of a mesoscopic *precursor* layer on the substrate; in the case $h_m \leq d$, one can speak of an adsorption layer on the solid surface.

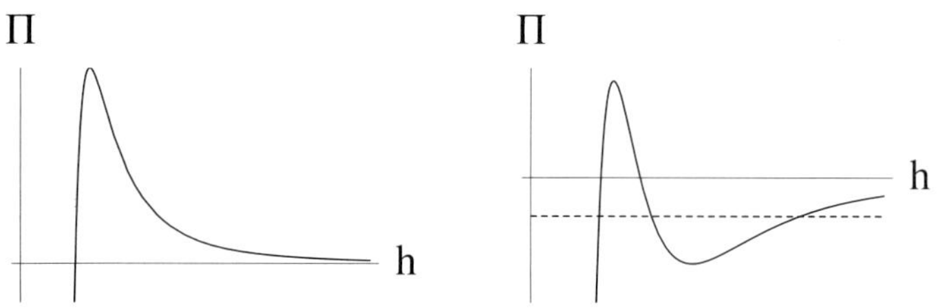

Figure 1. Dependence of disjoining pressure on layer thickness: left: nonwetting liquid with a precursor; right: two equilibrium thicknesses (the dashed line marks the Maxwell construction level).

A curve $\Pi(h)$ with a minimum, as in Figure 1 (right), may result from an interplay of different interactions with disparate decay laws, e.g. van der Waals and polar forces (Sharma, 1993). In this case, coexistence between films with different mesoscopic thicknesses is possible (more on this in Section 5.1).

1.3 Effective Mobility

In the standard case of Stokes flow with no slip, the h-dependent mobility function is

$$q(h) = \tfrac{1}{3}\eta^{-1}h^3, \tag{3}$$

where η is dynamic viscosity. This expression leads to the well-known stress singularity at the contact line and can be used in this unmodified only when a precursor layer exceeding a single molecular layer is formed ahead of the moving contact line. A widely used alternative way to alleviate the singularity is to apply the Navier slip condition on the solid surface (Hocking, 1977), leading to a modified mobility function

$$q(h) = \tfrac{1}{3}\eta^{-1}h^2\,(h + 3\lambda)\,, \tag{4}$$

where λ is the slip length, with values typically comparable with molecular dimensions but dependent on wetting properties of the substrate (as confirmed by experiment and molecular dynamics simulations). This expression relaxes to a quadratic function at $h \to 0$, thereby eliminating the singularity.

Another alternative is to replace viscous flow by activated slip in the first molecular layer adjacent to the solid substrate (Pismen and Rubinstein, 2001). The resulting expression is

$$q(h) = \eta^{-1}\left[\lambda^2 h + \tfrac{1}{3}(h-d)^3\right]. \tag{5}$$

The effective slip length λ is here not just a phenomenological constant as in (4) but is related to physical properties of the liquid:

$$\lambda = \sqrt{\frac{D_l \eta}{n_l T}}, \tag{6}$$

where D_l is self-diffusivity in the liquid, n_l is the liquid number density, and T is temperature measured in energy units. An estimate using an approximate relation between viscosity and diffusivity, $D_l\eta \approx 10^2 T/3\pi d$ and a density estimate $n_l \approx d^{-3}$ yields again values comparable with molecular dimensions. The mobility function becomes linear at $h \to 0$, eliminating the singularity as above.

1.4 Contact Angle

A flat layer of a "nonwetting" liquid is unstable, and tends to break into droplets separated by either "dry" substrate or patches covered by a precursor layer. The three-phase boundary is a *contact line*, and the boundary condition required to find the shape of a macroscopic liquid volume is given by the *contact angle* defined formally by the Young–Laplace formula

$$\frac{\gamma_{gs} - \gamma_{ls}}{\gamma_0} = \cos\theta \approx \frac{\theta^2}{2}, \tag{7}$$

where γ_{gs}, γ_{ls} are, respectively, the energies per surface area of the gas-solid and liquid-solid interfaces. With both disjoining pressure and external forces neglected, the shape of a finite droplet is just a spherical cap, or a parabolic cap of a radius R in lubrication approximation valid at $\theta \ll 1$:

$$h_0(x) = \frac{R\theta}{2}\left[1 - \left(\frac{x}{R}\right)^2\right], \tag{8}$$

where x is the distance from the droplet's center of symmetry. Take note that the contact angle falls out if R, as well as x, is measured on the scale extended by the factor θ^{-1} compared to the "vertical" scale of h.

A function $\Pi(h)$ of the kind shown in Figure 1 (left) contains the information determining the contact angle, which can be computed by finding a stationary solution of Eq. (2) with appropriate asymptotic conditions at $x \to \pm\infty$. Under equilibrium conditions in the absence of external forces, Eq. (2) reduces to

$$\gamma_0 \nabla^2 h - \Pi(h) = 0. \tag{9}$$

For a straight contact line normal to the x axis, it is convenient to use a "phase plane" representation of this equation obtained by taking the layer thickness h as the independent, and the squared slope $y = (h'(x))^2$ as the dependent variable:

$$\frac{\mathrm{d}y}{\mathrm{d}h} = \frac{2}{\gamma_0}\Pi(h), \qquad y(h_m) = 0. \tag{10}$$

Integrating Eq. (10) from h_m to infinity yields the asymptotic equilibrium contact angle

$$\theta = \sqrt{y(\infty)} = \left[\frac{2}{\gamma_0}\int_{h_m}^{\infty} \Pi(h)\mathrm{d}h\right]^{1/2}. \tag{11}$$

The correspondence with the Young–Laplace formula (7) can be seen by using the relation $\Pi = \mathrm{d}\gamma/\mathrm{d}h$ to evaluate the integral explicitly, and observing that $\gamma(h_m)$ equals to the gas-solid surface tension γ_{gs}, while $\gamma(\infty) = \gamma_0 + \gamma_{ls}$ is the sum of the gas-liquid and liquid-solid surface tensions.

For a finite droplet, the contact line can be nominally defined as the locus of the inflection point of the profile $h(x)$. The slope at the inflection point coincides with the "apparent" contact angle one would measure in a macroscopic experiment, which does not resolve a strongly curved transitional region separating the bulk fluid and the precursor layer. The profile in the transitional region can be easily matched to a macroscopic solution defining the equilibrium shape of the droplet bulk, e.g. Eq. (8).

2 Quasistationary Motion

2.1 Perturbation Expansion

We shall now concentrate on the case when the droplet is set into slow motion due to either external forces or long-range interactions. We assume that the deviation from equilibrium shape remains weak and can be treated as a small perturbation everywhere. The droplet mobility can be deduced then from integral conditions based on an equilibrium solution (Pismen and Pomeau, 2004). This allows us to avoid solving dynamic equations explicitly and computing a perturbed shape. We shall consider motion of a "droplet" defined as a compact bulk fluid region with a macroscopic thickness h residing on solid substrate, which is, possibly, covered by an ultrathin "precursor layer" with a thickness $h_m(\boldsymbol{x})$. The droplet is stationary in the absence of external forces when $\Pi(h)$ does not depend on position explicitly or the contact angle is constant around its periphery; slow motion with a velocity $\epsilon \boldsymbol{U} \ll 1$ ensues when the symmetry is weakly broken (e.g., due to substrate inhomogeneities or external forces) or if boundary conditions on the contact line are slightly perturbed.

For a droplet moving without change of volume, the time dependence in Eq. (1) is reduced in the leading order to translation with a velocity $\epsilon \boldsymbol{U}$, as yet unknown. Thus, the time derivative contributes to the expansion of Eq. (1) an inhomogeneity proportional to the translation velocity. This yields the quasistationary equation

$$\epsilon \boldsymbol{U} \cdot \nabla h = \nabla \cdot \left[q(h) \nabla \left(\gamma_0 \nabla^2 h - \Pi(h, \boldsymbol{x}) \right) \right] . \tag{12}$$

Disjoining pressure (or external potential if one is present) can be separated into a symmetric part $\Pi_0(h)$, which has no dynamic effect, and a small asymmetric part $\epsilon \Pi_1(h, \boldsymbol{x})$, which may set the droplet into motion:

$$\Pi(h, \boldsymbol{x}) = \Pi_0(h) + \epsilon \Pi_1(h, \boldsymbol{x}). \tag{13}$$

We are looking for a perturbed solution in the form

$$h(\boldsymbol{x}) = h_0(\boldsymbol{x}) + \epsilon \widetilde{h}(\boldsymbol{x}, t) + \ldots . \tag{14}$$

where $h_0(\boldsymbol{x})$ is a stationary solution verifying Eq. (9), now rewritten as

$$\gamma_0 \nabla^2 h_0 - \Pi_0(h_0) = 0. \tag{15}$$

The first-order equation has a general form

$$\mathcal{L} \widetilde{h} + \Psi(\boldsymbol{x}) = 0, \tag{16}$$

which contains the linear operator

$$\mathcal{L} = -\nabla \cdot \left\{ q(h_0) \nabla \left[\gamma_0 \nabla^2 - \Pi_0'(h_0) \right] \right\} \tag{17}$$

and the inhomogeneity

$$\Psi(\boldsymbol{x}) = \boldsymbol{U} \cdot \nabla h_0 + \nabla \cdot \left[q(h_0) \nabla \Pi_1(h_0) \right]. \tag{18}$$

The operator $\mathcal{L}$ is not self-adjoint. The adjoint operator is

$$\mathcal{L}^\dagger = -\left[\gamma_0 \nabla^2 - \Pi_0'(h_0) \right] \nabla \cdot q(h_0)\, \nabla. \tag{19}$$

The unknown velocity is obtained by computing a solvability condition of Eq. (16); there is no need to actually solve it, unless weak deformation expressed by the first-order function $\widetilde{h}$ is of interest.

The operator $\mathcal{L}$ has a translational Goldstone mode $\varphi(\boldsymbol{x})$ that verifies

$$\nabla \varphi = \boldsymbol{I}\, \frac{h_0 - \widehat{h}}{q(h_0)}, \tag{20}$$

where $\boldsymbol{I}$ is an arbitrarily directed unit vector and $\widehat{h}$ is a constant to be specified later. Taking two vectors $\boldsymbol{I}$ along the two Cartesian axes gives two Goldstone modes corresponding to two translational degrees of freedom in the plane. The identity $\mathcal{L}^\dagger \varphi = 0$ can be checked directly by plugging Eq. (20) in Eq. (19) and comparing the resulting expression,

$$-\left[\gamma_0 \nabla^2 - \Pi_0'(h_0) \right] \boldsymbol{I} \cdot \nabla h_0, \tag{21}$$

with the identical vanishing expression obtained by applying $\boldsymbol{I} \cdot \nabla$ to Eq. (15). In some cases, there is no need in an explicit integral form of φ; otherwise, it can be obtained by solving Eq. (20) and depends on appropriate boundary conditions.

Another Goldstone mode of $\mathcal{L}$ is $\varphi = \text{const}$, which corresponds to volume-changing symmetry and is relevant for spreading and evaporation processes. Since the two Goldstone modes should be mutually orthogonal, φ should be antisymmetric for a symmetric droplet. It can be therefore conveniently computed by setting $\varphi = 0$ on a symmetry axis and integrating Eq. (20) in two opposite directions along $\boldsymbol{I}$.

2.2 Translational Solvability Condition

The solvability condition of Eq. (16) defining the translation speed is given by orthogonality of the inhomogeneity $\Psi(\boldsymbol{x})$ to the translational Goldstone mode $\varphi(\boldsymbol{x})$. Generally, the solvability condition may be applicable in a finite region $\mathcal{R}$ bounded by a contour Γ, and includes both the area integral and contour integrals. The area integral is evaluated using integration by parts:

$$\begin{aligned}\mathcal{I}_{\mathcal{R}} &= \int_{\mathcal{R}} \varphi(\boldsymbol{x})\left\{\boldsymbol{U}\cdot\nabla h_0 + \nabla\cdot[q(h_0)\nabla\Pi_1(h_0)]\right\}\mathrm{d}\boldsymbol{x} \\ &= -U\mathcal{J} + \mathcal{F}_{\mathcal{R}} + \oint_{\Gamma}\varphi(s)\left\{h_0\,\boldsymbol{n}\cdot\boldsymbol{U} + q(h_0)\,\boldsymbol{n}\cdot\nabla\Pi_1(h_0)\right\}\mathrm{d}s.\end{aligned} \tag{22}$$

Here $\boldsymbol{I} = \boldsymbol{U}/U$ is chosen to be directed along $\boldsymbol{U}$ and $U = |\boldsymbol{U}|$, $\boldsymbol{n}$ is the outer normal to Γ, and s is the arc length. The transformed expression contains two area integrals: the *bulk force*

$$\mathcal{F}_{\mathcal{R}} = -\int_{\mathcal{R}}(h_0 - \widehat{h})\,\boldsymbol{I}\cdot\nabla\Pi_1(h_0)\,\mathrm{d}\boldsymbol{x} \tag{23}$$

and the *friction factor*

$$\mathcal{J} = \int_{\mathcal{R}} \frac{h_0(h_0 - \widehat{h})}{q(h_0)}\,\mathrm{d}\boldsymbol{x}. \tag{24}$$

Additional contour integrals dependent on the unknown first-order function $\widetilde{h}$ are contributed by the operator $\mathcal{L}$ when it is multiplied by φ and integrated by parts:

$$\begin{aligned}\mathcal{I}_{\Gamma} &= -\oint_{\Gamma}\varphi(s)\,q(h_0)\,\boldsymbol{n}\cdot\nabla\left[\gamma_0\nabla^2 - \Pi_0'(h_0)\right]\widetilde{h}\,\mathrm{d}s \\ &+ \oint_{\Gamma}(\boldsymbol{n}\cdot\boldsymbol{I})\,(h_0 - \widehat{h})\left[\gamma_0\nabla^2 - \Pi_0'(h_0)\right]\widetilde{h}\,\mathrm{d}s \\ &- \gamma_0\oint_{\Gamma}\left[(\boldsymbol{I}\cdot\nabla h_0)\,\boldsymbol{n}\cdot\nabla\widetilde{h} - (\boldsymbol{n}\cdot\boldsymbol{I})\,\nabla^2 h_0\,\widetilde{h}\right]\mathrm{d}s.\end{aligned} \tag{25}$$

These integrals, combined with the contour integral in Eq. (22), represent the *boundary force*.

The perturbation $\widetilde{h}$ can be expanded in angular harmonics

$$\widetilde{h}(r,\phi) = \sum h_n(r)\mathrm{e}^{\mathrm{i}n\phi}. \tag{26}$$

Harmonic perturbations with $n > 1$ distort the circular shape of the droplet footprint (hence, of the contact line Γ) and can be accommodated by a perturbed droplet with the respective symmetry. The first harmonic (dipole $n = 1$) is, however, an exception, since the respective perturbation of Γ is equivalent to translation without change of form. Therefore it is the dipole asymmetry which is expected to set the droplet into motion without affecting the shape of its footprint (to the first order in ϵ), and it is sufficient to compute only the dipole components of all integrals in Eqs. (22), (25).

2.3 Motion due to Asymmetry of Contact Angles

We consider first a macroscopic droplet on "dry" substrate; then $\mathcal{R}$ is just the droplet footprint, $h_0(\Gamma) = 0$, and the constant $\widehat{h}$ should be set to zero. In the framework of a macroscopic description with fixed contact angles at each location around the droplet circumference, disjoining pressure is excluded, so that, in the absence of external forces, the only non-vanishing area integral is the friction factor $\mathcal{J}$; the contour integral in Eq. (22) also vanishes. The first integral in Eq. (25) vanishes, since at $h_0 \to 0$ $q(h_0) \propto h_0^2$, while φ, as follows from Eq. (20), is only logarithmically divergent. The second integral vanishes at $h_0(\Gamma) = 0$ as well. The third integral expresses the *boundary force* due to the variable part of the contact angle $\widetilde{\theta} = \theta - \theta_0 = -\epsilon^{-1}\boldsymbol{n} \cdot \nabla\widetilde{h}$. The last integral depends on droplet shape distortion when $\widetilde{h}(\Gamma) \neq 0$. We have noticed that the dipole component does not affect the shape of the droplet footprint, besides its displacement. One has to require therefore $h_1(\Gamma) = 0$; hence, this integral vanishes. The remaining boundary force integral is evaluated using the identity $\boldsymbol{I} \cdot \nabla h_0 = -\theta_0 \cos\phi$ as

$$\mathcal{I}_\Gamma = -\gamma_0\theta_0 R \int_{-\pi}^{\pi} \cos\phi\, \widetilde{\theta}(\phi)\, \mathrm{d}\phi = -\gamma_0\theta_0 R \;\mathrm{Re}\; \theta_1 \equiv -\mathcal{F}. \tag{27}$$

The friction factor $\mathcal{J}$ diverges if the classical mobility function (3) is used, which is a consequence of the notorious contact line singularity. This is the reason for introducing the slip length in Eq. (4) or (5). Since, however, this length is very small, being measured on molecular scale, the integral can be evaluated by separating it into two parts. Near the contact line, i.e. in a ring $R \leq x \leq R - l$ where $\lambda \ll l \ll R$, the integration can be carried out using the linearized profile $h_0 = \theta_0(R - r)$. Using Eq. (4), this yields, asymptotically at $l \gg \lambda$,

$$\mathcal{J}_1 = 6\pi R\eta \int_{R-l}^{R} [\theta_0(R-r) + 3\lambda]^{-1}\, \mathrm{d}r \asymp \frac{6\pi R\eta}{\theta_0} \ln \frac{\theta_0 l}{3\lambda}. \tag{28}$$

In the bulk region $r \leq R - l$, λ can be neglected, and the integration yields, asymptotically at $l \ll R$,

$$\mathcal{J}_2 = \frac{6\pi\eta}{R\theta_0} \int_0^{R-l} \left[1 - \left(\frac{r}{R}\right)^2\right]^{-1} r\, \mathrm{d}r \asymp \frac{6\pi R\eta}{\theta_0} \ln \frac{R}{2l}. \tag{29}$$

When both integrals add up, the auxiliary length l falls out, resulting in an expression containing the logarithm of the ratio of the macroscopic and microscopic scales:

$$\mathcal{J} = \frac{6\pi R\eta}{\theta_0} \ln \frac{\theta_0 R}{6\lambda}. \tag{30}$$

Thus, the dipole component of the solvability condition defining the droplet velocity reads

$$U = \frac{\mathcal{F}}{\mathcal{J}} = \frac{\gamma\theta_0^2}{6\pi\eta} \ln^{-1}\left(\frac{\theta_0 R}{6\lambda}\right) \int_{-\pi}^{\pi} \cos\phi\, \widetilde{\theta}(\phi) \mathrm{d}\phi. \tag{31}$$

Other ways of eliminating the singularity, e.g. using Eq. (5), yield similar results differing only by the numerical coefficient under the logarithm. The ratio of the macroscopic to microscopic length under the logarithm is a telltale sign of the resolved contact line singularity.

3 Interactions Mediated by the Precursor

3.1 Moving Droplet on a Precursor Film

For a droplet sitting on a precursor layer, the contour Γ should be chosen as a circle with a radius L sufficiently exceeding that of the droplet. Practically, it is sufficient to include a precursor ring large on a *molecular* scale, since this is the scale of exponential decay of h to its equilibrium value h_m near the contact line, so that L can be considered coincident with the droplet radius on the macroscopic scale. The constant $\widehat{h}$ can be identified with the average thickness of the precursor around Γ:

$$\widehat{h} = \langle h_m \rangle = (2\pi R)^{-1} \oint h_m \,\mathrm{d}s. \tag{32}$$

All contour integrals in Eq. (25) are negligible, since the precursor layer is both very thin and flat. The first term in the contour integral Eq. (22) is negligible as well, and its second term vanishes by symmetry when the principal part of $\nabla\Pi_1$ is directed along the normal. The driving force is determined by the area integral (23). If the motion is due to asymmetry of disjoining pressure caused by substrate inhomogeneity, we expect, however, that, in the case of a macroscopic droplet, it should be possible to approximate this "bulk" driving force by the boundary force (27) due to locally varying contact angles. This can be, indeed, proven in the following way. Expanding Eq. (11)

$$\theta = \theta_0 + \epsilon\widetilde{\theta} = \left[\frac{2}{\gamma_0}\int_{h_m}^{\infty} \Pi_0(h) dh\right]^{1/2} \left[1 + \frac{\epsilon}{2}\int_{h_m}^{\infty} \Pi_1(h,\boldsymbol{x}) \mathrm{d}h \Big/ \int_{h_m}^{\infty} \Pi_0(h) \mathrm{d}h\right] \tag{33}$$

relates the perturbation of the contact angle $\widetilde{\theta}$ to the perturbation of disjoining pressure $\Pi_1(h, \boldsymbol{x})$:

$$\widetilde{\theta} = \frac{1}{\gamma_0\theta_0}\int_{h_m}^{\infty} \Pi_1(h,\boldsymbol{x})\, \mathrm{d}h. \tag{34}$$

On the other hand, the integral (23) is transformed to a contour integral by integrating in the direction normal to the contact line, i.e. along the radius x:

$$\begin{aligned}\mathcal{F}_{\mathcal{R}} &= \oint_\Gamma F(s)\mathrm{d}s, \qquad F(s) = \int_0^L (h_0 - \widehat{h})\frac{\partial \Pi_1(x,s)}{\partial x}\,\mathrm{d}x \\ &= -\int_{h_m(s)}^{\infty} (h_0 - \widehat{h})\frac{\partial \Pi_1}{\partial h_0}\,\mathrm{d}h_0 = \int_{h_m(s)}^{\infty} \Pi_1\,\mathrm{d}h_0. \end{aligned} \tag{35}$$

The integrand of $\mathcal{F}_{\mathcal{R}}$ is transformed assuming that the radial dependence of the disjoining pressure is determined by the local thickness $h_0(x)$, while local substrate properties weakly dependent on the arc length s affect parameters of the function $\Pi_1(h_0)$. The last expression is obtained by integration by parts; the boundary values vanish because the disjoining pressure vanishes both at $h \to \infty$ and at $h = h_m$, by definition of the precursor layer thickness. In view of Eq. (34), the first harmonic of Eq. (35) coincides with Eq. (27).

The friction factor $\mathcal{J}$ is computed in the same manner as in Section 2.3. The difference is in evaluating the integral $\mathcal{J}_1$. With $\widehat{h}$ defined by Eq. (32), the integrand vanishes in the precursor. The singularity at $h \to 0$ is no longer present, so one can use the standard Stokes mobility function (3). The contribution of the contact line region, i.e. a ring with the width l satisfying the inequality $R \gg l \gg \widehat{h}$, to the dissipative integral should be evaluated using the actual stationary profile $h_0(x)$ computed by solving Eq. (15) or (10):

$$\mathcal{J}_1 = 6\pi R\eta \int_{\widehat{h}}^{l\theta_0} \frac{(h-\widehat{h})}{\sqrt{y(h)}h^2}\,\mathrm{d}h \asymp \frac{6\pi R\eta}{\theta_0}\ln\frac{l\theta_0}{b\widehat{h}}, \tag{36}$$

The integral is computed by extracting the part diverging at $l \to \infty$; for the remaining finite integral, the upper limit can be extended to infinity, yielding a numerical constant dependent on the form of function $\Pi(h)$:

$$\ln b = \int_{\widehat{h}}^{\infty} \left[\frac{1}{h\theta_0} - \frac{(h-\widehat{h})}{\sqrt{y(h)}h^2}\right]\mathrm{d}h. \tag{37}$$

Adding here the bulk contribution (29) removes the auxiliary length l yielding an expression where $\widehat{h}$ replaces λ in the role of a microscopic length:

$$\mathcal{J} = \frac{6\pi R\eta}{\theta_0}\ln\frac{\theta_0 R}{2b\widehat{h}}. \tag{38}$$

3.2 Mass Transport through the Precursor

Even when the substrate is homogeneous, motion can be caused by fluxes induced in the precursor due to interaction of droplets. We shall consider now an ensemble of droplets sitting on the precursor film, and neglect both external forces and evaporation. The droplets are assumed to be sufficiently far removed, so that a contour Γ lying in the precursor, as defined in the preceding subsection, could be drawn around each of them separately. This requirement is, in fact, not very stringent, since relaxation to a flat precursor outside the contact line region is exponential with a characteristic length of $O(h_m)$. The droplets might have been formed as a result of spinodal dewetting and subsequent coarsening. In this case, what we are going to describe is a late stage of evolution following the formation of well separated macroscopic droplets.

A single macroscopic droplet, which has the form of a parabolic cap defined by Eq. (8), has the curvature $2\theta/R$, and, according to Eq. (9), is in equilibrium with the precursor film with the thickness

$$h_m(R) \approx h_\infty \left(1 - \frac{\Delta}{R}\right), \qquad \Delta = \frac{2\gamma_0\theta}{\Pi'(h_\infty)}. \tag{39}$$

where h_∞ is the precursor thickness in equilibrium with a flat layer of bulk fluid. The precursor is *thinner* near larger droplets, which should therefore grow at the expense of smaller ones. This is an analog of the Gibbs–Thomson effect, where mass exchange with the precursor replaces evaporation.

The precursor evolution, as well as droplet motion, can be considered quasistationary when interaction is weak. Since the curvature of the precursor is negligible, Eq. (1) reduces to the diffusion equation with variable effective diffusivity $D(h)$:

$$\nabla \cdot (D\,\nabla h) = 0, \quad D(h) = q(h)\Pi'(h) > 0. \tag{40}$$

For a common strong thickness dependence of the disjoining pressure ($\Pi \sim h^{-3}$), diffusivity may *increase* in thinner layers, in spite of a sharp decrease of mobility $q(h)$. Deviations from h_∞ are expected be very small, as the dependence $\Pi(h)$ is typically very steep at short distances from the substrate. The diffusion equation (40) can be replaced therefore by the Laplace equation $D\nabla^2 h = 0$ with the effective diffusivity $D(h_\infty)$.

In 2D (i.e. for a 1D film), quasistationary fluxes ependent on droplet positions can be conveniently computed by considering interactions between the nearest neighbors (Glasner and Witelski, 2003), but in 3D (2D film) exact theory is very complicated when more than two droplets are involved.In the simplest case of a pair of droplets separated at a distance r much larger than their radii $R_\pm$, one can solve the Laplace equation using the bipolar coordinate system with the foci at the droplet centers. The coordinates σ, τ (respectively, circular and unbounded) are related to the Cartesian coordinates x, y as

$$x = \frac{r}{2}\,\frac{\sinh\tau}{\cosh\tau + \cos\sigma}, \qquad y = \frac{r}{2}\,\frac{\sin\sigma}{\cosh\tau + \cos\sigma}. \tag{41}$$

The droplets circumference corresponds to

$$\tau_\pm = \pm\ln\frac{r}{R_\pm} + O\left(\frac{R_j}{r}\right). \tag{42}$$

The Laplace equation reduces to $h_{\tau\tau} = 0$. At large separations, there is no σ-dependence in the leading order, and the solution is

$$h(\tau) = (\tau_+ - \tau_-)^{-1}[h_-\tau_+ - h_+\tau_- + (h_+ - h_-)\tau], \tag{43}$$

where $h_\pm = h_m(R_\pm)$. Using here Eqs. (39), (42) yields, to the leading order in $R_\pm/r \ll 1$, the total flux out of either droplet:

$$j_\pm = \pm\pi D\Delta h_\infty\,\frac{R_+^{-1} - R_-^{-1}}{\ln\left(r\,/\sqrt{R_+R_-}\,\right)}. \tag{44}$$

The flux j_+ is negative when $R_- < R_+$, which causes the larger droplet to grow at the expense of the smaller one.

3.3 Coarsening

Exchange through precursor sets off coarsening of the droplet size distribution. Dynamics of this process can be followed by considering pair interactions considered above as additive. This still leads to rather complicated equations, since each pair exchange depends on radii of both droplets as well as on their positions. An alternative is the effective medium approximation commonly used in theory of coarsening. The total flux out of the droplet can be defined then as exchange with the precursor film of some effective thickness h_c slowly changing with time. Thus, the flux out of kth droplet with the equilibrium precursor thickness $h_k = h_m(R_k)$ is, in this approximation

$$j_k = 2\pi D(h_k - h_c). \tag{45}$$

The value of h_c is defined by mass conservation requiring the sum of fluxes out of all droplets to vanish, which requires h_c to be equal to the average

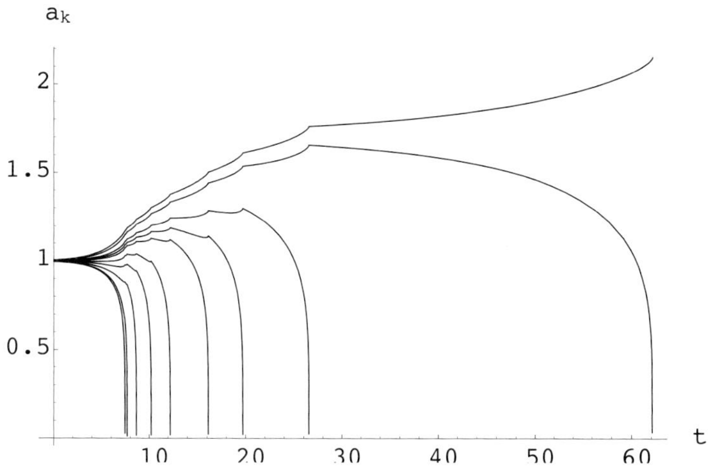

Figure 2. A typical coarsening sequence starting from ten droplets with dimensional radii $a_k = R_k/R_c(0)$; the initial size differences are less than 1%; the inverse time scale is $\frac{8}{3}D\Delta h_\infty R_c(0)^{-4}$.

$\langle h_k \rangle$. As in the standard coarsening theory (Lifshitz and Slyozov, 1961), h_c can be conveniently defined using Eq. (39) through the radius R_c of a critical droplet that neither grows nor decays:

$$h_c = h_m(R_c) = h_\infty \left(1 - \Delta/R_c\right), \tag{46}$$

where $R_c = 1/\langle R_k^{-1} \rangle$ is the harmonic average radius.

Since the volume of a parabolic cap is $V = \frac{1}{4}\pi\theta R^3$, the evolution equation of the droplet radii following from the mass conservation law $V_k'(t) = -j_k$ is

$$\frac{\mathrm{d}R_k}{\mathrm{d}t} = \frac{8}{3}\frac{D\Delta h_\infty}{\theta R_k^2}\left(\frac{1}{R_c(t)} - \frac{1}{R_k}\right). \tag{47}$$

Direct numerical computations using Eq. (47) with a moderate number of droplets are straightforward; an example is shown in Fig. 2. The results are, of course, not exact, as Eq. (47) is a mean field approximation that does not take into account fluctuations in spatial distribution of droplets of different size. The slight kinks visible on the curves for the extant droplets near extinction events are artifacts due to the quasistationarity assumption for the precursor. In the case of long-time evolution of a very large number of droplets, the Lifshitz–Slyozov theory can be applied to compute the asymptotics droplet size distribution. It corresponds to the universality class $n = 2$ in a generalized theory (Giron *et al*, 1998)), differing from the original case ($n = 1$) of diffusion-controlled ripening (Lifshitz and Slyozov, 1961). Deviations from the asymptotic theory are likely to arise due to migration and coalescence (see the next subsection).

3.4 Migration of Interacting Droplets

Unlike the standard coarsening problem, the droplets do not only grow or decay, but also move under the influence of the gradient of disjoining pressure in the precursor. The principal contribution to the driving force comes from the second contour integral in Eq. (25), where the Laplacian can be neglected in view of flatness of the precursor. The driving force stems from a small dipole component of $\widetilde{h}$ caused by mass fluxes from far removed droplets, which is proportional to the gradient $\nabla\widetilde{h}$. The gradient $\nabla\widetilde{h}_i$ at the location of a test droplet can be expressed in the effective medium approximation through superposition of the fluxes from the rest of the droplets:

$$\nabla\widetilde{h}_i = -D^{-1}\sum_{i\neq k}\boldsymbol{j}_k = \sum_{i\neq k}(h_k - h_c)\,\frac{\boldsymbol{r}_{ik}}{r_{ik}^2}, \tag{48}$$

where $\boldsymbol{r}_{ik}$ is the vector directed from the ith to kth droplet and $r_{ik} = |\boldsymbol{r}_{ik}|$ is the distance between their centers.

The constant $\widehat{h}$ in Eq. (25) should be set equal to h_∞. Using here Eqs. (39), (48), the driving force acting on the test droplet is computed in the leading order as

$$\mathcal{F}_i = -\mathcal{I}_\Gamma = -2\gamma_0\theta\oint_\Gamma(\boldsymbol{I}\cdot\nabla\widetilde{h}_i)\,\mathrm{d}s = -2\pi\gamma_0 h_\infty\Delta R_i\sum_{i\neq k}\left(R_k^{-1} - R_c^{-1}\right)\frac{\boldsymbol{I}\cdot\boldsymbol{r}_{ik}}{r_{ik}^2}. \tag{49}$$

The test droplet is, respectively, attracted and repelled by the droplets above and below the critical size. This is caused by the action of weak fluxes sucked in by larger and blown out by smaller droplets. In a droplet ensemble, the direction of motion $\boldsymbol{I}$ is determined by balancing fluxes induced by all droplets.

Using the expression for the friction factor (38), the velocity of the droplet is computed as

$$\boldsymbol{U}_i = -\frac{\gamma_0\theta_0 h_\infty\Delta}{3\eta}\ln^{-1}\left(\frac{\theta_0 R_i}{2bh_\infty}\right)\sum_{i\neq k}\left(R_k^{-1} - R_c^{-1}\right)\frac{\boldsymbol{I}\cdot\boldsymbol{r}_{ik}}{r_{ik}^2}. \tag{50}$$

A small droplet migrating in the direction of a large droplet with $R > R_c$ moves somewhat faster due to a smaller logarithmic factor, and therefore

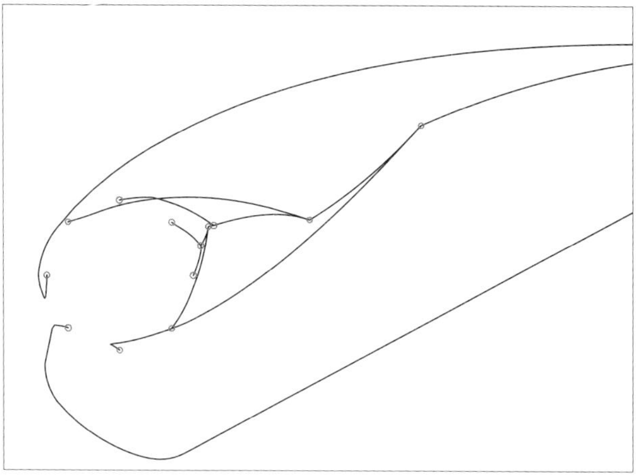

Figure 3. Typical trajectories of migrating droplets. Small circles mark the initial positions and positions at the coalescence events. Size changes due to mass exchange are neglected. Initially, eight droplets with dimensionless radii randomly distributed in the interval $0.9 < a < 1.1$ were placed symmetrically on a circle. The droplets proceed to the right to distant final coalescence events.

has some chance to catch up and coalesce before disappearing. Some typical trajectories computed using Eq. (50) with size changes due to mass exchange neglected are shown in Fig. 3.

Actual details of evolution involving both migration and mass exchange, which can be tracked numerically when the number of droplets is moderate, depend on many accidental factors. An example of evolution starting from a random initial arrangement of droplets with slightly different initial sizes is shown in Fig. 4. In this case, the droplets which have grown due to coalescence acquire decisive advantage, while the others become extinct. As a result, the coarsening process is greatly accelerated, and long runs seen in Fig. 3 are eliminated.

4 Chemical Self-Propulsion

4.1 Substrate Modification

Variation of the contact angle may be caused by substrate modification, e.g. dissolution of the surfactant adsorbed on the substrate in experiments of Sumino *et al* (2005). The surfactant adsorption and desorption is de-

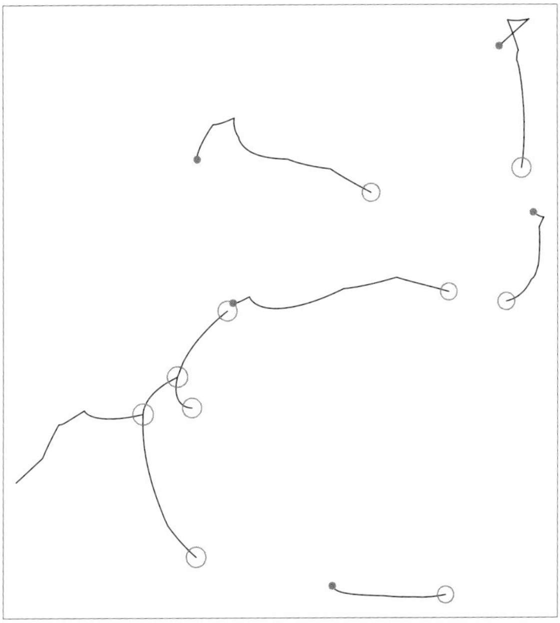

Figure 4. Trajectories of droplets interacting through the precursor, involving both migration and mass exchange and starting from a random distribution with a comparable mean separation. The ratio of characteristic migration and coalescence time is unity. Circles mark the initial positions and the coalescence events; dots mark extinction events.

scribed by a linear kinetic equation $\mathrm{d}c/\mathrm{d}t = k_a c_0 - kc$, where c_0 is the bulk concentration and k_a, k are kinetic constants; the latter is assumed for simplicity to be the same in the droplet and the continuous phase. Taking into account surface diffusion and assuming the bulk concentration in the droplet to be negligible, we write the adsorption-diffusion equation for the surfactant coverage on the substrate in the dimensionless form

$$c_t = \nabla^2 c - c + H(\boldsymbol{x}), \tag{51}$$

where $H(\boldsymbol{x})$ equals to 1 outside and 0 inside the droplet footprint. The surfactant coverage c is scaled by the coverage in equilibrium with the surfactant concentration in the continuous phase $k_a c_0/k$, time by the inverse desorption rate constant k, and length, by $\sqrt{D/k}$, where D is the surfactant diffusivity on the substrate. At later stages of the process when surfactant accumulates in the droplet bulk, one can use the same equation, taking into account that only the difference of adsorption rates outside and inside the droplet is important, and replacing $k_a c_0$ by the difference of the respective values.

The characteristic velocity scale is $U^* = \sqrt{Dk}$. If Eq. (12) is brought to a dimensionless form by replacing $\epsilon U = U^* v$, the advective term is recast as $\mathrm{Ca}\, v \cdot \nabla h$, where $\mathrm{Ca} = \sqrt{Dk}\eta/\gamma_0$ is the capillary number based on the characteristic "chemical" velocity U^*, or the ratio of the "chemical" velocity to the characteristic "hydrodynamic" velocity γ/η, which determines the influence of viscous stresses on the droplet shape sustained by surface tension. This implies that the capillary number is the true small parameter of the problem; deviations of the contact angle should be of $O(\mathrm{Ca})$ to justify the perturbation approach. This condition is likely to hold under realistic experimental conditions, since surface diffusion is slow, which insures that the shape of chemically propelled droplets is only weakly distorted by viscous stresses.

Equation (51) transformed to the frame moving with a dimensionless velocity v (scaled by $\sqrt{Dk}$) along the x axis is

$$vc_x + \nabla^2 c - c + H(\boldsymbol{x}) = 0. \tag{52}$$

The solution of this equation can be expressed with the help of an appropriate Green's function through an integral over the droplet footprint area and subsequently transformed into a contour integral with the help of the Gauss theorem (Pismen, 2001). For a circular contour with a dimensionless radius (Thiele modulus) $a = R\sqrt{k/D}$, the concentration on its boundary, which determines the propagation speed, is computed in this way as

$$\begin{aligned} c(\phi) = \ & 1 - \tfrac{a}{2\pi} \int_{-\pi}^{\pi} \mathrm{e}^{-\frac{1}{2} va(\cos\phi - \cos\xi)} \left[\tfrac{v}{2} \cos\xi \, K_0 \left(2a\sqrt{1 + \tfrac{v^2}{4}} \sin \tfrac{|\phi-\xi|}{2} \right) \right. \\ & \left. + \sqrt{1 + \tfrac{v^2}{4}} \sin \tfrac{|\phi-\xi|}{2} \, K_1 \left(2a\sqrt{1 + \tfrac{v^2}{4}} \sin \tfrac{|\phi-\xi|}{2} \right) \right] \mathrm{d}\xi, \end{aligned} \tag{53}$$

where K_n (and I_n below) are modified Bessel functions.

Assuming the contact angle to be a linear function of the surfactant coverage, $\epsilon\widetilde{\theta} = -\beta c$, the propagation velocity is computed by solving the equation obtained by combining Eqs. (31) and (53):

$$v = \frac{M}{\pi} \int_0^{\pi} \widetilde{c}(\phi; v) \cos\phi \, \mathrm{d}\phi, \tag{54}$$

where $\widetilde{c} = 1 - c$ and all relevant parameters, except the dimensionless droplet radius a in Eq. (53), are lumped into a single dimensionless combination,

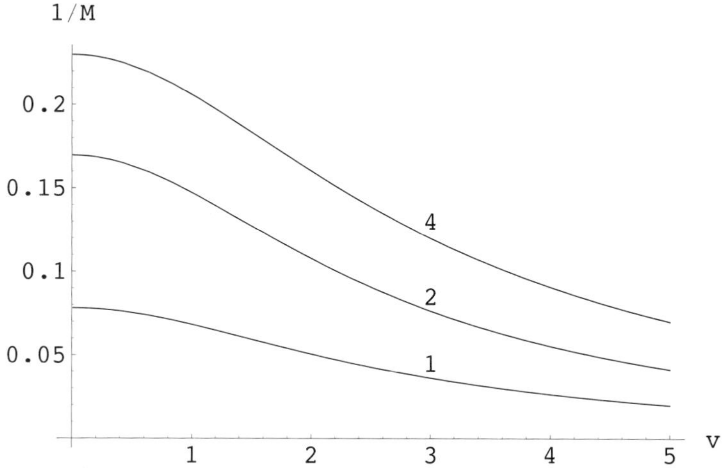

Figure 5. Plots of $G(a, v)$ defined by Eq. (56) as a function of v; the values of a are marked at the respective curves.

which can be called the *droplet mobility* parameter:

$$M = \frac{\theta_0^2 \gamma \beta}{6\eta\sqrt{Dk}} \ln^{-1} \frac{\theta_0 R}{6\lambda}. \tag{55}$$

The parameter M retains a weak logarithmic dependence on the droplet radius.

Equation (54) can be rewritten as $M^{-1} = G(a, v)$ and solved after computing numerically the double integral

$$G(a,v) = \frac{a}{\pi^2 v} \int_0^{\pi} \cos\phi \, \mathrm{d}\phi \int_{-\pi}^{\pi} \mathrm{e}^{-\frac{1}{2}va[\cos\phi - \cos(\phi+\zeta)]} \times \left[\frac{v}{2}\cos(\phi+\zeta)\, K_0\left(2a\sqrt{1+\frac{v^2}{4}}\sin\frac{|\zeta|}{2}\right) + \sqrt{1+\frac{v^2}{4}}\sin\frac{|\zeta|}{2}\, K_1\left(2a\sqrt{1+\frac{v^2}{4}}\sin\frac{|\zeta|}{2}\right)\right] \mathrm{d}\zeta. \tag{56}$$

The function $G(a, v)$ is plotted against v at several values of a in Fig. 5. The curves peak at the ordinate axis at the bifurcation value M_0^{-1} to be computed in the next subsection.

4.2 Traveling Bifurcation

Simplified expressions can be obtained in the limiting cases $v \ll 1$ and $v \gg 1$. The perturbation approach is still applicable also in latter case, since the capillary number is typically very small.

The limit $v \ll 1$ (i.e. fast diffusion) is analogous to the fast inhibitor limit in reaction-diffusion systems, which is conducive to formation of stationary patterns (see e.g. Pismen, 2006). Equation (53) is expanded in this limit to the first order in v as

$$\widetilde{c}(\phi) = \tfrac{a}{\pi} \int_0^\pi \left\{ \sin \tfrac{\zeta}{2} K_1 \left(2a \sin \tfrac{\zeta}{2} \right) \right. \tag{57}$$
$$\left. + v \cos\phi \left[\tfrac{1}{2} \cos\zeta \, K_0 \left(2a \sin \tfrac{\zeta}{2} \right) + a \sin^3 \tfrac{\zeta}{2} K_1 \left(2a \sin \tfrac{\zeta}{2} \right) \right] + O(v^2) \right\} \mathrm{d}\zeta,$$

where $\zeta = \xi - \phi$. In the leading $O(1)$ order, the surfactant distribution is circularly symmetric. The first-order dipole term in Eq. (57) is the only one contributing to the integral in Eq. (54) (another term vanishing upon integration is omitted). The angular integrals are evaluated using the identities

$$\Phi_k(a) = \int_0^\pi \sin^{2k} \frac{\phi}{2} K_0 \left(2a \sin \frac{\phi}{2} \right) \mathrm{d}\phi = -\frac{1}{2a} \frac{\mathrm{d}(a\Psi_{k-1})}{\mathrm{d}a},$$
$$\Psi_k(a) = \int_0^\pi \sin^{2k+1} \frac{\phi}{2} K_1 \left(2a \sin \frac{\phi}{2} \right) \mathrm{d}\phi = -\frac{1}{2} \frac{\mathrm{d}\Phi_k}{\mathrm{d}a},$$

starting from $\Phi_0(a) = \pi I_0(a) K_0(a)$. Plugging the resulting expressions in Eq. (54) yields the condition for the onset of motion

$$M_0^{-1} = \frac{a^2}{2} \left[I_1(a) K_2(a) - I_0(a) K_1(a) \right]. \tag{58}$$

The critical value M_0^{-1} as a function of a is plotted in Fig. 6. Since the radial dependence in Eq. (58) saturates when the droplet radius far exceeds the diffusional range, so that $M_0 \to 4$ at $a \to \infty$, no droplet can move below this limiting value.

Because of the logarithmic dependence of M on the droplet radius, the size dependence of the traveling threshold in Eq. (58) still remains implicit. An explicit dependence can be extracted after rewriting Eq. (55) as

$$M = \mu_d \ln^{-1} \frac{a}{\lambda_d}, \quad \mu_d = \frac{\theta_0^2 \gamma \beta}{6\eta\sqrt{Dk}}, \quad \lambda_d = \frac{6\lambda}{\theta_0} \sqrt{\frac{k}{D}}. \tag{59}$$

In the limit $a \gg 1$ when the critical value M_0 in Eq. (58) approaches the limit $M_0 = 4$, the droplet is mobile at $a < \lambda_d \exp(\mu_d/4)$. This suggests that

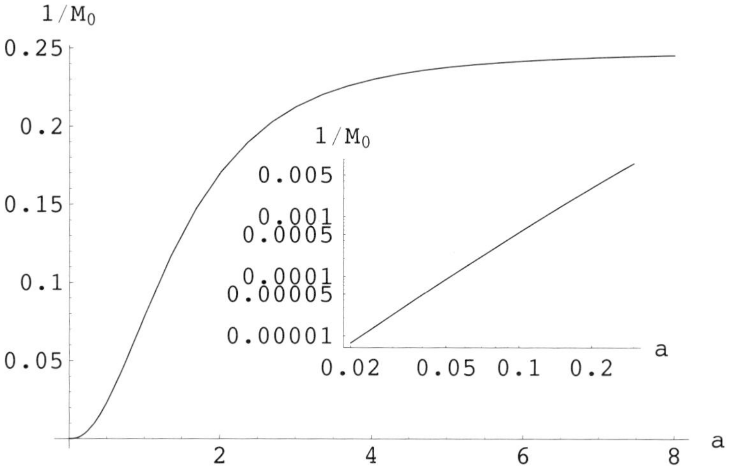

Figure 6. The critical value M_0^{-1} as a function of the dimensionless droplet radius a. Inset: blow-up near the origin on logarithmic scale. The droplets are mobile below this curve.

the radius of mobile droplets is bounded both from below and above, as it indeed follows from the existence of a maximum in the radial dependence of the critical value of the parameter μ_d, such as seen in Fig. 7.

Beyond the critical point, the velocity can be obtained using further terms in the expansion (57). The dipole component of the term quadratic in v vanishes, and therefore there is no contribution to motion in this order. The dipole component of the third-order term is $\alpha_3 v^3$ where

$$\alpha_3 = -\frac{a^3}{16}\left[I_0(a)K_1(a) - I_1(a)K_0(a) - a^{-1}I_1(a)K_1(a)\right]. \qquad (60)$$

This coefficient is negative; hence, the bifurcation is supercritical and propagation is possible at $M > M_0$. For small deviations $M_2 = M - M_0 > 0$, the velocity is $v = \sqrt{-M_2/\alpha_3}$.

4.3 Non-diffusive Limit

In the opposite limit when diffusion is negligible, the diffusional scaling of the droplet radius and velocity becomes invalid, and dimensionless relations should be constructed using the capillary number $\mathrm{Ca} = U\eta/\gamma$ and the rescaled dimensionless radius $\rho = Rk\eta/\gamma$. The surfactant concentration

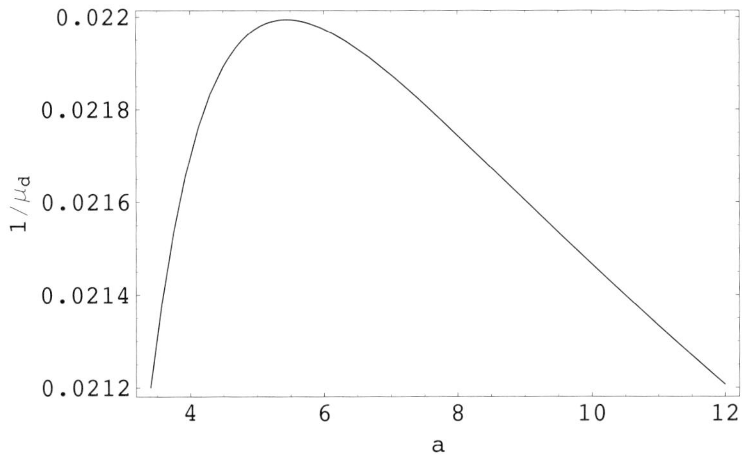

Figure 7. The critical value of μ_d^{-1} as a function of the dimensionless droplet radius a for $\lambda_d = 10^{-4}$.

can be obtained directly by integrating along the direction of motion with the initial condition $c = 1$ on the advancing contact line ($|\phi| < \pi/2$). The resulting concentration on the receding contact line,

$$c(\phi) = \mathrm{e}^{-2\tau \cos\phi}, \tag{61}$$

depends only on the ratio

$$\tau = \frac{a}{v} = \frac{kR}{U} = \frac{\rho}{\mathrm{Ca}}. \tag{62}$$

The equation for τ following from Eq. (31) is

$$P = \frac{\tau}{\pi} \int_0^{\pi/2} \cos\phi \left(1 - \mathrm{e}^{-2\tau\cos\phi}\right) \mathrm{d}\phi \equiv \tau F(\tau), \tag{63}$$

containing a single parameter

$$P = \frac{a}{M} = \frac{6\eta R k}{\theta_0^2 \gamma \beta} \ln \frac{\theta_0 R}{6\lambda} = \frac{\rho}{\mu_h} \ln \frac{\rho}{\lambda_h}, \tag{64}$$

where $\mu_h = \frac{1}{6}\beta\theta_0^2$, $\lambda_h = 6\lambda\gamma/(k\eta\theta_0)$. The function $F(\tau)$ is evaluated as

$$F(\tau) = \frac{1}{\pi} + \frac{1}{2}\left[I_1(2\tau) - \mathbf{L}_{-1}(2\tau)\right], \tag{65}$$

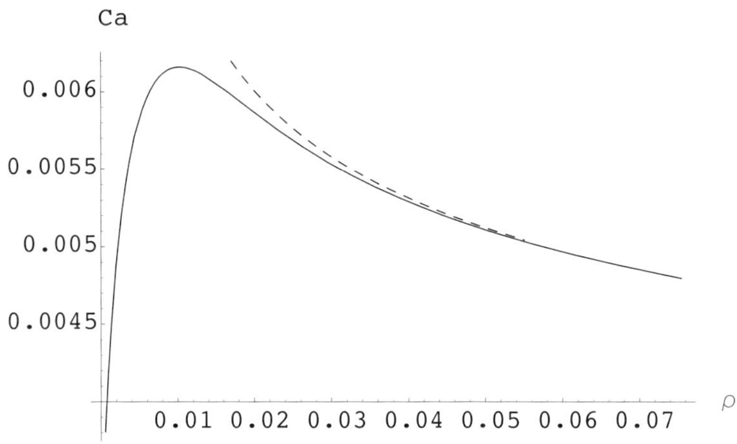

Figure 8. The dependence of the capillary number ϵ on the dimensionless radius ρ for $\lambda_h = 10^{-4}$, $\mu_h = 0.1$. The dashed line shows the asymptotic dependence (67).

where $\mathbf{L}_n(x)$ is a Struve function. The function $F(\tau)$ increases monotonically from 0 at $\tau = 0$ to $1/\pi$ at $\tau \to \infty$.

The dependence of the capillary number on the dimensionless droplet radius can be obtained analytically in two limiting cases corresponding to the unsaturated and saturated regimes, respectively, at small and large τ. In the former case, one can use the approximation $F(\tau) = \tau/2 + O(\tau^2)$ to obtain

$$\mathrm{Ca} \approx \left[\frac{\rho \mu_h}{2 \ln(\rho/\lambda_h)} \right]^{1/2}. \tag{66}$$

In the opposite limit $\tau \gg 1$, $F(\tau) \approx 1/\pi$ and

$$\frac{1}{\mathrm{Ca}} \approx \frac{\pi}{\mu_h} \ln \frac{\rho}{\lambda_h}. \tag{67}$$

Thus, the velocity increases with droplet size in the unsaturated and decreases in the saturated regime, in agreement with experiment (Lee *et al*, 2002) and numerical computations (Thiele *et al*, 2004). For intermediate values of τ, the dependence of velocity on radius obtained by solving Eq. (63) numerically is plotted in Fig. 8.

4.4 Relaxation to a Stationary Pattern

At high diffusivities when droplets do not travel spontaneously, they still can move under the influence of mutual interactions. The surfactant depletion in the far field of a stationary droplet (at distances far exceeding its radius) is well approximated by the stationary solution of Eq. (51) with c replaced by $\widetilde{c}$ and $H(\boldsymbol{x})$, by the delta-function multiplied by the droplet area:

$$\widetilde{c} = \frac{1}{2} a^2 K_0(r). \tag{68}$$

The circular symmetry of the surfactant distribution around a single droplet is perturbed by the far field of its neighbors. The resulting repelling interaction induces, according to Eq. (31), motion with the velocity proportional to the concentration gradient at the droplet location. If there is a number of droplets, their action is additive. This leads to the equation of motion for droplet centers $\boldsymbol{X}_j$

$$\frac{\mathrm{d}\boldsymbol{X}_j}{\mathrm{d}t} = M_j a_j \sum_{k \neq j} \frac{\boldsymbol{X}_j - \boldsymbol{X}_k}{|\boldsymbol{X}_j - \boldsymbol{X}_k|} \frac{a_k^2}{2} K_1 \left(|\boldsymbol{X}_j - \boldsymbol{X}_k| \right). \tag{69}$$

This is a gradient dynamical system

$$\frac{\mathrm{d}\boldsymbol{X}_j}{\mathrm{d}t} = -\frac{M_j}{a_j} \frac{\partial V}{\partial \boldsymbol{X}_j}, \tag{70}$$

evolving to minimize the potential

$$V = \frac{1}{2} \sum_{k \neq j} a_j^2 a_k^2 K_0 \left(|\boldsymbol{X}_j - \boldsymbol{X}_k| \right). \tag{71}$$

In a "thermodynamic" limit of an infinitely large region containing an infinite number of droplets, the potential is expected to be minimized by a regular hexagonal pattern with spacing dependent on the number density of droplets. The ordering tendency is demonstrated by an example of evolution shown in Fig. 9. The confinement is effected in this computation by adding to Eq. (71) a centripetal external potential $\Pi_0 = m \sum |\boldsymbol{X}_j|^2$ with a suitable constant m . One can see that evolution starting from a random arrangement of droplets evolves to regular pattern where circles mark final positions falling roughly on a hexagonal grid. This might be a practical way to arrange a regular dewetting pattern on a homogeneous substrate.

4.5 Scattering

The surfactant depletion in the far field of a steadily moving droplet can be obtained by solving Eq. (52) with c replaced by $\widetilde{c}$ and $H(\boldsymbol{x})$, by the

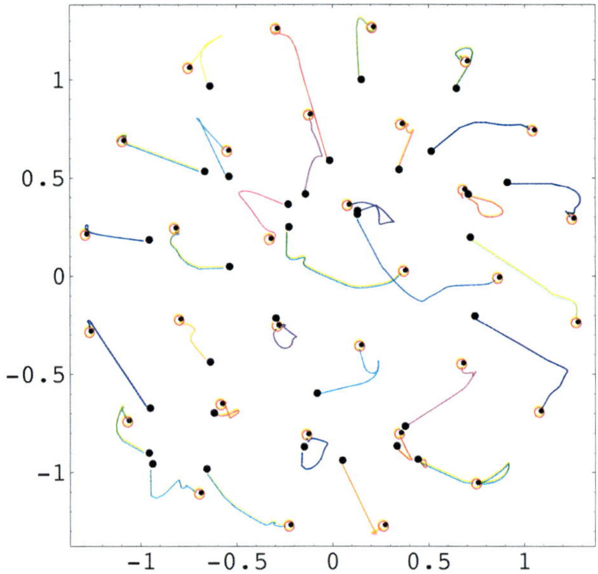

Figure 9. Trajectories of identical droplets moving according to Eq. (69) with added centripetal force. The dots mark the original random arrangement, and the circles, final positions.

delta-function multiplied by the droplet area. The solution is expressed in polar coordinates r, ϕ centered on the droplet as

$$\widetilde{c} = \frac{a^2}{2}\, \mathrm{e}^{-\frac{1}{2} v r \cos\phi} K_0 \left(r\sqrt{1 + \frac{v^2}{4}} \right). \tag{72}$$

The depletion field is strongly asymmetric, with a slower decay behind the droplet (Fig. 10).

Droplets moving one toward another are scattered by mutually repelling interaction created by the gradient of the far field. The problem remains tractable in the quasistationary approximation as long as velocity induced by interaction is much smaller than the speed of self-propelled motion. Otherwise, the far field becomes dependent on the entire history of motion, and solving the full non-stationary problem (51) is necessary.

Consider as an example two droplets of equal size propagating along

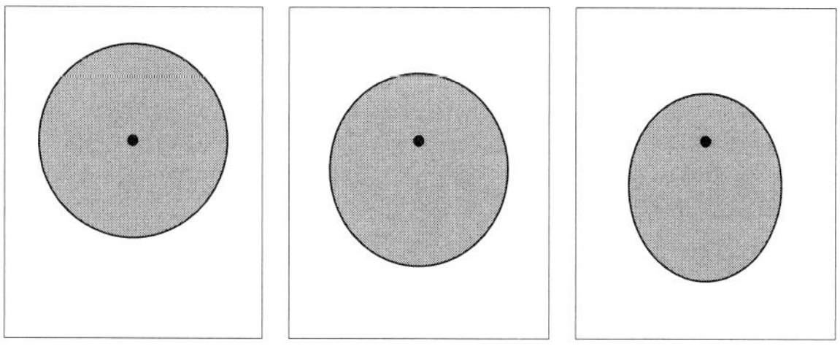

Figure 10. Surfactant depletion in the far field of a stationary droplet (left) and droplets propagating with the speed $v = 1$ (center) and $v = 2$ (right). The droplet position is marked by a dot. The area with surfactant depletion above the same level is shaded, showing the depletion in the tail region increasing at higher speed.

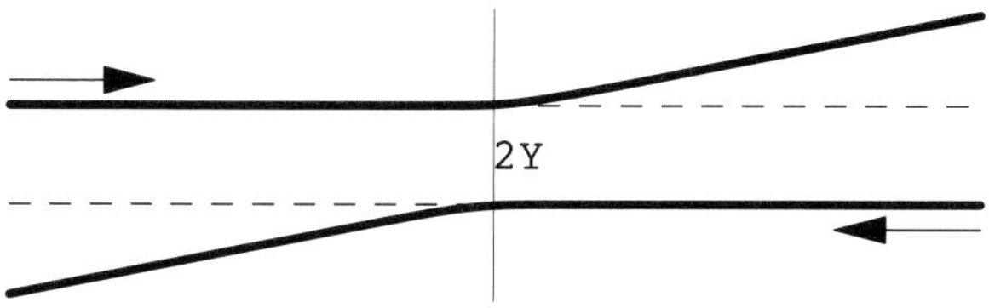

Figure 11. Scattering trajectories. The vertical line marks the location of the closest approach.

the x axis with identical speed v on antiparallel trajectories shifted by the interval $2Y$, as in Fig. 11. Since a deflected droplet keeps moving on a perturbed course, the scattering action is equivalent to acceleration in the direction normal to self-propelled motion. Restricting to the quasistationary approximation, the dynamic equation for the deviation $\widetilde{y}$ normal to the original trajectory is therefore

$$\frac{\mathrm{d}^2\widetilde{y}}{\mathrm{d}t^2} = -Ma\,\frac{\partial\widetilde{c}}{\partial y}\,, \tag{73}$$

where the derivative of the surfactant depletion given by Eq. (72) is computed at a current distance between the droplets at the moment t equal to

$$r(t) = 2\left[(vt)^2 + (Y + \widetilde{y})^2\right]^{1/2} \tag{74}$$

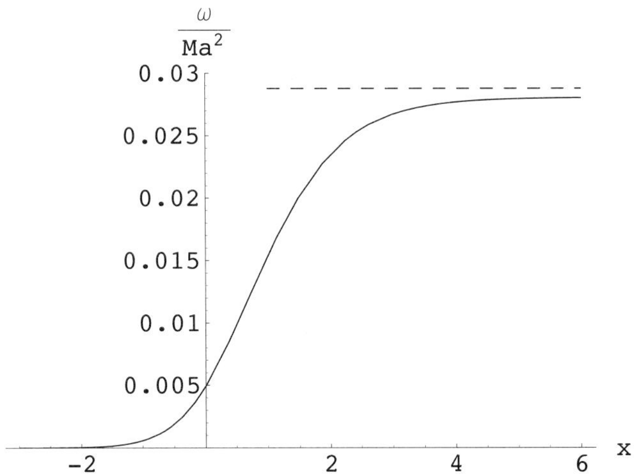

Figure 12. The change of the scattering angle with distance for $v = 1$, $Y = 2$. The dashed line shows the scattering angle at infinity computed with the help of Eq. (75).

where the moment of closest approach is taken as $t = 0$. Neglecting the change of the velocity component along the x-axis, time t in Eq. (73) can be replaced by x/v. A typical trajectory obtained by integration is shown in Fig. 11 and a more quantitative example of the change of the scattering angle with distance is shown in Fig. 12. Take note that, due to a faster decay of depletion ahead of the droplet, scattering largely accumulates already after the droplets have passed the point of closest approach.

For moderate deviations, a reasonable approximation for the scattering angle at infinity $\omega = \widetilde{y}'(\infty)$ gives the formula neglecting $\widetilde{y}$ compared to Y:

$$\omega = \frac{MYa^3}{2v}\sqrt{1+\frac{v^2}{4}}\int_{-\infty}^{\infty}\frac{e^{vx}}{\sqrt{x^2+Y^2}}K_1\left(r\sqrt{(4+v^2)(x^2+Y^2)}\right)\mathrm{d}x. \quad (75)$$

Scattering angle computed with the help of this formula only weakly depends on velocity (Fig. 13, above). A much stronger dependence on the separation interval is shown in logarithmic coordinates in Fig. 13 (below).

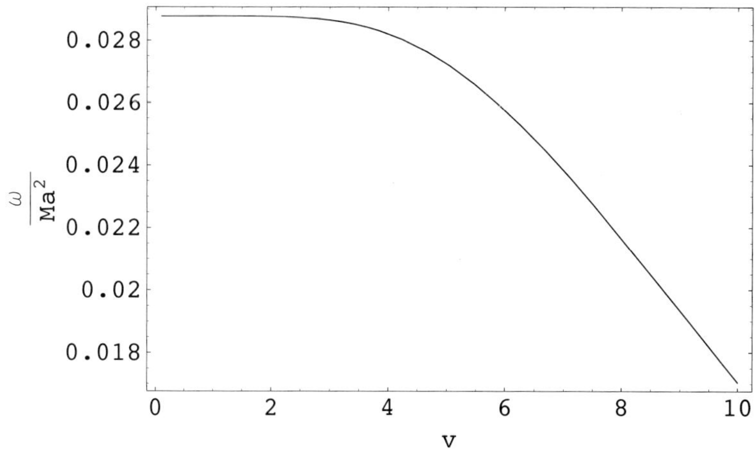

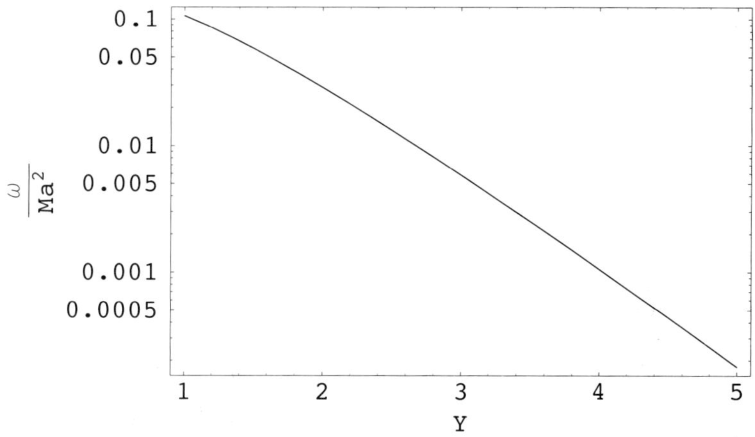

Figure 13. Dependence of the change of the scattering angle on velocity for $Y = 0$ (above) on the separation interval for $v = 1$ (below). Both curves are computed using Eq. (75).

5 Thickness Fronts

5.1 Static Thickness Fronts

For a dependence $\Pi(h)$ of the kind shown in Figure 1 (right), two stable roots of $\Pi(h) = \mu_0$, or minima of $\gamma(h) - h\mu_0$, may exist in within a certain

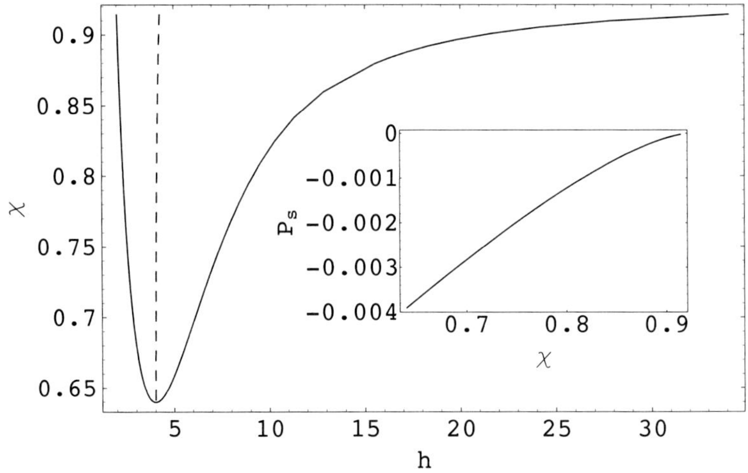

Figure 14. Dependence of the stationary film thickness on the parameter χ. The two branches of the solid curve show the values of $h_s^{\pm}$, and the dashed line shows the intermediate unstable state h_s^0. Insert: dependence $\Pi(\chi)$.

interval of μ_0, determined by the respective shift of chemical potential $\widetilde{\mu} = \mu_0/n_l$. A suitable form of $\Pi(h)$ is the Sharma function (Sharma, 1993) combining van der Waals and polar interactions:

$$\Pi(h) = -Q_s h^{-3} + Q_p \mathrm{e}^{-h/d}. \tag{76}$$

The dimensionless form using d as the thickness scale and Q_s/d^3 as the energy density scale is

$$\Pi(h) = -h^{-3} + \chi \mathrm{e}^{-h}, \qquad \chi = d^3 Q_p/Q_s. \tag{77}$$

Two stable values of the film thickness at a fixed level of μ_0 are possible when the curve $\Pi(h)$ is S-shaped, i.e. in the interval $3(\mathrm{e}/4)^4 < \chi < \mathrm{e}^2/8$, or approximately $0.6398 < \chi < 0.9236$. The alternative stationary states $h_s^{\pm}$ merge at the lower end of this interval, while at the upper end h_s^+ diverges (see Fig. 14).

Under these conditions, the liquid film may separate into domains with alternative stable values $h = h_{\pm}$. The domain boundaries (thickness fronts) are set into motion during evaporation, condensation, or coarsening, as we shall see below. In the absence of bulk fluid, there is no asymptotic contact angle in the Young–Laplace sense, and the "droplets", i.e. "thick" film domains, are shaped like pancakes rather than spherical caps. Both thin and "thick" films may fall into a nanoscopic range. When the liquid volume is large, both thin and thick film domains may be in equilibrium with bulk fluid, i.e. a layer of macroscopic thickness $h \to \infty$.

A straight-line stationary front separating domains with two alternative stable values $h = h_s^{\pm}$ verifies the stationary equation (2) with the asymptotic conditions $h = h_s^{\pm}$ at $x \to \pm\infty$. We shall write this equation in the dimensionless form

$$h''(x) - \Pi(h) - \mu_0 = 0, \tag{78}$$

where the "horizontal" coordinate x is scaled by $l = \sqrt{\gamma_0 d/Q_s}$, and $\mu_0 = Pd^3/Q_s$. When the liquid mass is conserved, a single planar interface cannot propagate; therefore Maxwell construction should be reached by adjusting μ_0. The equilibrium value $\mu_0 = \mu_s$ is obtained by multiplying (78) by $h'(x)$ and integrating across the front. The differential term vanishes upon integration, while the integral of the algebraic part yields the Maxwell condition, which can be written, recalling that $\Pi = \gamma'(h)$, as

$$\mu_s = -\frac{\gamma(h_s^+) - \gamma(h_s^-)}{h_s^+ - h_s^-}. \tag{79}$$

This, together with $\mu_s = \Pi(h_s^{\pm})$, defines the three unknowns μ_s, $h_s^{\pm}$.

The stationary front profile is computed most easily in the "phase plane" representation, using h as an independent, and $y = [h'(x)]^2$ as a dependent variable. Integrating Eq. (78) yields

$$y(h) = 2\left[\gamma(h) - \gamma(h_s^-) + \mu_s(h - h_s^-)\right], \tag{80}$$

and the front profile is expressed in an implicit form

$$x(h) = \frac{1}{\sqrt{2}} \int_{h_s^0}^{h} \left[\gamma(h) - \gamma(h_s^-) + \mu_s(h - h_s^-)\right]^{-1/2} \mathrm{d}h. \tag{81}$$

The origin $x = 0$ may be arbitrary, but, to be definite, we have chosen it to coincide with the unstable intermediate solution h_s^0. Two typical front profiles are shown in Fig. 15.

5.2 Evaporation and Condensation

Thickness fronts can be set into motion when the liquid is *volatile*. In this case, the volume is not conserved, and Eq. (1) is modified by adding a

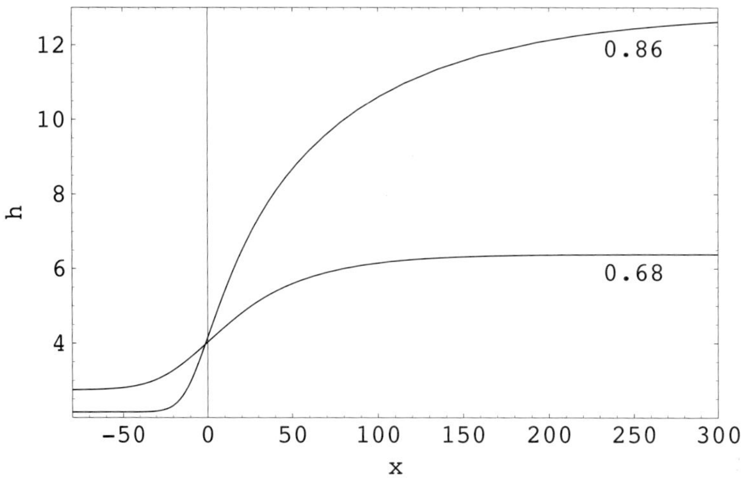

Figure 15. The stationary front profiles for $\chi = 0.86$ and $\chi = 0.68$.

source (sink) term. The evaporation (condensation) rate is proportional to the difference of chemical potentials in the liquid and gas phases. The shift of chemical potential in the liquid is related to pressure defined by Eq. (2), $\mu = P/n_l$, where n_l is the liquid number density. Thus, we can write

$$\partial_t h = n_l \nabla \cdot [q(h) \nabla \mu] - \beta(\mu - \mu_0), \tag{82}$$

where μ_0 is chemical potential of the ambient vapor phase, β is an effective rate constant, and the zero level of μ corresponds to equilibrium with saturated vapor over a flat interface sufficiently removed from the substrate and not subject to external forces.

The characteristic evaporation scale $L = \sqrt{d^3 n_l / \beta\eta}$ is determined by the balance between advective mass transport along the film (with $q(h) \propto d^3$) and evaporation. We shall consider the case when evaporation is sufficiently slow, so that this scale is *long*, compared with the internal scale l that determines the extent of a region where the interface may be strongly curved due to interaction with the substrate. The respective time scales are

$$T_l = \frac{l^2 \eta}{d^2 n_l Q_s} = \frac{\gamma_0 \eta}{d n_l Q_s^2}, \qquad T_L = \frac{L^2 \eta}{d^2 n_l Q_s} = \frac{d}{\beta Q_s}.$$

Evolution is *quasistationary* if it proceeds on the slower time scale.

The scale ratio

$$\epsilon = \frac{l}{L} = \sqrt{\frac{T_l}{T_L}} = \frac{1}{d}\sqrt{\frac{\beta\gamma_0\eta}{n_l Q}} \tag{83}$$

is the small parameter of the problem. We shall distinguish between the short-scale dimensionless coordinates and time, $\boldsymbol{x}$, t and the long-scale variables $\boldsymbol{X}$, T, as well as between the respective gradient operators ∇ and $\widehat{\nabla}$. In a scalar notation, x or X will denote coordinates normal to the front, and y or Y, the transverse coordinates. Using the "inner" (short) and the "outer" (long) scales, we rewrite Eq. (82) as

$$\partial_t h = \nabla \cdot [q(h)\nabla\mu] - \epsilon^2(\mu - \mu^0), \qquad \mu = -\nabla^2 h + \Pi(h). \tag{84}$$

$$\partial_T h = \widehat{\nabla} \cdot [q(h)\widehat{\nabla}\mu] - (\mu - \mu^0), \qquad \mu = -\epsilon^2\widehat{\nabla}^2 h + \Pi(h). \tag{85}$$

5.3 Fluxes and Mobility of the Front

Across a narrow front, the film thickness switches between the two alternative values corresponding to the same constant value of chemical potential. The latter is constant everywhere when the two "phases" are at equilibrium, while under non-equilibrium conditions it varies on a longer $O(\epsilon^{-1})$ scale. Problems containing widely separated scales should be solved by matching expansions in the *inner* region (localized at the front), and the *outer* regions spreading out to the domains where the film thickness is almost constant. Evaporation takes place in the outer domains and can be neglected in a narrow front region. The front may be set into motion under the action of a weak gradient of chemical potential developing as a result of evaporation. The multiscale expansion approach is suitable to track slow motion of the front at times far exceeding the characteristic relaxation time T_l to a stationary front profile.

We shall allow the front to be weakly curved with a curvature radius of $O(\epsilon^{-1})$. The equations in the inner region can be written then in the *aligned comoving frame*. By convention, the x axis is directed normally to the nominal front position in such a way that the film thickness is lower to the left. The characteristic scale along this direction is the $O(1)$ short scale, while the coordinate y parallel to the front is scaled by ϵ^{-1}. We suppose that the curvature radius is of $O(\epsilon^{-1})$; then the curvature is written as $\epsilon\kappa$ when measured on the short inner scale. The curvature is positive when the thin film domain is convex. Inasmuch as the front region is assumed to be locally at equilibrium, the front is expected to move under the influence of long-scale changes of chemical potential. Therefore the propagation speed should be measurable on a long scale L/T_L, and can be written as ϵc. The chemical potential within the front region should differ from the Maxwell construction by $O(\epsilon)$, and expressed as $\mu = \mu_s + \epsilon\mu_1$. Using this in Eq. (84) and expanding also the film thickness, $h = h_0 + \epsilon h_1 + \ldots$, we obtain in the first-order

$$\mathrm{d}_x \left[q(h_0)\mu_1'(x)\right] + ch_0'(x) = 0. \tag{86}$$

$$-\mu_1 = h_1''(x) + \kappa h_0'(x) - \Pi'(h_0)h_1. \tag{87}$$

The solutions of the inner equations should be matched at $x \to \pm\infty$ with the outer solutions, which we denote as $\mu^{\pm}(X)$, $h^{\pm}(X)$. The matching point should lie at a distance from the front that is large on the inner but small on the outer scale; the result must be independent of a precise matching position within this range.

Integrating Eq. (86) and using the matching conditions

$$\mu_1'(x) = \epsilon^{-1}\partial_x\mu^{\pm} = \boldsymbol{n}\cdot\widehat{\nabla}\mu^{\pm} \equiv -j^{\pm} \tag{88}$$

yields the material balance relation

$$c(h^+ - h^-) = -q(h^+)\boldsymbol{n}\cdot\widehat{\nabla}\mu^+ + q(h^-)\boldsymbol{n}\cdot\widehat{\nabla}\mu^- \equiv j^+ - j^-. \tag{89}$$

Here $\boldsymbol{n}$ is the normal to the front (directed in the same way as the x axis). The r.h.s. of Eq. (89) is the difference of the fluxes on the two sides $j^{\pm}$ driven by the gradients of chemical potential in the outer regions. This integral condition defines therefore the speed of local interface displacement required to ensure mass conservation.

Since the variable part of μ is restricted in the front region to $O(\epsilon)$, the film thickness in the matching regions may deviate from the equilibrium values $h_s^{\pm}(\mu_s)$ by no more than $O(\epsilon)$. Linearizing $\Pi(h)$, the fluxes $j^{\pm}$ can be expressed therefore as $j^{\pm} = -D^{\pm}\boldsymbol{n}\cdot\widehat{\nabla}h^{\pm}$, where $D^{\pm} = q(h_s^{\pm})\Pi'(h_s^{\pm})$ are effective diffusivities. It is notable that, in spite of a strong dependence of mobility on h, the effective diffusivity turns out to be larger in the *thin* film domain (see Fig. 16. The mass flux through the front is constant, and is given by either of the equivalent expressions

$$-j_1 = ch_s^{\pm} + q(h_s^{\pm})\boldsymbol{n}\cdot\widehat{\nabla}\mu^{\pm} = ch_s^{\pm} - j^{\pm} = \frac{j^+h_s^- - j^-h_s^+}{h_s^+ - h_s^-}. \tag{90}$$

5.4 Solvability Condition

It remains to determine the first-order correction to the interfacial chemical potential and the related values of the film thickness $h^{\pm}$ at the matching locations on both sides of the interface. For this purpose, Eq. (86) is integrated twice to yield

$$\mu_1(x) = \overline{\mu}_1 - \int_{x_0}^{x} \frac{j_1 + ch_0(x)}{q(h_0(x))}\,\mathrm{d}x. \tag{91}$$

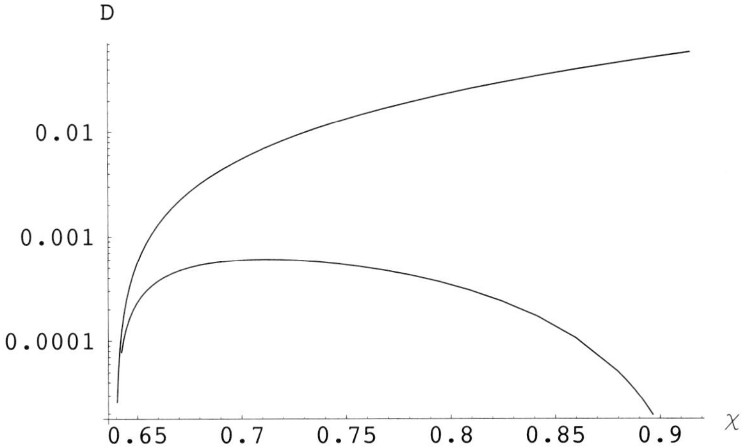

Figure 16. Dependence of effective diffusivities in thin (upper curve) and "thick" (lower curve) film domains on the parameter χ.

The integration constant $\overline{\mu}_1 = \mu_1(x_0)$ has to be determined by using the above expression in Eq. (87) and computing the solvability condition of this linear inhomogeneous equation. The position x_0 is arbitrary, and should fall out of the final result.

The linear operator $\mathrm{d}_x^2 - \Pi'(h_0(x))$ in Eq. (87) is self-adjoint and has a zero eigenvalue corresponding to the translational Goldstone mode $h_0'(x)$. Multiplying Eq. (87) by $h_0'(x)$ and integrating yields

$$\overline{\mu}_1(h_s^+ - h_s^-) = j_1\mathcal{J}_0 + c\mathcal{J}_1 - \kappa\mathcal{I}, \tag{92}$$

where the integrals $\mathcal{I}$, $\mathcal{J}_k$ depend on the stationary front profile only:

$$\mathcal{I} = \int_{-\infty}^{\infty} [h_0'(x)]^2 \mathrm{d}x = \int_{h_s^-}^{h_s^+} p(h)\mathrm{d}h, \tag{93}$$

$$\begin{aligned} \mathcal{J}_k &= \int_{-\infty}^{\infty} h_0'(x)\mathrm{d}x \int_{x_0}^{x} \frac{h_0^q(\widetilde{x})}{q(h_0(\widetilde{x}))}\mathrm{d}\widetilde{x} = \int_{h_s^-}^{h_s^+} \mathrm{d}h \int_{h^0}^{h} \frac{\widetilde{h}^k}{p(\widetilde{h})q(\widetilde{h})}\mathrm{d}\widetilde{h} \\ &= \int_{h^0}^{h_s^+} \frac{\widetilde{h}^q(h_s^+ - \widetilde{h})}{p(\widetilde{h})q(\widetilde{h})}\mathrm{d}\widetilde{h} - \int_{h_s^-}^{h^0} \frac{\widetilde{h}^k(\widetilde{h} - h_s^-)}{p(\widetilde{h})k(\widetilde{h})}\mathrm{d}\widetilde{h} \equiv \mathcal{J}_k^+ - \mathcal{J}_k^-. \end{aligned} \tag{94}$$

The value $h^0 = h(x_0)$ can be chosen to coincide with the unstable intermediate stationary solution h_s^0, but this is not necessary, as any value within the interval $h_s^- < h^0 < h_s^+$ will fit as well.

The integral $\mathcal{I}$ is recognized as dimensionless line tension, or energy per unit length of the front. The dependence of the excess chemical potential on curvature in Eq. (92) expresses a 2D analog of the Gibbs–Thomson relation, while the other two terms give dynamic corrections due to the flux through the interface.

The two values of chemical potential $\mu^\pm$ or film thickness $h^\pm$ to be used as interfacial boundary conditions for the outer equation are obtained by matching the outer limit of the inner solution with the inner limit of the outer solution. The limit of the inner solution (91) at $x \to \pm\infty$ is

$$\lim_{x\to\pm\infty} \mu_1(x) = \overline{\mu}_1 + c\mathcal{J}_0^\pm - j^\pm \mathcal{K}^\pm + (x - x_0)\,\partial_x \mu^\pm, \tag{95}$$

where we have used Eq. (90) and separated the converging integrals $\mathcal{J}_0^\pm$, defined by Eq. (94), and

$$\mathcal{K}^\pm = \int_{x_0}^{\pm\infty} \left[\frac{1}{q(h_0(x))} - \frac{1}{q(h_s^\pm)}\right] \mathrm{d}x = \int_{h_s^0}^{h_s^\pm} \frac{1}{p(h)} \left[\frac{1}{q(h)} - \frac{1}{q(h_s^\pm)}\right] \mathrm{d}h. \tag{96}$$

This formula, together with Eqs. (89), (90), and (92), yields local relations between the values of chemical potential and fluxes on both sides of the front, which can serve as the boundary conditions for outer equations. One can check by differentiating the integrals $\mathcal{J}_k^\pm$, $\mathcal{K}^\pm$ with respect to the variable limit that the limiting value of $\mu(x)$ is indeed independent of x_0 (which can be now set to zero) or h^0 . The respective limits $h^\pm$ can be obtained using the near-equilibrium linearized relation

$$\mu^\pm = \mu_s + \Pi'(h_s^\pm)(h^\pm - h_s^\pm). \tag{97}$$

6 Evaporative Patterns

6.1 Straight-line front

The film evaporates when the ambient chemical potential μ^0 drops below the prevailing chemical potential in the film, i.e. the Maxwell construction level μ_s. Evaporation taking place in the outer regions is described by Eq. (85) written in the long-scale coordinates. Assuming $\Delta\mu = \mu^0 - \mu_s = O(\epsilon)$, the changes of chemical potential are restricted to $O(\epsilon)$ also in the outer domains, so that $h^\pm$ remain close to the two stationary values $h_s^\pm$. When $|\Delta\mu|$ is sufficiently small, evaporation is slowed down to such a level that evolution is quasistationary. Then the time derivative in l.h.s of Eq. (85) is negligible, and it reduces, in the leading order, to the inhomogeneous Helmholtz equation

$$q^{\pm}\widehat{\nabla}^2\mu^{\pm} + \mu^0 - \mu^{\pm} = 0, \tag{98}$$

where $q^{\pm} = q(h_s^{\pm})$. This equation has to be solved in both outer domains to find a relation between the fluxes $j^{\pm}$ and the values of chemical potential at the front. The problem is closed by applying the matching condition (95).

Equation (98) can be solved as a Dirichlet problem, setting $\mu^{\pm}(\Gamma) = \overline{\mu}^{\pm}$, where Γ denotes the instantaneous position of the front, and $\overline{\mu}^{\pm}$ is as yet unknown chemical potential on either side of the front. Once the solution is found, the fluxes $j^{\pm}$ are computed. The chemical potential close to the front, i.e. at a distance $X \ll 1$ along the normal $\boldsymbol{n}$ to Γ, which is small on the outer scale, is computed by expanding the solution in Taylor series, and the first-order expansion should match the last term in Eq. (95). The remaining constant terms yield two matching conditions for computing $\overline{\mu}^{\pm}$:

$$\begin{aligned} \overline{\mu}^{\pm} - \mu_s = \; & \pm j^{\pm}\left[\frac{\mathcal{J}_1 - h_s^{\mp}\mathcal{J}_0}{(h_s^+ - h_s^-)^2} + \frac{\mathcal{J}_0^{\pm}}{h_s^+ - h_s^-} - \mathcal{K}^{\pm}\right] \\ & \mp j^{\mp}\,\frac{\mathcal{J}_1 + h_s^+\mathcal{J}_0^- - h_s^-\mathcal{J}_0^+}{(h_s^+ - h_s^-)^2} - \frac{\kappa\mathcal{I}}{h_s^+ - h_s^-}. \end{aligned} \tag{99}$$

If the fluxes $j^{\pm}$ are expressed through $\overline{\mu}^{\pm}$ using an appropriate Green's function, Eq. (99) reduces to an integral equation defining the the local values of chemical potential. Once the fluxes on the front are known, the local propagation speed is computed with the help of Eq. (89). Successive positions of the front can be tracked by shifting it in accordance to the computed values. Following the evolution of the front in this way requires, of course, solving the integral equations at each time step. Simpler explicit solutions, which can be resolved analytically to the end, are obtained for symmetric arrangements further in this Section.

The simplest example is a stationary solution of Eq. (98) corresponding to a straight-line front at $X = 0$. The general solution is

$$\mu^{\pm} = \mu^0 + C^{\pm}\exp\left(\mp X/\sqrt{q^{\pm}}\right). \tag{100}$$

The constants $C^{\pm}$ should be obtained by matching the inner limit of this expression with the outer limit (95) of the inner solution, where the fluxes $j^{\pm} = \pm\sqrt{q^{\pm}}C^{\pm}$ are evaluated using Eq. (100). The last term in Eq. (95) matches the first-order expansion of Eq. (100):

$$x\,\partial_x\mu^{\pm} = X\,\partial_X\mu^{\pm} = \mp C^{\pm}X/\sqrt{q^{\pm}}.$$

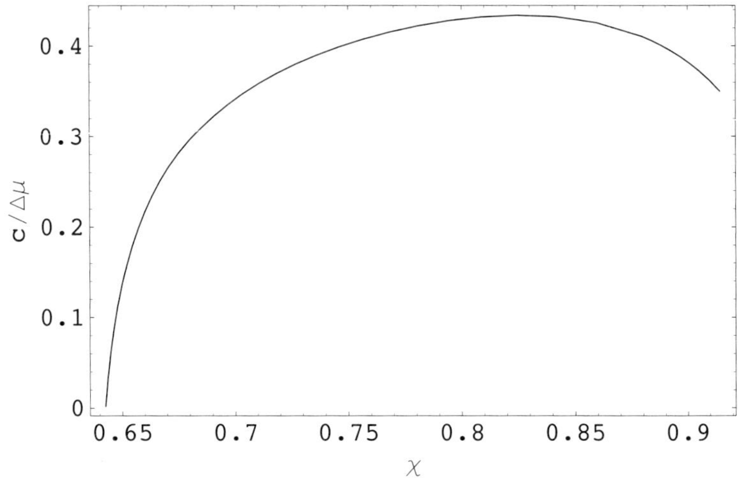

Figure 17. Dependence of the speed of a straight-line front per unit chemical potential difference, $c/\Delta\mu$, on the parameter χ.

The remaining constant terms yield two matching conditions for computing $C^{\pm}$:

$$\begin{aligned}\Delta\mu + C^{\pm} = \; & C^{\pm}\sqrt{q^{\pm}}\left[\frac{\mathcal{J}_1 - h_s^{\mp}\mathcal{J}_0}{(h_s^+ - h_s^-)^2} + \frac{\mathcal{J}_0^{\pm}}{h_s^+ - h_s^-} - \mathcal{K}^{\pm}\right] \\ & + C^{\mp}\sqrt{q^{\mp}}\,\frac{\mathcal{J}_1 + h_s^+\mathcal{J}_0^- - h_s^-\mathcal{J}_0^+}{(h_s^+ - h_s^-)^2}.\end{aligned} \tag{101}$$

The dependence of the ratio $c/\Delta\mu$ on the parameter χ obtained by solving these two linear equations is shown in Fig 17. The speed is positive, i.e. the thin film advances, when $\Delta\mu < 0$. A slow-down at χ close to the upper limit is due to increasing capacity of the thick film. Since the equilibrium thickness decreases on the thick side and increases on the thin side when μ becomes more negative, the film profile develops a bump on the thick side a dip on the thin side during evaporation. When $\Delta\mu > 0$, i.e. during condensation, the profile near the front is monotonic.

6.2 "Pancake" and "hole"

Circularly symmetric solutions are applicable to isolated droplets, "pancakes" or holes removed from other similar objects, as well as from any boundaries, at a distance far exceeding the characteristic horizontal scale

$L = d^{3/2}/\sqrt{\beta\eta}$, i.e. unity in the long-scale dimensionless units of Eq. (98). The stationary solution of Eq. (98) corresponding to a circular thick-film "pancake" of a radius a immersed in an infinite thin layer is

$$\mu^+ = \mu^0 + C^+ I_0\left(r/\sqrt{q^+}\right), \quad \mu^- = \mu^0 + C^- K_0\left(r/\sqrt{q^-}\right), \tag{102}$$

where r is the radial coordinate and I_0, K_0 are modified Bessel functions. The constants $C^\pm$ should be obtained in the same way as in the preceding subsection using the matching condition (101) at the front $r = a$, where one should set

$$j^+ = \sqrt{q^+} I_1\left(a/\sqrt{q^+}\right), \quad j^- = -\sqrt{q^-} K_1\left(a/\sqrt{q^-}\right), \quad \kappa = -a^{-1}.$$

The problem is solved analytically, but the resulting expressions are too cumbersome and the results can better assessed graphically. The front speed is expressed by a linear relation

$$c = f_0(a) + f_1(a)\Delta\mu. \tag{103}$$

Notably, the front speed *increases* with growing radius. The dependence is monotonic at large $|\Delta\mu|$, while at smaller $|\Delta\mu|$ the speed passes a maximum at a certain radius (see Fig. 18).

The increase of the front propagation speed means that "large droplets evaporate faster", in accordance to experimental observations (Leizerson *et al*, 2003). This is, in essence, a consequence of the "bulk" evaporation from the interior of the droplet, which has no analog in conventional evaporation of 3D droplets. The effect disappears when evaporations when flux through the boundary plays a larger role in evaporation of "thick" domains. One can seen from a simple calculation that, if evaporation rate per unit area were constant, the shrinkage rate following from the material balance would be $da/dt \propto a$. The actual radial dependence is weaker than linear, which can be attributed to slower evaporation from larger droplets, which are closer to equilibrium with the ambient vapor phase in their interior. The total evaporation time *increases* with radius, as it must do, since even a large droplet becomes small before disappearing altogether. Nevertheless, smaller droplets are less disadvantaged here than in a usual coarsening process when the front speed increases with increasing curvature.

For an opposite arrangement – a thin "hole" of a radius a immersed in an infinite thick layer, the stationary solution of Eq. (98) is

$$\mu^+ = \mu^0 + C^+ K_0\left(r/\sqrt{q^+}\right), \quad \mu^- = \mu^0 + C^- I_0\left(r/\sqrt{q^-}\right). \tag{104}$$

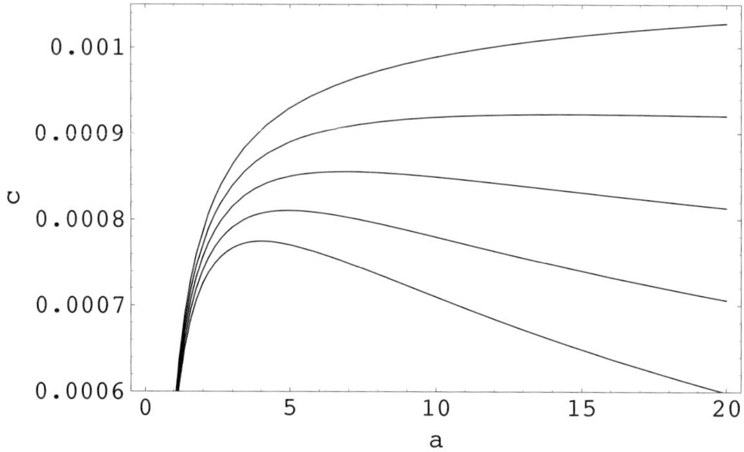

Figure 18. The dependence of the speed of the circular boundary on the "pancake" radius at $\chi = 0.9$ and different values of $\Delta\mu$. The latter change from $\Delta\mu = -0.5 \times 10^{-3}$ for the lower curve to $\Delta\mu = -2 \times 10^{-3}$ for the uper curve with the increment 0.5×10^{-3}.

The constants $C^{\pm}$ are obtained using the matching condition (101) at the front $r = a$, where one should set

$$j^{+} = \sqrt{q^{+}} K_1\left(a/\sqrt{q^{+}}\right), \quad j^{-} = -\sqrt{q^{-}} I_1\left(a/\sqrt{q^{-}}\right), \quad \kappa = a^{-1}.$$

The dependence of the critical radius a_c of an incipient hole nucleating during evaporation on $\Delta\mu$ is shown in Fig 19.

6.3 Solution in a comoving frame

The quasistationary approximation becomes inadequate when the front propagation speed increases. Then the non-stationary equation (85) has to be solved in the outer regions. Assuming as before that $h^{\pm}$ remain close to the two stationary values $h_s^{\pm}$, and using the equilibrium relation (97), the outer equations can be rewritten as

$$[\Pi'(h^{\pm})]^{-1}\partial_t \mu^{\pm} = q^{\pm}\widehat{\nabla}^2\mu^{\pm} + \mu^0 - \mu^{\pm}. \tag{105}$$

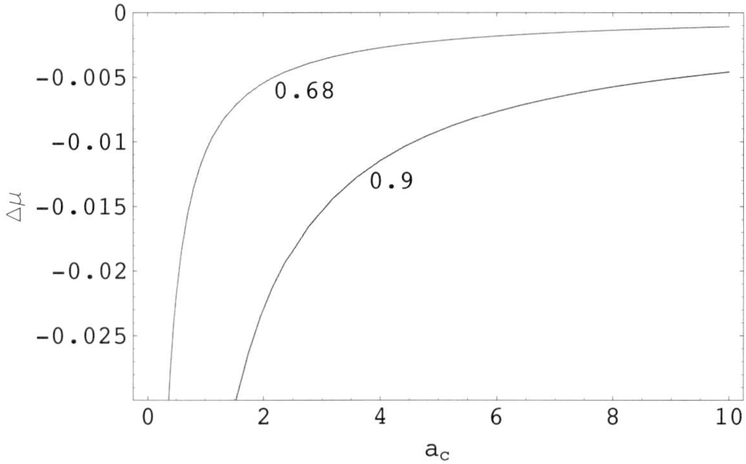

Figure 19. The dependence of the critical radius a_c of a hole nucleating during on $\Delta\mu$ at two different values of the parameter χ.

For a straight-line front, this equation can be solved in the comoving frame propagating with an as yet unknown speed c. It is rewritten then as

$$\widehat{c}\mu_X^{\pm} + q^{\pm}\widehat{\mu}_{XX}^{\pm} + \mu^0 - \mu^{\pm} = 0, \tag{106}$$

where $\widehat{c} = c/\Pi'(h_s^{\pm})$. The general solution is

$$\mu^{\pm} = \mu^0 + C^{\pm}\mathrm{e}^{\lambda^{\pm}X}, \quad \lambda^{\pm} = -\frac{\widehat{c}}{2q^{\pm}}\left(1 \pm \sqrt{1 + \frac{4q^{\pm}}{\widehat{c}^2}}\right). \tag{107}$$

The fluxes $j^{\pm} = -q^{\pm}\lambda^{\pm}C^{\pm}$ are now velocity-dependent, so that these expressions have to be solved together with Eq. (89), and the algebraic structure of the matching conditions (99) becomes very cumbersome. Fortunately, the dependence on $\Delta\mu$ remains linear, and the implicit dependence $\Delta\mu(c)$ can be obtained analytically, albeit in a form not fit for a human eye. Two typical curves are shown in Fig. 20. A graphical comparison with the quasistationary solution in the inset shows that the difference between the propagation speed computed in the comoving frame and that computed in the quasistationary approximation remains within a few percentage points when $\Delta\mu$ is within the same range as the difference between μ_s and the equilibrium chemical potential for the bulk fluid. This can serve as an indication

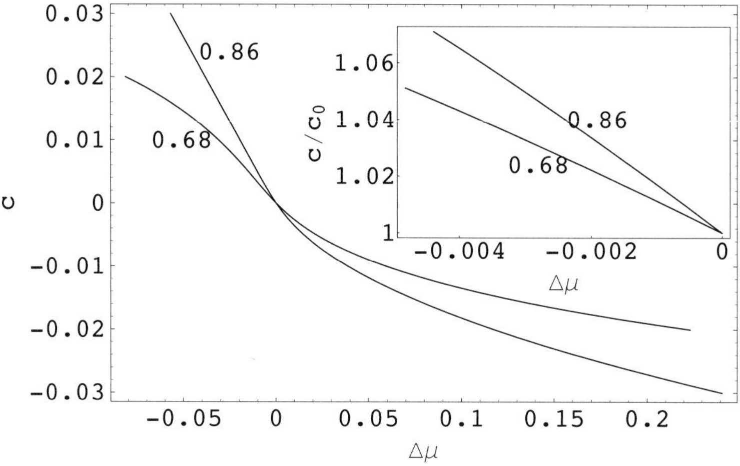

Figure 20. The propagation speed computed in comoving frame as a function of $\Delta\mu$ at two different values of the parameter χ. Inset: the ratio of propagation speed computed in comoving frame to that computed in the quasistationary approximation.

of reliability of the above results for the "pancake" and "hole" configurations, which do not admit a comoving formulation of the outer problem, and leave full dynamic simulation as the only alternative to the quasistationary solution.

The nonlinear character of the $c(\Delta\mu)$ dependence , which is characteristic to the comoving frame computations, becomes qualitatively important at high speeds. There is a maximum propagation speed at $\Delta\mu < 0$, but it is achieved at very high potential differences, far beyond the instability threshold to be computed in the next subsection. The nonlinearities are interrelated with changes in the flux through the interface j_1. At relevant small propagation speeds, up to and beyond the instability limit, the flux is directed during evaporation towards the thin film domain, implying a higher efficiency of evaporation in the thin layer.

6.4 Zigzag instability

A straight-line front may become unstable to transverse perturbations. Since the front is neutrally stable to translations, the onset of instability is most likely to occur in a *long-scale* mode. We consider a straight-line front parametrized by the coordinate Y and denote the instantaneous displacement of the front from its unperturbed position along the X axis directed towards the region occupied by the thick film as $\zeta(Y,T)$. The inner front solution is assumed to remain quasistationary, while instability is developing or decaying in the outer domains; therefore the outer (long) scale is suitable for stability analysis. As long as the amplitude of the perturbation is much smaller than its wavelength, the normal vector defining the direction of the front propagation, is almost parallel to the X axis, and the propagation speed is expressed as $\zeta_T = c$. In the same approximation, the curvature is given by $\kappa = -\zeta_{YY}$.

The long-scale instability can be studied in the comoving frame using Eq. (106) corrected by taking into account front curvature, which is assumed to be small *even on the extended scale*:

$$\widehat{c}\mu_X^{\pm} + q^{\pm}(\widehat{\mu}_{XX}^{\pm} - \kappa\mu^{\pm}) + \mu^0 - \mu^{\pm} = 0. \tag{108}$$

The local front propagation speed, as well as $\mu^{\pm}$ and $j^{\pm}$, are expanded in $\kappa \ll 1$. The curvature correction to the front propagation speed is presented as $c = c_0 + \widetilde{c}\kappa$, where c_0 is the speed of a straight-line front evaluated in the preceding subsection. The first-order correction $\widetilde{c}$ is computed by using the first-order solution of Eq. (108) in the matching conditions at the front (89), (99) expanded likewise to the first order in κ. The problem is solved analytically along the same lines as in the preceding subsection, producing awesome expressions handled by computer algebra.

The front is unstable when the curvature correction $\widetilde{c}$ is positive. The instability is observed at $\Delta\mu < 0$ (i.e. during evaporation) in an interval limited both from below and from above; at $\Delta\mu > 0$ (condensation) the front is always stable. The curves $\widetilde{c}(\Delta\mu)$ for two chosen values of χ are shown in Fig. 21. The onset of instability is observed at rather low propagation speeds, and the front stabilizes again at higher speeds or $|\Delta\mu|$, as, apparently, perturbations are swept under by the propagating front. It is possible, however, that other instabilities, which we do not investigate here, become relevant at high speeds. The values of $|\Delta\mu|$ and c_0 at the lower limit of zigzag instability are plotted as functions of χ in Fig. 22. A full dispersion relation, including both transverse instabilities at finite wavelength and oscillatory instabilities, can be obtained using the same approach as described above, but rather extensive computations are needed to determine the instability limits.

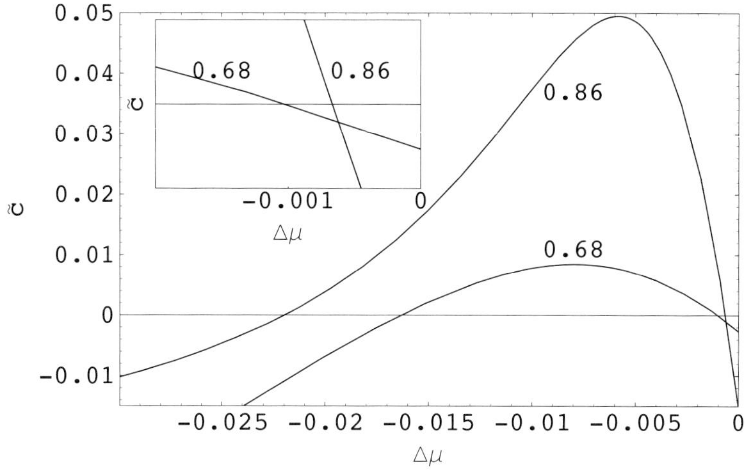

Figure 21. The curvature correction $\widetilde{c}$ to the front propagation speed computed in comoving frame as a function of $\Delta\mu$ at two different values of the parameter χ. Inset: blow-up near the origin. The front is unstable when $\widetilde{c} > 0$.

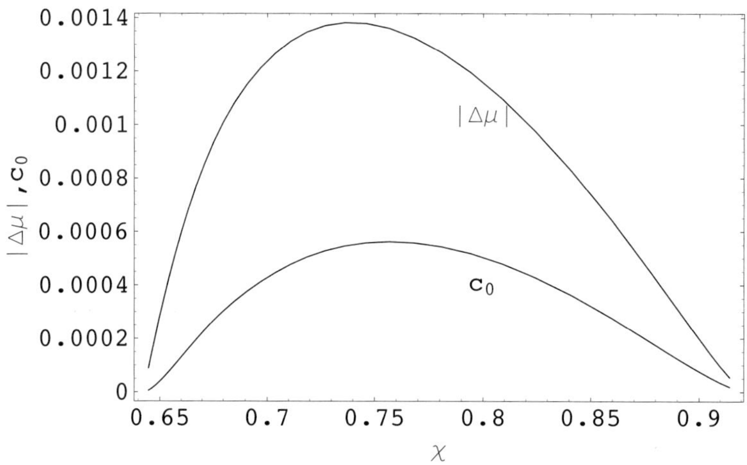

Figure 22. The values of the chemical potential difference $|\Delta\mu|$ and propagation speed of the straight-line front c_0 at the lower limit of zigzag instability as functions of χ.

Acknowledgement This work has been supported by Israel Science Foundation (grant 55/02).

Bibliography

B.V. Derjaguin, N.V. Churaev and V.M. Muller, *Surface Forces*, Consultants Bureau, New York, 1987.
A. Oron, S.G. Bankoff, and S.H. Davis, Rev. Mod. Phys. **6**9, 931 (1997).
L.M. Pismen, Colloids and Surfaces A, **20**6 11 (2002).
M. Bestehorn, this volume.
A. Sharma, Langmuir **9**, 3580 (1993).
L.M. Hocking, J. Fluid Mech. **7**9, 209 (1977).
L.M. Pismen and B.Y. Rubinstein, Langmuir, **17** 5265 (2001).
L.M. Pismen and Y. Pomeau, Phys. Fluids **1**6, 2604 (2004).
K.B. Glasner and T.P. Witelski, Phys. Rev. E **6**7, 016302 (2003).
I.M. Lifshitz and V.V. Slyozov, J. Phys. Chem. Solids **1**9, 35 (1961).
B. Giron, B. Meerson, and P.V. Sasorov, Phys. Rev. E **5**8, 4213 (1998).
Y. Sumino, H. Kitahata, and K. Yoshikawa, Phys. Rev. E **7**2, 041603 (2005).
L. M. Pismen, Phys. Rev. Lett. **8**6, 548 (2001).
L. M. Pismen, *Patterns and Interfaces in Dissipative Dynamics*, Springer, Berlin (2006).
S. W. Lee, D. Y. Kwok, and P. E. Laibinis, Phys. Rev. E **65**, 051602 (2002).
U. Thiele, K. John, and M. Bär, Phys. Rev. Lett. **93**, 027802 (2004).
I. Leizerson, S.G. Lipson, and A.V. Lyushnin, Phys. Rev. E **6**8, 051601 (2003).

Interfacial Phenomena in Materials Science

Alexander A. Golovin
Department of Engineering Sciences and Applied Mathematics
Northwestern University, USA

Abstract The present lecture notes discuss several important problems in materials science where interfacial phenomena play a crucial role, namely: equilibrium crystal shapes and Wulff construction, growth of a crystal nucleus from an undercooled melt, Mullins-Sekerka instability, growth of a dendrite and Ivantsov's solution, surface-diffusion phenomena and grain boundary grooving as well as the Asaro-Tiller-Grinfeld instability of solid epitaxial films that leads to the formation of quantum dots. Mathematical models of the corresponding phenomena are presented in detail. The lecture notes are based on the books of Pimpinelli and Villain (2001), Davis (2001), Godreche (1992), Landau and Lifshits (1978) and original papers cited in the references.

Contents

1 Equilibrium crystal shape

One of the oldest interfacial problems in materials science is associated with the following question: what determines the equilibrium crystal shape? The answer to this question is the same as to another one: why liquid drops are spherical? It is common knowledge that liquid drops are spherical because a sphere is a body that has minimal surface area at a given constant volume. Since a free liquid surface is characterized by excess, surface energy per unit area, a spherical shape minimizes the drop energy and therefore corresponds to thermodynamic equilibrium.

The similar argument works for equilibrium crystal shapes. The difference is that, unlike simple liquids, crystals are not isotropic: they have distinguished directions determined by the structure of the crystal lattice. As a result, surface energy of a free surface of a crystal depends on the surface orientation with respect to the crystal lattice. Thus, the question of the equilibrium shape of a crystal can be mathematically formulated in the following way: given the dependence of the crystal surface energy on the surface orientation, find the surface shape that minimizes the total surface energy, provided the crystal volume is constant.

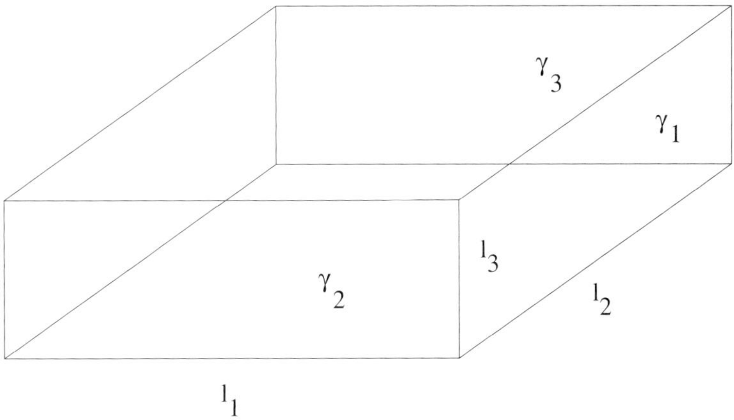

Figure 1. A crystal parallelepiped with anisotropic surface energy.

Consider first the following simple problem. Consider a crystal that has the shape of a parallelepiped with the edges having the lengths l_1, l_2 and l_3 (see Fig.1). Let the faces normal to the edges l_i have the surface energy per unit area γ_i, $i = 1, 2, 3$, see Fig.1. The total surface energy is, therefore,

$$E_s = \gamma_1 l_2 l_3 + \gamma_2 l_3 l_1 + \gamma_3 l_1 l_2, \tag{1}$$

and the volume of the crystal is

$$V = l_1 l_2 l_3. \tag{2}$$

Thus we have the following problem: find $l_{1,2,3}$ such that $E_s = min$ and $V = const$. This is a problem of a conditional extremum that should be solved using a Lagrange multiplier. Consider $\Phi = E_s - \lambda V$, where λ is the Lagrange multiplier. The condition of an extremum,

$$\frac{\partial \Phi}{\partial l_1} = \frac{\partial \Phi}{\partial l_2} = \frac{\partial \Phi}{\partial l_3} = 0 \tag{3}$$

gives the system of equations

$$\begin{aligned} \gamma_3 l_2 + \gamma_2 l_3 - \lambda l_2 l_3 = 0, \\ \gamma_1 l_3 + \gamma_3 l_1 - \lambda l_1 l_3 = 0, \\ \gamma_1 l_2 + \gamma_2 l_1 - \lambda l_1 l_2 = 0, \end{aligned} \tag{4}$$

from which one obtains

$$\frac{\gamma_3}{l_3} + \frac{\gamma_2}{l_2} = \frac{\gamma_1}{l_1} + \frac{\gamma_3}{l_3} = \frac{\gamma_1}{l_1} + \frac{\gamma_2}{l_2} = \lambda, \tag{5}$$

or

$$\frac{\gamma_1}{l_1} = \frac{\gamma_2}{l_2} = \frac{\gamma_3}{l_3}. \tag{6}$$

Thus, one can see that such a crystal will have an equilibrium shape if the length of each edge is proportional to the surface energy of the crystal face normal to this edge. Equivalently, one can say that in equilibrium, the distance to each face from the crystal center must be proportional to the face surface energy. The lengths of the edges, $l_{1,2,3}$, can be found from the system of four equations, (4) and (2) for four unknowns, $l_{1,2,3}$ and λ.

Now consider a general case. Suppose that the crystal surface shape is given in Descartes coordinates by a function $z(x, y)$, and the surface energy per unit area, γ, depends on the surface orientation, i.e. $\gamma = \gamma(p, q)$, where $p = z_x$, $q = z_y$. Thus we have the following variational problem: find $z(x, y)$ that gives minimum value to the surface energy functional,

$$E_s = \int \gamma(p, q) \sqrt{1 + p^2 + q^2}\, dxdy = min, \tag{7}$$

under condition that the crystal volume is constant,

$$V = \int z\, dxdy = const. \tag{8}$$

This is a problem for a conditional extremum of a functional. Therefore, one should use the Euler-Lagrange equation for the functional

$$F = E_s - \lambda V = \int (f - \lambda z)\, dxdy, \tag{9}$$

where $f = \gamma(p, q)\sqrt{p^2 + q^2}$ is the so-called weighted surface energy, and λ is the Lagrange multiplier. We obtain

$$\frac{\partial F}{\partial z} - \frac{\partial}{\partial x}\frac{\partial F}{\partial p} - \frac{\partial}{\partial y}\frac{\partial F}{\partial q} = 0, \tag{10}$$

or

$$\frac{\partial}{\partial x}\frac{\partial f}{\partial p} + \frac{\partial}{\partial y}\frac{\partial f}{\partial q} = -\lambda. \tag{11}$$

Eq. (11) is a nonlinear partial differential equation for an unknown function of the crystal shape, $z(x, y)$. Remarkably, this second-order nonlinear partial differential equation has a very elegant geometric solution.

Let us introduce a new function,

$$\zeta = px + qy - z = z_x x + z_y y - z. \tag{12}$$

This function is called a *Legendre transform* of the function $z(x, y)$. Since $dz = z_x dx + z_y dy$, then

$$d\zeta = xdp + pdx + ydq + qdy - dz = xdp + ydq. \tag{13}$$

Thus, ζ is a function of the variables p and q, and

$$x = \frac{\partial \zeta}{\partial p}, \qquad y = \frac{\partial \zeta}{\partial q}. \tag{14}$$

Now let us use the properties of the Jacobian of two functions, ϕ and ψ,

$$\frac{\partial(\phi, \psi)}{\partial(x, y)} = \begin{vmatrix} \phi_x & \phi_y \\ \psi_x & \psi_y \end{vmatrix}, \tag{15}$$

and write

$$\frac{\partial}{\partial x}\frac{\partial f}{\partial p} = \frac{\partial(f_p, y)}{\partial(x, y)} = \frac{\partial(f_p, \zeta_q)}{\partial(x, y)}, \quad \frac{\partial}{\partial y}\frac{\partial f}{\partial q} = \frac{\partial(x, f_q)}{\partial(x, y)} = \frac{\partial(\zeta_p, f_q)}{\partial(x, y)}. \tag{16}$$

Using (16), we can rewrite eq.(11) as

$$\frac{\partial(f_p, \zeta_q)}{\partial(x, y)} + \frac{\partial(\zeta_p, f_q)}{\partial(x, y)} = -\lambda = const. \tag{17}$$

Now multiply eq.(17) by $\partial(x,y)/\partial(p,q)$ and use the property of a Jacobian,

$$\frac{\partial(\phi,\psi)}{\partial(x,y)} \cdot \frac{\partial(x,y)}{\partial(u,v)} = \frac{\partial(\phi,\psi)}{\partial(u,v)}, \tag{18}$$

to obtain

$$\frac{\partial(f_p,\zeta_q)}{\partial(p,q)} + \frac{\partial(\zeta_p,f_q)}{\partial(p,q)} = -\lambda\frac{\partial(\zeta_p,\zeta_q)}{\partial(p,q)}. \tag{19}$$

By simple inspection we can see that

$$f = -\frac{\lambda}{2}\zeta \tag{20}$$

satisfies eq.(19). This gives the following relation between the surface energy and the surface shape:

$$\gamma(p,q)\sqrt{1+z_x^2+z_y^2} = \frac{\lambda}{2}(z-px-qy),$$

or

$$\frac{2}{\lambda}\gamma(p,q) = \frac{z-xz_x-yz_y}{\sqrt{1+z_x^2+z_y^2}} = (x,y,z)\cdot(-z_x,-z_y,1)\frac{1}{\sqrt{1+z_x^2+z_y^2}}. \tag{21}$$

Eq.(21) has a clear geometrical meaning shown in Fig.2. It implies that the projection of a radius-vector, drawn from the origin to a point at the crystal surface, on the direction normal to the surface at this point is proportional to the surface energy at this point. This, in turn, implies the following method of constructing an equilibrium crystal shape determined by an anisotropic surface energy, see Fig.2. Make a *polar plot* of surface energy such that the distance from the origin to a point on this plot is proportional to the surface energy at this point. At each point on this polar plot draw a straight line perpendicular to the radius-vector from the origin to this point. Then draw an *envelope* of this family of straight lines. This envelope will give you the equilibrium crystal shape. This is called *Wulff construction*, or *Wulff plot*. It was discovered by Wulff in 1901 (Wulff , 1901). The derivation of (21) presented in this chapter closely follows Landau and Lifshits (1978).

One can see that the shape of a crystal that has a shape of a rectangular parallelepiped is a particular case of the general Wulff construction. A spherical equilibrium shape corresponding to a constant surface energy is, obviously, a particular case of the Wulff construction, too.

The Wulff construction can explain why corners and edges appear on equilibrium crystal shape. For simplicity, consider a two-dimensional crystal whose one-dimensional surface is described by a function $h(x)$; a 1D

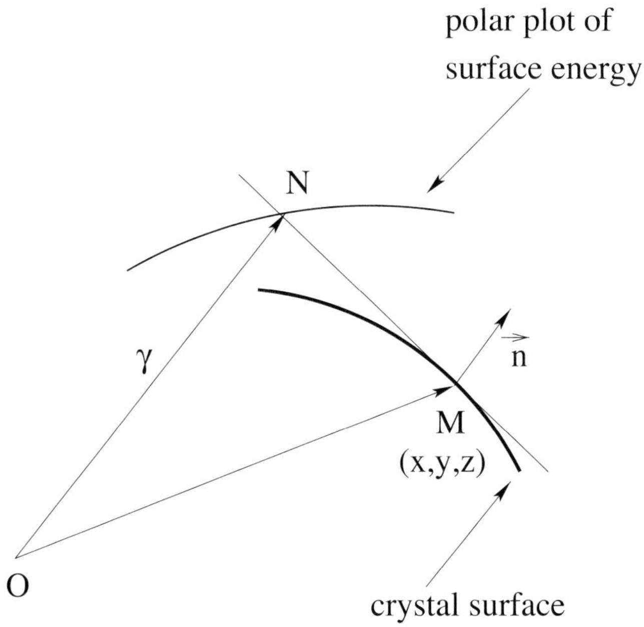

Figure 2. Wulff construction.

surface of a 2D crystal is often called a 1+1 surface. Let us assume that the surface energy depends on the surface orientation (slope), $\gamma = \gamma(h_x)$. Then, the equilibrium shape of such a crystal will be described by the following problem:

$$E_s = \int \gamma(h_x)\sqrt{1+h_x^2}\,dx = min, \quad V = \int h\,dx = const. \qquad (22)$$

It is convenient to introduce the surface slope angle, θ, so that $h_x = \tan\theta$. Then, after some algebraic manipulations, one can obtain that the Euler-Lagrange equation for the conditional extremum problem (22) gives the following condition for the equilibrium crystal shape:

$$K(\gamma + \gamma_{\theta\theta}) = const, \qquad (23)$$

where K is the surface curvature. The quantity $\gamma + \gamma_{\theta\theta}$ is called the *surface stiffness*.

For $\gamma = const$ (isotropic surface energy) eq.(23) describes a circle. If the surface energy is anisotropic and $\gamma + \gamma_{\theta\theta} > 0$ the surface parts having larger surface stiffness have smaller curvature, and vice versa. For $\gamma+\gamma_{\theta\theta} \to 0$ $K \to \infty$ and the corner forms. And what if $\gamma+\gamma_{\theta\theta} < 0$? Note that in the isotropic case the left hand side of eq.(23) is nothing but the capillary pressure, or the surface *chemical potential*: energy needed to add one particle to the system. Anisotropic system with negative surface stiffness is similar to a system with negative pressure, which is thermodynamically unstable. In other words, for positive surface stiffness the total energy is increased if the surface area is locally increased but for a negative surface stiffness the total energy decreases with the local increase of the surface area. This makes the surface with this particular surface orientation thermodynamically unstable: no equilibrium crystal shape with the surface orientation corresponding to a negative surface stiffness will be observed in experiment. The orientations for which the surface stiffness is negative are called *forbidden* or *missing* orientations.

This situation can be clearly illustrated for the case of a model, two-dimensional crystal with a cubic (four-fold) symmetry, whose surface energy depends on the surface orientation as

$$\gamma = \gamma_0(1 + \epsilon \cos 4\theta), \tag{24}$$

where θ is the surface orientation angle, and $0 < \epsilon < 1$ is the anisotropy coefficient. In this case the surface stiffness is

$$\gamma + \gamma_{\theta\theta} = \gamma_0(1 - 15\epsilon \cos 4\theta). \tag{25}$$

For $\epsilon < 1/15$, the surface stiffness is positive for all θ and the equilibrium crystal shape, although not spherical, will have no corners. For $\epsilon > 1/15$ the surface stiffness will be negative for some surface orientations. These are forbidden orientations that cannot exist at the equilibrium crystal shape. In this case the equilibrium crystal shape will develop *corners*. This situation is illustrated in Fig.3.

An interesting question is the following: if an equilibrium crystal shape develops a corner – what is the corner angle? The answer to this question can be obtained by analyzing the Euler-Lagrange equation for the equilibrium 1+1 crystal surface shape $h(x)$:

$$\frac{d}{dx}\frac{df}{dh_x} = \lambda = const, \quad f = \gamma(h_x)\sqrt{1 + h_x^2}. \tag{26}$$

indeed, from (26) one obtains

$$f = \lambda \int x \, dh_x = \lambda(xh_x - h). \tag{27}$$

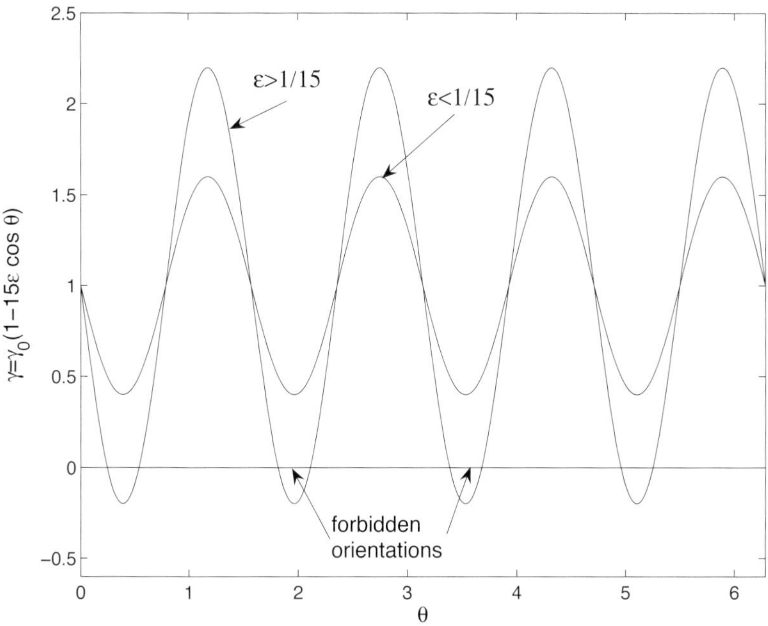

Figure 3. Forbidden orientations.

At the corner, $x = x_c$, $h = h_c$, and at the two sides of the corner $h_x^{(1)} \neq h_x^{(2)}$, $f^{(1)} \neq f^{(2)}$, $(df/dh_x)_1 = (df/dh_x)_2$, so that $f_{1,2} = \lambda(x_c h_x^{(1,2)} - h_c)$ and

$$\frac{f_1 - f_2}{h_x^{(1)} - h_x^{(2)}} = \lambda x_c = \left(\frac{df}{dh_x}\right)_1 = \left(\frac{df}{dh_x}\right)_2 . \tag{28}$$

The relations (28) mean that the two slopes at the corner correspond to the points of a *double tangent* to the graph of the weighted surface energy, $f = \gamma(h_x)\sqrt{1 + h_x^2}$, as a function of the surface slope, h_x. This is shown in Fig.4.

2 Growth of a spherical crystal nucleus

In this lecture we shall consider the growth of a spherical crystal nucleus from melt. We shall demonstrate that interfacial effects play a key role in this process.

Consider an undercooled melt whose temperature T_0 is less than the liquid-solid equilibrium (melting) temperature, T_m. Let us assume that a

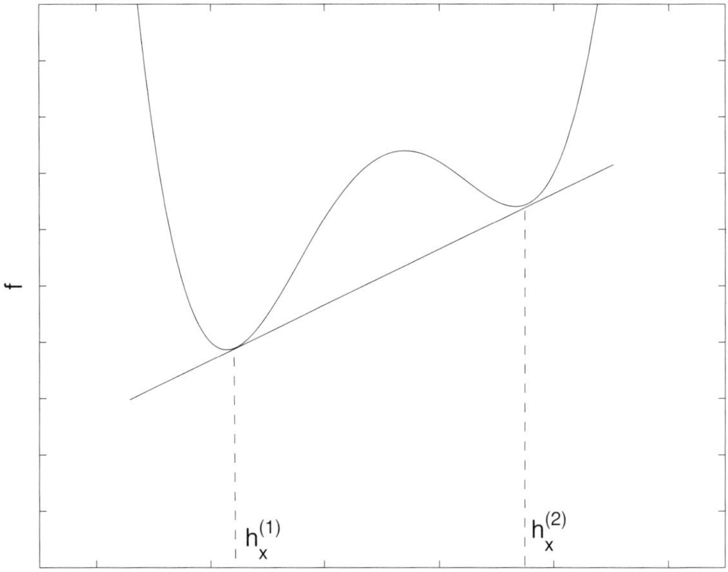

Figure 4. Double-tangent construction for corner angles.

spherical crystal nucleus of radius R has formed in the melt, and compute the corresponding change of free energy in the system, ΔG. The total free energy change has two components: the bulk component, ΔG_V, and the surface component, ΔG_S. Since the melt is undercooled,

$$\Delta G_V = -\Delta\mu \frac{4}{3}\pi R^3 < 0, \tag{29}$$

where $\Delta\mu = \mu_{melt} - \mu_{crystal} \propto T_m - T_0 > 0$ is the difference between the chemical potentials of the melt and the crystal at the ambient temperature $T_0 < T_m$. Eq.(29) shows that the bulk component of the free energy decreases since the melt temperature is less than the equilibrium freezing temperature.

At the same time, when a spherical crystal nucleus appears, a liquid-melt interface is created which has an excess free energy,

$$\Delta G_S = 4\pi R^2 \gamma > 0, \tag{30}$$

where γ is the crystal-melt interfacial energy per unit area; we assume for now that it is isotropic. Thus, the total free energy change is

$$\Delta G = -\Delta\mu \frac{4}{3}\pi R^3 + \gamma\, 4\pi R^2. \tag{31}$$

The function $\Delta G(R)$ is shown in Fig.5. One can see that when the nucleus is small enough the input of the surface energy is dominant and the free energy grows with the radius. When the radius of the nucleus becomes larger than R_c, corresponding to the maximum of the function $\Delta G(R)$, the free energy decreases with the radius. At this point, for $R > R_c$, the growth of the nucleus becomes thermodynamically preferable. However, for $R < R_c$ the nucleus growth is thermodynamically forbidden. One can say that the growth of a spherical crystal nucleus has a free energy barrier corresponding to R_c (see Fig.5). The critical value R_c is called a *critical nucleation radius* and can be easily computed as a point of maximum of the function $\Delta G(R)$,

$$R_c = \frac{2\gamma}{\Delta\mu}. \tag{32}$$

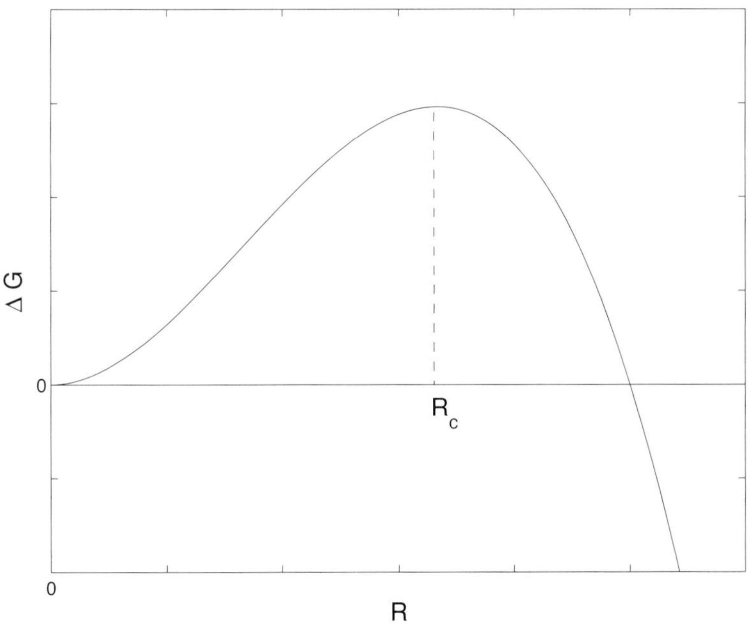

Figure 5. Critical nucleation radius.

For $T_0 = T_m - \Delta T$ and at constant pressure p, the difference of chemical potentials of the liquid and the solid phases in eq.(32), $\Delta\mu$, can be computed as

$$\begin{aligned}\Delta\mu &= \mu_l(T_m - \Delta T) - \mu_s(T_m - \Delta T) \\ &\approx \mu_l(T_m) - \left(\frac{\partial \mu_l}{\partial T}\right)_p \Delta T - \mu_s(T_m) + \left(\frac{\partial \mu_s}{\partial T}\right)_p \Delta T.\end{aligned} \tag{33}$$

At the equilibrium melting temperature T_m, corresponding to a given constant pressure and a planar liquid-solid interface, the chemical potentials of the liquid and solid phases are equal, $\mu_l(T_m) = \mu_s(T_m)$. Then, using the thermodynamic relation

$$\left(\frac{\partial \mu}{\partial T}\right)_p = -s,$$

where s is the specific entropy (here per unit volume), we obtain

$$\Delta\mu = (s_l - s_s)\Delta T = \frac{L_v}{T_m}\Delta T, \tag{34}$$

where L_v is the latent heat of melting per unit volume. This gives the following expression for the critical nucleation radius:

$$R_c = \frac{2\gamma T_m}{L_v \Delta T}. \tag{35}$$

One can see that the smaller the undercooling the larger the critical nucleation radius (and vice versa). When $\Delta T \to 0$, $R_c \to \infty$, and no crystal can form.

Eq.(35) can be interpreted in a different way: it determines an equilibrium temperature T_e at which a spherical crystal is in thermodynamic equilibrium with the surrounding melt. One obtains from (35)

$$T_e = T_m\left(1 - \frac{2\gamma}{L_v}\mathcal{K}\right), \tag{36}$$

where $\mathcal{K} = 1/R$ is the mean curvature of the sphere. Formula (36) is known as the Gibbs-Thomson relation and it applies for any curved interface: it gives the shift of the equilibrium temperature at any curved phase boundary.

Let us now consider the growth of a spherical crystal nucleus in an undercooled melt. The growth is governed by the evolution of the temperature field in the melt and the crystal. Let a spherical crystal have the initial radius R_0 at $t = 0$, and the radius $R(t)$ at later time t. We have the following initial-value problem:

$$\frac{\partial T_l}{\partial t} = \chi_l \nabla^2 T_l, \quad r > R(t), \tag{37}$$

$$\frac{\partial T_s}{\partial t} = \chi_s \nabla^2 T_s, \quad r < R(t), \tag{38}$$

$$T_l = T_s = T_m \left(1 - \frac{2\gamma}{L_v R}\right), \quad r = R(t), \tag{39}$$

$$L_v \frac{dR}{dt} = \lambda_s \frac{\partial T_s}{\partial r} - \lambda_l \frac{\partial T_l}{\partial r}, \quad r = R(t), \tag{40}$$

$$T_l \to T_0 \text{ for } r \to \infty, \quad T_s < \infty \text{ for } r \to 0, \tag{41}$$

where $\chi_{l,s}$ and $\lambda_{l,s}$ are the thermal diffusivities and heat conductivities of the liquid and the solid, respectively, L_v is the latent heat of melting per unit volume, γ is the surface energy, T_m is the equilibrium melting temperature, and $T_0 < T_m$ is the temperature in the melt far from the nucleus. The boundary condition (39) is the Gibbs-Thomson relation for the equilibrium temperature at the spherical surface of the nucleus, and the boundary condition (40) describes the balance of the heat flux that takes into account the latent heat of the phase transition.

Let us introduce the following dimensionless variables:

$$\hat{r} = \frac{r}{R_0}, \quad \hat{t} = \frac{\lambda_l \Delta T}{R_0^2 L_v} t, \quad \theta = \frac{T - T_m}{\Delta T}, \quad \Delta T = T_m - T_0. \tag{42}$$

In the new variables the heat conduction equations (37) and (38) read

$$S^{-1} \frac{\partial \theta_l}{\partial t} = \nabla^2 \theta_l, \quad S^{-1} \frac{\partial \theta_s}{\partial t} = \chi \nabla^2 \theta_s, \tag{43}$$

where $\chi = \chi_s / \chi_l$, and

$$S = \frac{L_v}{\rho c_p \Delta T}$$

is the Stefan number. Usually, the Stefan number is very large so that S^{-1} is very small; for example, for copper $S^{-1} \approx \Delta T / 682\text{K} \ll 1$. Therefore, we can neglect the terms with time derivatives in (43) and consider the following problem in *quasistationary approximation*:

$$\frac{1}{r^2} \frac{d}{dr} \left(r^2 \frac{d\theta_{l,s}}{dr} \right) = 0, \tag{44}$$

$$r \to \infty : \ \theta_l \to -1, \quad r \to 0, \ |\theta_s| < \infty, \tag{45}$$

and

$$r = R(t): \quad \theta_l = \theta_s = -\frac{\Gamma}{R}, \quad \frac{dR}{dt} = \lambda\frac{d\theta_s}{dr} - \frac{d\theta_l}{dr}, \tag{46}$$

where

$$\Gamma = \frac{2\gamma T_m}{L_v R_0 \Delta T} \equiv \frac{R_c}{R_0}, \quad \lambda = \frac{\lambda_s}{\lambda_l}$$

and $\lambda_{l,s}$ are heat conductivities of the melt and the crystal, respectively.

The problem (44)-(46) has the following solution that describes a growing spherical crystal:

$$\theta_{l0} = \frac{R}{r}\left(1 - \frac{\Gamma}{R}\right) - 1, \quad \theta_{s0} = -\frac{\Gamma}{R}, \tag{47}$$

where the sphere radius is described by the following differential equation:

$$\frac{dR}{dt} = \frac{1}{R}\left(1 - \frac{\Gamma}{R}\right). \tag{48}$$

One can see from eq.(48) that if $R < \Gamma$, i.e. when the crystal radius is less than the critical nucleation radius, the radius of the sphere decreases with time and the crystal shrinks. If $R > \Gamma$, the radius of the sphere increases and the crystal grows. For $R \gg 1$ $R(t) \sim \sqrt{t}$. For $R \ll \Gamma$, $R \sim (R_0^3 - \Gamma t)^{1/3}$, which means that a crystal, whose radius is less than the critical nucleation radius, shrinks in a finite time.

3 Mullins-Sekerka instability

One might ask the following question: will a growing spherical crystal remain spherical during all the time of its growth? In order to answer this question we shall perform a linear stability analysis of a spherical crystal growing into an undercooled melt.

Consider a growing spherical crystal at a certain moment of time when its radius is equal to R. We introduce the same dimensionless variables as in the previous Section, and use the quasi-stationary approximation. Spherically-symmetric temperature profiles in the melt and the solid, θ_{0l} and θ_{0s}, are described by eq.(47). Consider perturbed temperature profiles

$$\theta_l = \theta_{l0}(r) + \tilde{\theta}_l(r,\theta,t), \quad \theta_s = \theta_{s0}(r) + \tilde{\theta}_s(r,\theta,t), \tag{49}$$

where the temperature perturbations $\tilde{\theta}_{ls}(r,\theta,t)$ depend on the radial coordinate and the polar angle, and on time; the dependence of the perturbations on the azimuthal angle can be neglected. We also consider the shape of the growing crystal to be perturbed as

$$r = R + \tilde{R}(t,\theta). \tag{50}$$

In the quasistationary approximation the perturbations of the temperature fields are described by the Laplace equation,

$$\nabla^2 \tilde{\theta}_{l,s} = 0, \tag{51}$$

where

$$\nabla^2 = \frac{1}{r^2}\frac{\partial}{\partial r}\left(r^2 \frac{\partial}{\partial r}\right) + \frac{1}{r^2}\mathcal{L}_\theta, \quad \mathcal{L}_\theta = \frac{\partial}{\partial \theta}\left(\sin\theta \frac{\partial}{\partial \theta}\right). \tag{52}$$

The linearized boundary conditions (46) for the perturbations $\tilde{\theta}_{l,s}$ and $\tilde{R}$ read

$$\tilde{\theta}_l + \frac{d\theta_{0l}}{dr}\tilde{R} = \tilde{\theta}_s + \frac{d\theta_{0s}}{dr}\tilde{R} = -\Gamma\tilde{\mathcal{K}} \quad \text{at} \quad r = R, \tag{53}$$

$$\frac{d\tilde{R}}{dt} = \lambda\frac{d\tilde{\theta}_s}{dr} - \frac{d\tilde{\theta}_l}{dr} + \left(\lambda\frac{d^2\tilde{\theta}_s}{dr^2} - \frac{d^2\tilde{\theta}_l}{dr^2}\right)\tilde{R} \quad \text{at} \quad r = R. \tag{54}$$

Here, $\tilde{\mathcal{K}}$ is the perturbation of the mean curvature corresponding to the perturbed shape of the crystal $R = \bar{R} + \tilde{R}(\theta)$, which is computed as

$$\tilde{\mathcal{K}} = -\frac{1}{2R^2}(\mathcal{L}_\theta + 1)\tilde{R}(\theta). \tag{55}$$

Solutions of (51), bounded at infinity and in the center of the crystal, have the following form

$$\theta_l = \sum_{l=1}^{\infty} A_l(t)\, r^{-l-2} P_l(\cos\theta), \quad \theta_s = \sum_{l=1}^{\infty} B_l(t)\, r^l P_l(\cos\theta), \tag{56}$$

where $P_l(\cos\theta)$ are Legendre polynomials (here we consider perturbations that do not change the volume of the crystal; the mode corresponding to $l = 0$ is therefore absent). Consider the perturbation of the crystal spherical shape in the form

$$\tilde{R} = \sum_{l=1}^{\infty} C_l(t) P_l(\cos\theta). \tag{57}$$

From the boundary condition (53) one finds

$$A_l = \left[R^l - \frac{\Gamma}{2} l(l+1) R^{l-1}\right] C_l, \quad B_l = \Gamma R^{-l-2}\left[1 - \frac{l(l+1)}{2}\right] C_l. \tag{58}$$

Eqs.(58) connect perturbations of the temperature fields to the perturbation of the crystal shape. Then, using (58) and the heat flux balance boundary condition (54) one obtains

$$\frac{dC_l}{dt} = C_l \frac{(l-1)}{R^2}\left\{1 - \frac{\Gamma}{2R}\left[\lambda\, l(l+2) + 4 + 3l + l^2\right]\right\}. \tag{59}$$

Eq.(59) gives the evolution equations for the coefficients C_l describing the perturbations of the crystal-melt interface. One can see that the mode corresponding to $l = 1$ neither grows nor decays since it corresponds to a homogeneous translation of the spherical crystal and the system is translationally invariant. The modes that describe the perturbations of the crystal spherical shape correspond to $l > 2$. They grow in time if $dC_l/dt > 0$. For $l = 2$ we find

$$\frac{dC_l}{dt} = C_l \frac{1}{R^2}\left[1 - \frac{\Gamma}{2R}(8\lambda + 14)\right]. \tag{60}$$

Thus, for $\lambda = 0$, the spherical shape of a growing crystal becomes unstable when $R > R_* = 7\Gamma$, i.e. when the crystal radius becomes 7 times larger than the critical nucleation radius, R_c. For $\lambda = 1$ the instability occurs when the crystal radius exceeds R_c 11 times, and for $\lambda = 2$ the instability occurs when the crystal radius exceeds R_c 15 times. Also, with the increase of the crystal radius, higher harmonics, corresponding to $l > 2$ become unstable.

This instability was first studied by Mullins and Sekerka (Mullins and Sekerka , 1963) and it is called Mullins-Sekerka instability. It is responsible for the formation of snowflakes. The qualitative physical mechanism of the Mullins-Sekerka instability is shown in Fig.6. When the surface of a crystal growing into an undercooled melt is deformed as shown in Fig.6, the latent heat flows away easier from the protrusion parts and is accumulated in the troughs, which leads to faster freezing of the protrusions and slower freezing of the troughs; this asymmetry causes surface perturbations to grow. This mechanism is universal in that it applies also to a planar surface of a crystal freezing into an undercooled melt. More information about Mullins-Sekerka instability in various systems with a moving crystal-melt interface can be found in Davis (2001).

4 Dendrites

The Mullins-Sekerka instability of a spherical crystal growing into an undercooled melt discussed in the previous section results in the formation of complex shapes, similar to the shape of a snowflake, called dendrites due to their branched structure. An example of a dendrite crystal is shown in Fig.7. Indeed, when perturbations of a crystal-melt interface grow and develop they "forget" about the initial shape of the unstable interface. After some time, the newly formed interface becomes unstable itself, by the same mechanism, thus leading to the growth of the secondary structure. This process continues towards the formation of tertiary and quartiary structures, etc. Each structure, however, has a common feature: a parabolic-like tip of a new branch advancing into the melt. In this section we shall discuss the

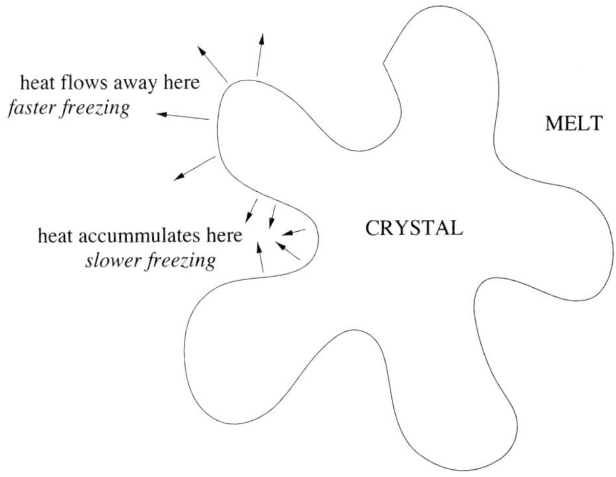

Figure 6. Mechanism of the Mullins-Sekerka instability.

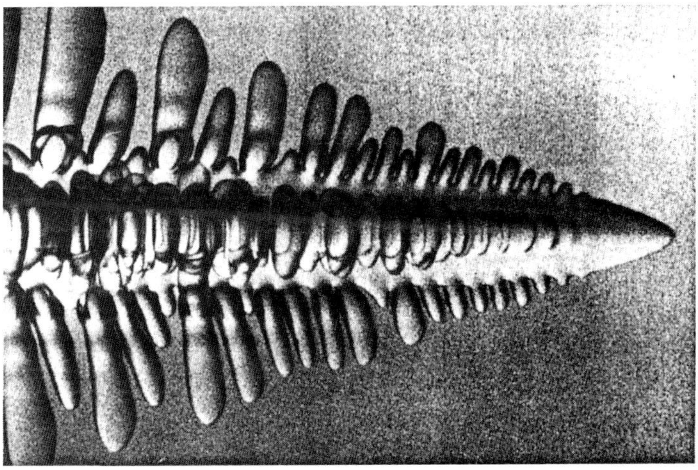

Figure 7. A dendrite crystal of succinonitrile. From Gliksman and Marsh (1993). (Reprinted with permission of Elsevier Science).

formation and propagation of this dendrite tip that forms the basis of the whole dendritic structure.

Let us generalize the problem of a growing spherical nucleus discussed in Section 2 and consider the growth of a crystal of an arbitrary shape into an undercooled melt. Let the crystal shape be described by a function $S(\mathbf{r}, t) = 0$. The melt temperature far away from the crystal, T_0, is less than the equilibrium melting temperature $T_m > T_0$. For simplicity, let us neglect the Gibbs-Thomson effect of the surface energy on the equilibrium temperature at the crystal-melt interface. In this case the temperature at the liquid-solid interface will be equal to the equilibrium melting temperature T_m and moreover, the temperature *inside* the crystal will be *constant* and equal to T_m, since otherwise, heat sources or sinks would be present inside the crystal. The problem can therefore be described completely by considering the temperature field $T(\mathbf{r},\, t)$ in the melt, and the boundary conditions at the crystal-melt interface describing the condition of thermodynamic equilibrium and the balance of heat flux:

$$\frac{\partial T}{\partial t} = \chi \nabla^2 T, \tag{61}$$

$$|\mathbf{r}| \to \infty : \; T = T_0; \tag{62}$$

$$\mathbf{r} \in S : \quad T = T_m, \tag{63}$$

$$L_v v_n = -\lambda \mathbf{n} \cdot \nabla T \tag{64}$$

Here χ is the thermal diffusivity, λ is the thermal conductivity, L_v is the latent heat of melting per unit volume, v_n is the normal velocity of the growing crystal surface, and $\mathbf{n}$ is the surface normal.

If an evolving surface in three-dimensional space is given by $S(\mathbf{r}, t) = 0$, then

$$\mathbf{n} = \frac{\nabla S}{|\nabla S|}, \quad v_n = -\frac{\partial S/\partial t}{|\nabla S|}. \tag{65}$$

Since in our problem the surface of a growing crystal is an *isotherm*, i.e. it is defined by the condition $T(\mathbf{r}, t) - T_m = 0$, one can rewrite the boundary condition (64) as

$$(\nabla T)^2 = \frac{L_v}{\lambda} \frac{\partial T}{\partial t} \quad \text{on the surface} \quad T = T_m. \tag{66}$$

Let us now use the following trick invented by Ivantsov (Ivantsov , 1947). Consider a *modified* eq.(66),

$$(\nabla T)^2 = f(T) \frac{\partial T}{\partial t}, \tag{67}$$

as a *first-order differential equation* for T; solve it and *then* find $f(T)$ to satisfy eq.(61).

Let us introduce a new function,

$$F(T) = \int^T \frac{dT}{f(T)}.$$

Then, $\nabla F = f^{-1}\nabla T$, $\partial F/\partial t = f^{-1}(\partial T/\partial t)$, and we obtain the following first-order partial differential equation for F:

$$\frac{\partial F}{\partial t} = (\nabla F)^2 \equiv \left(\frac{\partial F}{\partial x}\right)^2 + \left(\frac{\partial F}{\partial y}\right)^2 + \left(\frac{\partial F}{\partial z}\right)^2. \tag{68}$$

Let us consider the following general solution of eq.(68):

$$F = \frac{C_1 C_4\, x + C_2 C_4\, y + C_3 C_4\, z + C_4^2\, t}{C_1^2 + C_2^2 + C_3^2}, \tag{69}$$

where $C_{1,2,3,4}$ are arbitrary constants. At the crystal surface $T = T_m$ and $F(T) = F(T_m) = const$, therefore, the crystal surface is described by the *envelope* of the following *family of moving planes*:

$$C_1\, x + C_2\, y + C_3\, z + C_4\, t = \frac{C_1^2 + C_2^2 + C_3^2}{C_4}\, F = const. \tag{70}$$

Let us look for a crystal shape uniformly traveling with as yet unknown speed v in the z-direction, so that $C_4 = -vC_3$. Such shape is described by the envelope of the following family of moving planes:

$$\Psi(x, y, z, t, C_1, C_2, C_3) \equiv C_1\, x + C_2\, y + C_3\, (z - vt) + \frac{C_1^2 + C_2^2 + C_3^2}{vC_3}\, F = 0. \tag{71}$$

The shape of the envelope of (71) is determined by the conditions

$$\Psi = \frac{\partial \Psi}{\partial C_1} = \frac{\partial \Psi}{\partial C_2} = \frac{\partial \Psi}{\partial C_3} = 0,$$

from which we find

$$\frac{C_1}{C_3} = -\frac{xv}{2F}, \quad \frac{C_2}{C_3} = -\frac{yv}{2F},$$

and

$$z = -\frac{F}{v} + vt + \frac{v}{4F}(x^2 + y^2). \tag{72}$$

Eq.(72) shows that a uniformly moving surface of a growing crystal has the form of a paraboloid. For a paraboloid tip moving in positive z-direction with $v > 0$ we should choose $F < 0$. Solving eq.(72) for F we obtain

$$F = -\frac{v}{2}\left[z - vt + \sqrt{(z - vt)^2 + r^2}\right], \tag{73}$$

where $r^2 = x^2 + y^2$.

Thus, since $F = F(T)$, we can look for a *self-similar* solution of the problem (61)-(64) in the form

$$T = \Phi(\eta), \quad \eta = z - vt + \sqrt{(z - vt)^2 + r^2}, \tag{74}$$

where η is the *self-similar variable.* Substitute (74) in eq.(61) to obtain the following problem for Φ:

$$\eta\frac{d^2\Phi}{d\eta^2} + \frac{d\Phi}{d\eta}\left(1 + \frac{v}{2\chi}\eta\right) = 0, \quad \Phi(\infty) = T_0, \quad \Phi(\eta_0) = T_m, \tag{75}$$

where η_0 corresponds to the tip of the moving paraboloid, $z = vt$, and therefore it is equal to the tip radius of curvature, R_0. The solution of (75) is

$$\Phi = T_0 + (T_m - T_0)\,\frac{\mathrm{Ei}(v\eta/2\chi)}{\mathrm{Ei}(v\eta_0/2\chi)}, \quad \mathrm{Ei}(x) = \int_x^{\infty} \frac{e^{-x}}{x}\, dx. \tag{76}$$

Finally, we use the heat flux balance boundary condition (66), introduce the dimensionless Peclet number, $P = vR_0/2\chi$, and the Stefan number, $S = L_v/(\rho c_p \Delta T)$, and obtain the following equation:

$$P\,\mathrm{Ei}(P)\exp(P) = S^{-1}. \tag{77}$$

Eq.(77) allows to find the Peclet number of a moving paraboloid tip if the Stefan number (i.e. the undercooling) is given. The solution discussed in this section was first obtained by Ivantsov (Ivantsov , 1947), and a solid-liquid interface in the form of a paraboloid described by (74)-(77), uniformly moving into an undercooled melt, is called Ivantsov's paraboloid. The function $\mathrm{Iv}(x) \equiv x\mathrm{Ei}(x)\exp(x)$ is sometimes called the Ivantsov's function; it is shown in Fig. 8. One can see that the solution in the form of Ivantsov paraboloid exists only for $S > 1$, i.e. for small enough undercooling.

Note that eq.(77) allows one to find only the Peclet number and thus to relate the paraboloid speed, v, to the radius of curvature of the paraboloid tip, R_0. The discussed theory does not determine the speed v and the radius R_0 separately. This represents the so-called selection problem which is discussed in detail in Davis (2001). Note also that self-similar solutions

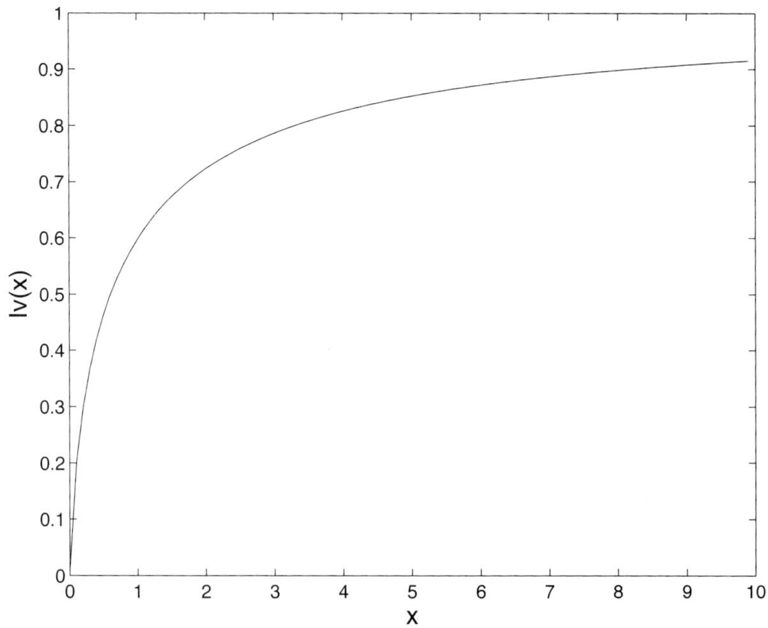

Figure 8. Ivantsov function Iv(x).

for a solid-liquid interface freezing into an undercooled melt other than a moving paraboloid are possible. These are discussed in Horvay and Cahn (1961).

5 Surface diffusion and surface-diffusion-controlled interface shape

On a microscopic level, the evolution of a crystal-melt interface described in Sections 2-4 is governed by the attachment-detachment of atoms, or, as it is often called, by the evaporation-condensation mechanism. There is a large class of interfacial phenomena in materials science (for example, molecular beam epitaxy or chemical vapor deposition) that are characterized by a different mechanism of surface evolution, namely, by diffusion of atoms along the surface they adsorbed at. This mechanism is called surface diffusion, and it leads to different dynamics of interface evolution. In the course of surface diffusion, atoms jump along the surface from a high-energy place to a low-energy place, thus reducing the total energy of the system. In the simplest case the surface energy of atoms is determined by the surface structure of the crystal and is associated with different number of neighbors that atoms on the surface have in comparison with atoms in the bulk, and the total energy of the surface is proportional to the surface area. As for usual diffusion, on average, atoms move along the surface against the gradient of the surface chemical potential, and the the average velocity of an atom, $\mathbf{v}_s$, is proportional to the surface gradient of the surface chemical potential, μ_s,

$$\mathbf{v}_s = -M_s \nabla_s \mu_s, \tag{78}$$

where M_s is the mobility coefficient connected to the surface diffusion coefficient, D_s, by the Einstein's relation,

$$M_s = \frac{D_s}{k_B T}, \tag{79}$$

where T is the absolute temperature and k_B is the Boltzmann constant. The surface flux of atoms, $\mathbf{j}_s$, is proportional to the average velocity of a single atom,

$$\mathbf{j}_s = \nu \mathbf{v}_s, \tag{80}$$

where ν is the number of atoms per unit area on the crystal surface.

If $\mathbf{j}_s$ is known, one can easily write an evolution equation for the surface shape, $h(x, y, t)$, see Fig. 9. Indeed, from the conservation of mass (number of atoms) we have

$$v_n = -\Omega \nabla_s \cdot \mathbf{j}_s, \tag{81}$$

where v_n is the normal velocity of the interface motion due to redistribution of atoms on its surface, and Ω is the atomic volume. The normal velocity is expressed in terms of the shape of an evolving interface as

$$v_n = \frac{h_t}{\sqrt{1 + |\nabla h|^2}}. \tag{82}$$

Eqs.(78)-(82) give the following general equation for the crystal surface evolution by surface diffusion,

$$\frac{h_t}{\sqrt{1 + |\nabla h|^2}} = \frac{D_s \nu \Omega}{k_B T} \nabla_s^2 \mu_s, \tag{83}$$

where we have assumed that all the parameters on the right hand side of eq.(83), except μ_s, are constant.

Thus, in order to derive an evolution equation for the surface shape, we need to know the connection between the surface shape and the surface

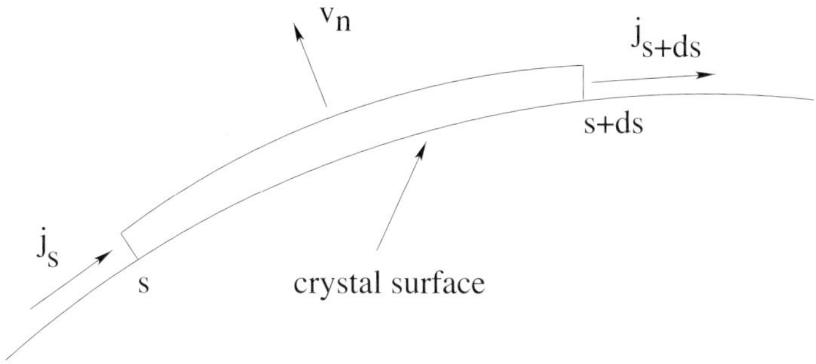

Figure 9. Surface diffusion.

chemical potential. Consider a spherical crystal nucleus with radius R. In equilibrium, the chemical potential of atoms inside the nucleus is constant, μ_0. Let a small amount of the crystalline material accumulate at the surface so that the radius of the nucleus increases by dR and the volume increases by $dV = 4\pi R^2 dR$. The number of atoms accumulated is

$$dN = \frac{dV}{\Omega} = \frac{4\pi R^2 dR}{\Omega}, \tag{84}$$

where Ω is the atomic volume. The energy of the nucleus changes due to the change of the total number of particles in the system as well as due to the creation of a new surface that has an excess surface energy. Thus, the energy change

$$dE = \mu_0 dN + \gamma dS, \tag{85}$$

where γ is the surface energy per unit area and dS is the change of the total area,

$$dS = 4\pi(R + dR)^2 - 4\pi R^2 = 8\pi R\, dR. \tag{86}$$

Thus, for the chemical potential we have

$$\mu = \frac{dE}{dN} = \frac{\mu_0 dN + \gamma\, dS}{dN} = \mu_0 + \frac{2\gamma\Omega}{R}. \tag{87}$$

Thus, we see that the chemical potential of a spherical nucleus consists of two parts: bulk part, μ_0, and the additional, surface part, $\mu_s = 2\gamma\Omega/R$. In the general case of a curved surface, using similar arguments, one obtains that the surface chemical potential

$$\mu_s = 2\gamma\Omega\mathcal{K}, \tag{88}$$

where $\mathcal{K}$ is the mean curvature. Eq.(88) has an important consequence: any curved surface, which has a constant (isotropic) surface energy, will relax by surface diffusion to either a sphere or a plane, depending on the initial conditions.

Eq.(83) for the surface evolution then becomes

$$\frac{h_t}{\sqrt{1+|\nabla h|^2}} = 2B\nabla_s^2\mathcal{K}, \tag{89}$$

where

$$B = \frac{D_s \nu \gamma \Omega^2}{k_B T},$$

and $\mathcal{K}$ is the mean curvature that can be expressed in terms of the surface shape function $h(x, y, t)$.

Eq.(89) is the evolution equation for the surface shape. In the model case of a one-dimensional surface, $h(x, t)$, of a two-dimensional crystal, using

$$\frac{\partial}{\partial s} = \frac{1}{\sqrt{1+h_x^2}}\frac{\partial}{\partial x}, \quad 2\mathcal{K} = -\frac{h_{xx}}{(1+h_x^2)^{3/2}},$$

we obtain

$$h_t = B\frac{\partial}{\partial x}\left[\frac{1}{\sqrt{1+h_x^2}}\frac{\partial}{\partial x}\left[-\frac{h_{xx}}{(1+h_x^2)^{3/2}}\right]\right]. \tag{90}$$

In the important case when the surface slope is small, $|h_x| \ll 1$, eq.(90) is reduced to the following linear equation:

$$h_t = -Bh_{xxxx}, \tag{91}$$

which is similar to the diffusion equation, but of the higher spatial order.

Now, using eq.(91), let us consider the following interesting problem: the evolution of a *grain-boundary groove*. Consider a surface of a semi-infinite crystal that consists of two semi-infinite grains having two different but symmetric crystallographic orientations, see Fig. 10. The planar interface between the two grains touches the crystal surface. The balance of surface and interfacial tensions at the point of the triple phase contact prescribes the slopes of the crystal surface at this point to be constant, $\pm m$, thus forming a grain boundary groove. Thus, if initially the crystal surface is planar, the next instant a grain boundary groove will form. However, due to surface diffusion, the surface curvature should decrease in time so that the groove should become deeper. Our main goal will be to determine the time dependence of the groove depth and the shape of the interface around the groove.

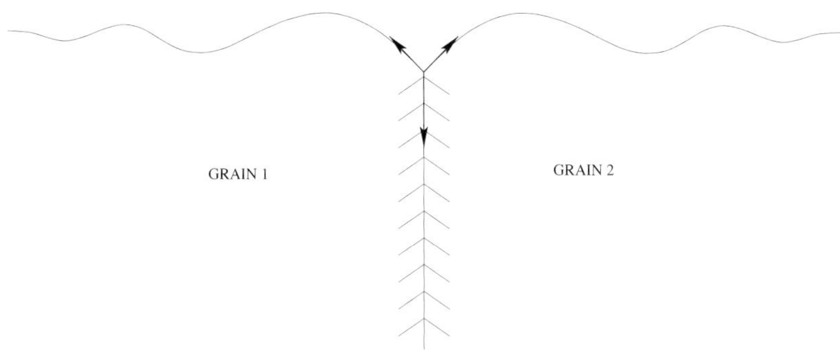

Figure 10. Grain boundary groove.

Due to the problem symmetry with respect to the planar interface between the two grains, we have the following problem for the surface shape $y(x,t)$:

$$\begin{gathered} y_t = -By_{xxxx},\ x > 0; \\ y(\infty,t) = 0,\ y(x,0) = 0,\ \ y_x(0,t) = m,\ \ y_{xxx}(0,t) = 0. \end{gathered} \tag{92}$$

The last boundary condition in (92) is the symmetry condition that describes zero surface diffusion flux at the triple contact point.

It is convenient to solve the problem (92) using the Laplace transform. Applying the Laplace transform to (92) we obtain the following problem for the image $Y(x,p) \equiv \int_0^\infty y(x,t)\, e^{-pt}\, dt$:

$$\begin{gathered} BY_{xxxx} + pY = 0,\ \ x > 0; \\ Y(\infty,p) = 0,\ Y_x(0,p) = \frac{m}{p},\ \ Y_{xxx}(0,p) = 0. \end{gathered} \tag{93}$$

Now we look for a solution of (93) in the form $Y = e^{\lambda x}$ and find that the solution bounded at $x \to \infty$ is

$$\begin{aligned} Y(x,p) = {} & \exp\left[-\left(\frac{p}{B}\right)^{1/4} \frac{x}{\sqrt{2}}\right] \times \\ & \left\{C_1 \cos\left[\left(\frac{p}{B}\right)^{1/4} \frac{x}{\sqrt{2}}\right] + C_2 \sin\left[\left(\frac{p}{B}\right)^{1/4} \frac{x}{\sqrt{2}}\right]\right\}. \end{aligned} \tag{94}$$

The constants C_1 and C_2 are found from the two boundary conditions at $x = 0$, so finally we obtain

$$Y(x,p) = -\frac{mB^{1/4}}{p^{5/4}} \exp\left[-\left(\frac{p}{B}\right)^{1/4} \frac{x}{\sqrt{2}}\right] \cos\left[\left(\frac{p}{B}\right)^{1/4} \frac{x}{\sqrt{2}} + \frac{\pi}{4}\right]. \quad (95)$$

Solution (95) allows one to obtain the dependence of the grain boundary groove depth on time. Indeed, for $x = 0$ we have

$$Y(0,p) = -\frac{mB^{1/4}}{\sqrt{2}p^{5/4}},$$

and taking the inverse Laplace transform we obtain

$$y(0,t) = -\frac{m}{\sqrt{2}\,\Gamma(5/4)}\,(Bt)^{1/4}, \quad (96)$$

where $\Gamma(s)$ is the Euler Gamma-function. Thus, we see that the depth of the grain-boundary groove increases as $t^{1/4}$.

Taking the inverse Laplace transform of (95) in order to find the shape of the crystal surface around the grain-boundary groove is quite cumbersome. Rather, we shall look for a self-similar solution of the problem (92) in the following form:

$$y(x,t) = m(Bt)^{1/4}\,\Phi(\eta), \quad \eta = \frac{x}{(Bt)^{1/4}}. \quad (97)$$

The boundary conditions for $\Phi(\eta)$ can be found from (96) and the following relations for the derivatives of $y(x,t)$ at $x = 0$:

$$y_x(0,t) = m,\ y_{xx}(0,t) = -\frac{m}{\sqrt{2}\Gamma(3/4)}\,(Bt)^{-1/4},\ y_{xxx} = 0.$$

Substitute (97) in (92) to find the following problem for $\Phi(\eta)$:

$$\Phi'''' - \frac{1}{4}\eta\,\Phi' + \frac{1}{4}\Phi = 0, \quad (98)$$

$$\Phi(0) = -\frac{1}{\sqrt{2}\,\Gamma(5/4)}, \quad \Phi'(0) = 1, \quad \Phi''(0) = -\frac{1}{\sqrt{2}\Gamma(3/4)}, \quad \Phi'''(0) = 0. \quad (99)$$

Solution of (98)-(99) can be found in terms of hypergeometric function. However, it is more convenient to look for the solution by means of Frobenius method, in the form

$$\Phi(\eta) = \sum_{n=0}^{\infty} a_n \eta^n. \quad (100)$$

Substitute (100) in (98)-(99) to obtain the following recurrent relation for the coefficients a_n,

$$(n+4)(n+3)(n+2)(n+1)\,a_{n+4} - \frac{1}{4}(n-1)\,a_n = 0, \qquad (101)$$

with

$$a_0 = -\frac{1}{\sqrt{2}\,\Gamma(5/4)}, \quad a_1 = 1, \quad a_2 = -\frac{1}{\sqrt{2}\Gamma(3/4)}, \quad a_3 = 0.$$

We find that the coefficients a_n tend to zero quite quickly, $a_0 = -0.7803$, $a_1 = 1.0000$, $a_2 = -0.2866$, $a_3 = 0$, $a_4 = 0.00813$, $a_5 = 0$, $a_6 = -0.0002, \ldots$. The resulting surface shape has the form shown in Fig. 10. It exhibits *oscillations* with a decaying amplitude and the wavelength growing as $\sim 4.6(Bt)^{1/4}$. The presence of surface oscillations is the characteristic feature of surface-diffusion-driven surface relaxation that follows from the fourth spatial derivative corresponding to the Laplacian of the mean curvature.

The derivation of the surface-diffusion equation and the solution of the grain-boundary groove problem was first obtained in the classical paper by Mullins (Mullins , 1957), which the derivation presented in this section closely follows.

6 Elastic instability of solid epitaxial films and self-assembly of quantum dots

In this section we shall discuss an important interfacial instability that is related to the field of nanotechnology and responsible for the formation of interesting nano-scale surface structures called "quantum dots". Consider a film of solid crystalline material deposited on a surface of a substrate made of another crystalline material. Solid-on-solid deposition is called epitaxy, and when the deposited material is different from that of the substrate it is called hetero-epitaxy. The deposition process is often done by means of a beam of atoms or molecules hitting the substrate, the so-called molecular beam epitaxy (MBE). When atoms are deposited on a substrate they stick to the atoms of the substrate so that the atoms inside the growing epitaxial film are arranged accordingly. Since the crystal lattice spacings in two different materials usually differ from one another (*lattice mismatch*), the film grows in an elastically stressed state, and has elastic energy stored in it. If, as a result of some fluctuation, a small perturbation in the form of a small hill appears on the film surface, there will be more room inside this hill for the atoms of the film to adjust their interatomic spacing to the normal equilibrium value, which will lead to the decrease of the elastic energy. This process of the hill formation will therefore be energetically preferable, which will result in an instability of a planar film that will decompose into a system of islands. The transport mechanism by which the film surface reconstructs itself into a system of islands is surface diffusion discussed in the previous section. This mechanism of the island formation is schematically shown in Fig.11.

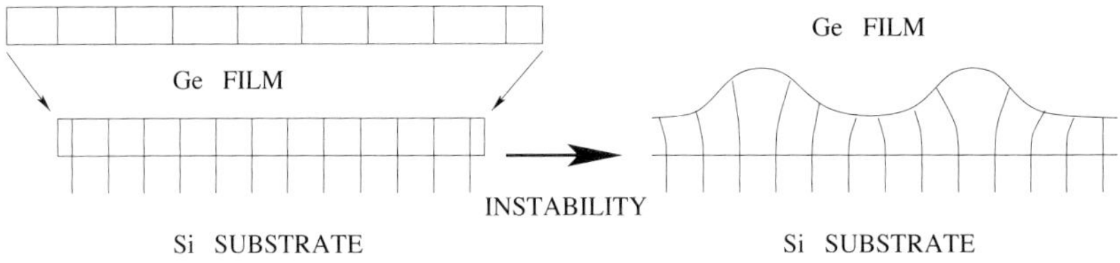

Figure 11. Instability of an epitaxially-strained solid film.

Examples of such islands are shown in Fig.12. One can see that they can have different shapes, faceted or rounded. A typical island size of the order of nanometers, which is of the same order of magnitude as de Broglie wavelength of electrons in semiconductors. Confined within such surface structures, electrons exhibit interesting quantum properties that can be used for fabrication of new electronic devices. For this reason such islands are called *quantum dots* (QDs). For reviews of the formation and properties of QDs see Shchukin and Bimberg (1999); Bimberg e.a. (2004); Stangl e.a. (2004).

In order to describe the dynamics of the formation of QDs in epitaxial films one should use the surface diffusion equation derived in the previous section. However, along with the surface energy, the surface chemical potential should include also the elastic energy accumulated in the film. Thus we have

$$\frac{h_t}{\sqrt{1+|\nabla h|^2}} = \mathcal{D}\nabla_s^2(\mu_e + \gamma\mathcal{K}), \tag{102}$$

where $\mathcal{D}$ is a constant proportional to the surface diffusivity of atoms, γ is the surface energy, $\mathcal{K}$ is the mean curvature, and μ_e is the part of surface chemical potential corresponding to the elastic energy stored in the epitaxial film. It is computed as (Voorhees and Johnson , 2004)

$$\mu_e = \frac{1}{2}E_{ij}\sigma_{ij}|_{z=h}, \tag{103}$$

where E_{ij} is the strain tensor, σ_{ij} is the stress tensor (Landau and Lifshits , 1981), and the usual convention of the summation over the repeated indices

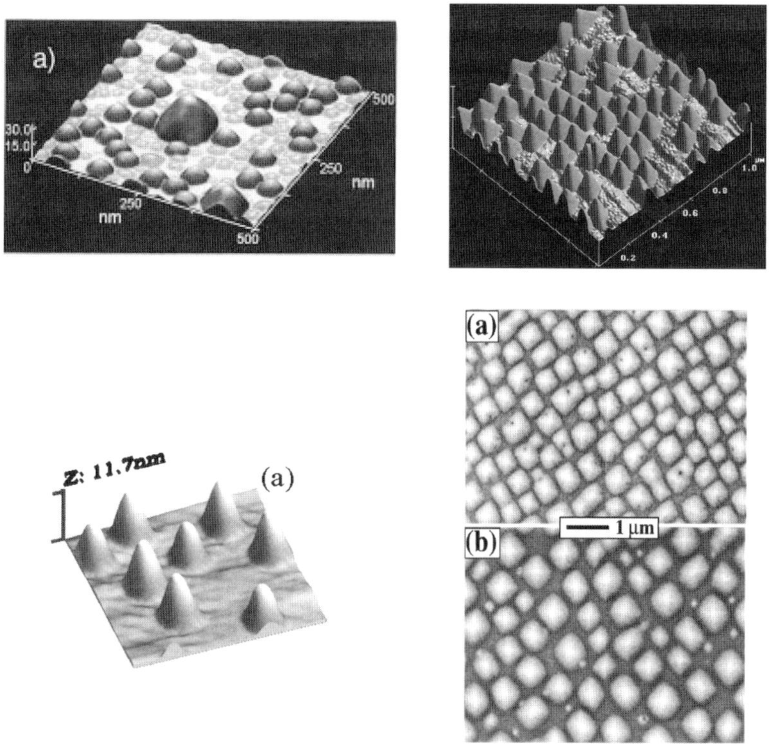

Figure 12. Examples of quantum dots: upper left: Ge QDs on Si(001) Williams e.a. (1998) (reprinted with permission from Williams e.a. (1998), ©1998 by the American Chemical Society); upper right: PbSe QDs on PbTe-on-Si(111) Alchalabi e.a. (2003) (reprinted with permission from Alchalabi e.a. (2003), ©2003 by the American Physical Society); lower left: InAs QDs on GaAs(001) Granados e.a. (2003) (reprinted with permission from Granados e.a. (2003), ©2003 by the American Institute of Physics); lower right: coarsening of SiGe QDs on Si(001) – figure b) corresponds to later time than a) Floro e.a. (2000) (reprinted with permission from Floro e.a. (2000), ©2000 by the American Physical Society). (Reprinted with permission from Springer).

in used. In the approximation of linear elasticity theory, σ_{ij} and E_{ij} are connected by the following relation (Landau and Lifshits , 1981):

$$\sigma_{ij} = 2\mu \left[\left(\frac{\nu}{1+\nu} \right) \delta_{ij} E_{kk} + E_{ij} \right], \tag{104}$$

where μ is the shear modulus, ν is the Poisson's ratio, and δ_{ij} is the Kroneker delta. The strain tensor is determined by the component of the displacement vector u_i that characterizes small deformations of a solid body,

$$E_{ij} = \frac{1}{2}(\partial_j u_i + \partial_i u_j). \tag{105}$$

Here ∂_i denotes differentiation with respect to the coordinate x_i, $i = 1, 2, 3$.

Thus, in order to find the elastic component of the film surface chemical potential we should find the displacement field in the film which is determined by the epitaxial stress caused by the lattice mismatch between the film and the substrate. Consider a thin solid film deposited on a semi-infinite solid substrate. The interface between the film and the substrate is planar and the z-axis is normal to it, with $z < 0$ corresponding to the substrate and $0 < z < h(x, y, t)$ corresponding to the film; $z = h(x, y, t)$ describes the shape of the film free surface, x and y are the coordinates in the plane parallel to the substrate. Since surface diffusion processes are much slower than the speed of sound in a crystalline material that determines the characteristic time of elastic relaxation, we can assume that at each moment of time there is mechanical equilibrium in the film and in the substrate which is described by the equation

$$\partial_j \sigma_{ij} = 0. \tag{106}$$

We should also set the following boundary conditions:

(i) stress-free film surface,

$$\sigma_{ij} n_j = 0, \tag{107}$$

where n_j are the components of the normal to the film surface;

(ii) continuity of displacement field in the film and the substrate at the planar substrate-film interface:

$$u_i^f = u_i^s + \epsilon \begin{pmatrix} x \\ y \\ 0 \end{pmatrix}; \tag{108}$$

here

$$\epsilon = \frac{a_s - a_f}{a_f} \tag{109}$$

is the lattice mismatch between the substrate and the film determined by the crystal lattice spacings of the corresponding materials, a_s and a_f;

(iii) stress balance at the planar substrate-film interface,

$$\sigma_{i3}^f = \sigma_{i3}^s; \tag{110}$$

(iv) decay of all deformations deep into the semi-infinite substrate far from the film-substrate interface:

$$u_i^s \to 0 \quad \text{for} \quad z \to \infty. \tag{111}$$

Let us first consider the "basic state" of the epitaxial film, in which the stresses and strains are constant and the film free surface is planar. Obviously, in this case the stress and strain in the substrate must be zero since a thin film cannot deform an infinitely deep substrate. The horizontal components of the strain tensor in the film in the basic state are determined by the lattice mismatch,

$$E_{xx}^{(0)} = E_{yy}^{(0)} = \epsilon. \tag{112}$$

Since the film free surface is stress-free and the stress is constant, then the vertical stress component must be zero, $\sigma_{zz}^{(0)} = 0$. Then from the stress-strain relation (104) we find the vertical component of the strain in the film,

$$E_{zz}^{(0)} = -\frac{2\epsilon\nu_f}{1-\nu_f}. \tag{113}$$

Then, again from (104) we find the horizontal components of the strain tensor,

$$\sigma_{xx}^{(0)} = \sigma_{yy}^{(0)} = 2\epsilon\mu_f \left(\frac{1+\nu_f}{1-\nu_f}\right). \tag{114}$$

Expressions (112)-(113) are called epitaxial strain and (114) is called epitaxial stress. We can see that both are proportional to the lattice mismatch ϵ. Now we can compute the elastic energy stored in an epitaxial film:

$$\mathcal{E}^{(0)} = \frac{1}{2}\sigma_{ij}^{(0)} E_{ij}^{(0)} = 2\epsilon^2\mu_f \left(\frac{1+\nu_f}{1-\nu_f}\right). \tag{115}$$

Note that this energy does not depend on the sign of the lattice mismatch, i.e. it does not depend on whether the epitaxial strain is compressive or tensile.

As discussed above, a planar, stressed epitaxial film should be unstable with respect to the undulations of the film free surface that can result in the formation of mounds and in the decrease of the elastic energy stored in the film. In order to show that we need to perform a linear stability analysis of the basic state described above. This analysis is quite cumbersome and the details can be found in Spencer e.a. (1991, 1993). Here we shall describe the main steps of this analysis.

We consider infinitesimal perturbations of the displacement field and the film free surface from the basic state in the form

$$\tilde{\mathbf{u}} = \mathbf{u} - \mathbf{u}^{(0)} = \mathbf{U}(z)e^{i\mathbf{k}\cdot\mathbf{x}+\omega t}, \quad \tilde{h} = h - h_0 = He^{i\mathbf{k}\cdot\mathbf{x}+\omega t}, \tag{116}$$

where $\mathbf{x} = (x, y)$, $\mathbf{k}$ is the perturbation wavevector and ω is the perturbation growth rate, and linearize the system (102)-(111). We then obtain a system of linear differential equations for the components of the vector-function $\mathbf{U}(z)$ and solve it using the linearized boundary conditions; $\mathbf{U}(z)$ will be obtained in terms of exponential functions of z with coefficients depending on H, $\mathbf{k}$, ω as well as elastic constants of the film and the substrate and the lattice mismatch ϵ. Then, using the linearized surface diffusion equation (102) we shall obtain

$$H\omega = HD(|\mathbf{k}|, \epsilon, \text{elastic constants}), \tag{117}$$

where D is some cumbersome function of its variables. From (117), as the condition to have a non-trivial H, we readily obtain the dispersion relation $\omega = D(|\mathbf{k}|, \omega, \epsilon, \text{elastic constants})$ between the perturbation growth rate and the wavenumber. In the long-wave limit, $|\mathbf{k}|\, h_0 \ll 1$, the dispersion relation will read

$$\omega = a|\mathbf{k}|^3 - b|\mathbf{k}|^4, \tag{118}$$

where $a \propto \mu_f \epsilon^2 > 0$ is proportional to the total elastic energy stored in the film, and $b \sim \gamma$ linearly depends on the surface energy γ (in the case when the elastic constants of the substrate and the film are equal, $b = \gamma$).

Thus, from the dispersion relation (118) we can see that epitaxially stressed solid film is always unstable with respect to surface undulations. The dispersion relation (118) is shown in Fig.13. The elastic energy has destabilizing effect whereas the surface energy has stabilizing effect. Destabilizing effect is larger than the stabilizing one for long-wave perturbations and smaller for short-wave perturbations. The competition of the two effects leads to the selection of a characteristic wavelength of the growing structure, $\lambda_c = 2\pi/k_c$, where k_c corresponds to the maximum of the dispersion relation $\omega(|\mathbf{k}|)$.

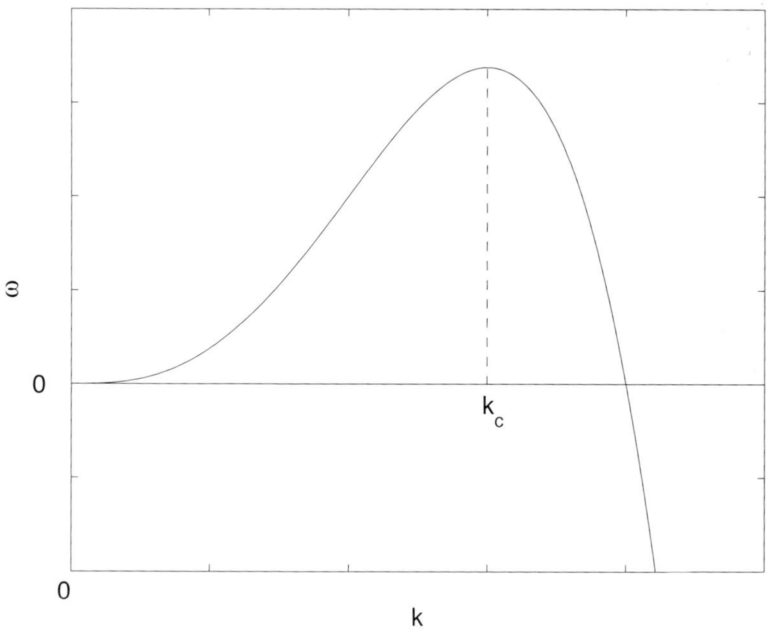

Figure 13. Dispersion relation (118).

The instability of an epitaxial film caused by epitaxial stress due to lattice mismatch between the film and the substrate is called Asaro-Tiller-Grinfeld (ATG) instability that was first shown to occur for a stressed semi-infinite solid with a free surface (Asaro and Tiller , 1972; Grinfeld , 1986). The stability analysis described in this section was first done in (Spencer e.a. , 1991, 1993). It describes the instability that results in the formation of quantum dots on the surface of an epitaxial solid film. There are other effects that can be important for the formation of QDs, for example, wetting interactions between the film and the substrate and surface-energy anisotropy. One can read about these effects as well as about nonlinear evolution of the arrays of QDs in (Golovin e.a. , 2006, 2004; Levine e.a. , 2007) and in references therein.

Bibliography

K. Alchalabi, D. Zimin, G. Kostorz, and H. Zogg. Self-assembled semiconductor quantum dots with nearly uniform sizes. *Phys. Rev. Lett.*, 90:026104, 2003.

R.J. Asaro and W.A. Tiller. Interface morphology development during stress corrosion cracking. *Metall. Trans.*, 3:1789–1796, 1972.

D. Bimberg, M. Grundmann, and N.N. Ledentsov. *Quantum Dot Heterostructures*, Chichester: Wiley, 1999.

S.H. Davis. *Theory of Solidification*, Cambridge University Press, 2001.

J.A. Floro, M.B. Sinclair, E. Chason, L.B. Freund, R.D. Twesten, R.Q. Hwang, and G.A. Lucadamo. Novel SiGe island coarsening kinetics: Ostwald ripening and elastic interactions. *Phys. Rev. Lett.*, 84:701–704, 2000.

M.E. Gliksman and S.P. Marsh. The dendrite. In D.T.J. Hurle, editor, *Handbook of Crystal Growth*, Elsevier, 1993.

C. Godreche, editor. *Solids far from Equilibrium*. Cambridge University Press, 1992.

A.A. Golovin, P.W. Voorhees, and S.H. Davis. Self-assembly of quantum dots from thin solid films. In A.A. Golovin and A.A. Nepomnyashchy, editors. *Self-Assembly, Pattern Formation and Growth Phenomena in Nano-Systems*. Springer, 2006.

A.A. Golovin, M.S. Levine, T.V.Savina, and S.H.Davis. Faceting instability in the presence of wetting interactions: a mechanism for the formation of quantum dots. *Phys. Rev. B*, 70:#235342, 2004.

D. Granados and J.M. Garcia. In(Ga)As self-assembled quantum ring formation by molecular beam epitaxy. *Appl. Phys. Lett.*, 82:2401–2403, 2003.

M.Ya. Grinfeld. Instability of the separation boundary between a nonhydrostatically stressed elastic body and a melt. *Sov. Phys. Dokl.*, 31:831–834, 1986.

G. Horvay and J.W. Cahn. Dendritic and spheroidal growth. *Acta Metall*, 9:695–705, 1961.

G.P. Ivantsov. Temperature field around a spherical, cylindrical and acircular crystal growing in a supercooled melt. *Dokl. Akad. Nauk SSSR* 58:567–569, 1947. English translation in P. Pelce, editor, *Dynamics of Curved Fronts*, Academic Press, 1988.

L.D. Landau and E.M. Lifshits. *Statistical Physics*, pt.1, Chap. 15, Sec. 155. Pergamon Press, Oxford, 1978-[1981].

L.D. Landau and E.M. Lifshits. *Elasticity Theory*. Pergamon Press, Oxford, 1981.

M.S. Levine, A.A. Golovin, S.H. Davis, and P.W. Voorhees. Self-Assembly of Quantum Dots in a Thin Epitaxial Film Wetting an Elastic Substrate. *Phys. Rev. B*. 75:#205312, 2007.

W.W. Mullins. Theory of thermal grooving. *J. Appl. Phys.*, 28:333–339, 1957.

W.W. Mullins and R. Sekerka. Morphological stability of a particle growing by diffusion or heat flow. *J. Appl. Phys.*, 34:323-329, 1963.

A. Pimpinelli and J. Villain. *Physics of Crystal Growth*, Cambridge University Press, 1998.

V.A. Shchukin and D. Bimberg. Spontaneous ordering of nanostructures on crystal surfaces. *Rev. Mod. Phys.*, 71:1125–1171, 1999.

B.J. Spencer, P.W. Voorhees, and S.H. Davis. Morphological instability in epitaxially-strained dislocation-free solid films. *Phys. Rev. Lett.*, 67:3696–3699, 1991.

B.J. Spencer, S.H. Davis, and P.W. Voorhees. Morphological instability in epitaxially-strained dislocation-free solid films: linear stability theory. *J. Appl. Phys.*, 73:4955–4970, 1993.

J. Stangl, V. Holy and G. Bauer. Structural properties of self-organized semiconductor nanostructures. *Rev. Mod. Phys.*, 76:725–783, 2004.

P.W. Voorhees and W.C. Johnson. The thermodynamics of elastically stressed crystals. *Sol. St. Phys. Adv. Res. Appl.* 59:1–201, 2004.

R.S. Williams, G. Medeiros-Ribeiro, T.I.Kamins, and D.A.A. Ohlberg. Equilibrium shape diagram for strained Ge nanocrystals on Si(001). *J. Phys. Chem. B.*, 102:9605–9609, 1998.

G. Wulff. Zur Frage der Geschwindigkeit des Wachstums und der Aufslösung der Krystallflächen. *Z. Kristallogr.*, 34:449-530, 1901.

The Physics and Analyses of Interfacial Instabilities that Arise from Phase Change

Ranga Narayanan
Department of Chemical Engineering, University of Florida, FL 32611, USA

Abstract In the three sections of this chapter the reader is given an idea of instabilities that occur when a phase change takes place. When a liquid turns into a solid or into a vapor, energy is either removed or added to the interface between both phases. The direction of motion of the surface, the sign of the potential gradients there and the influence of surface energy on pressure or temperature are factors that affect the morphology of the interface. We show the reader how such problems may be viewed and how they should be analyzed. Two examples, one from solidification and the other from evaporation, are given. The major references from which this material has been adapted are also given.

Contents

1 The Physics and Analysis of Instability During the Solidification of a Pure Material

In this section[1] of the volume we discuss the instability at an interface that arises when it divides two phases of a pure material: a low-temperature phase and a high-temperature one. Imagine an advancing solidification front undergoing a displacement. Our objective is to determine whether or not a solidification front, if it is planar, can regain planarity in the face of a small disturbance.

Here one phase turns into the other, and as liquid turns into solid, the speed of the moving front at any point depends on how fast the latent heat released there can be conducted through either phase before being rejected to the far surroundings. As in most front moving problems, the speed, the direction of the front and the sign of the potential gradients at the front determine its stability.

Figure 1 illustrates two ways of rejecting this heat. In both cases, a liquid lies in a container whose walls are held at a temperature below the freezing point. Heat is removed from the liquid at these walls. In the left-hand illustration, the liquid is above its freezing point and so a solid phase begins to form at the walls and starts to grow into the liquid. Here the latent heat is conducted through the solid and rejected to the walls. It turns out that this is a stable configuration; the solidification front moves smoothly toward the center of the container. In the right-hand illustration the liquid is subcooled[2] to a temperature below its freezing point. The latent heat is now conducted through the liquid before being rejected to the walls. This is an unstable configuration. Instead of moving smoothly outward, the solidification front takes on a snowflake-like shape with dendrites growing

1.1 The Physics of the Instability

To get an idea why rejecting the latent heat in one way should be stable, but rejecting it in the other may not look at Figure 2. Let the solid lie to the left and the liquid to the right and also let the two phases be divided by a planar interface. The solid advances to the right as liquid turns into solid, releasing its latent heat. In the top two sketches, this heat is conducted through the solid; in the lower two sketches, it is conducted through the liq-

[1] The reader will find the following reference helpful: "Interfacial Instability" by L.E. Johns and R. Narayanan, pub. Springer-Verlag, New York, (2002). Large parts of this section are adapted from this book.

[2] The term *subcooled liquid* denotes liquid cooled below its freezing point. Likewise the term *superheated solid* will denote solid heated above its melting point.

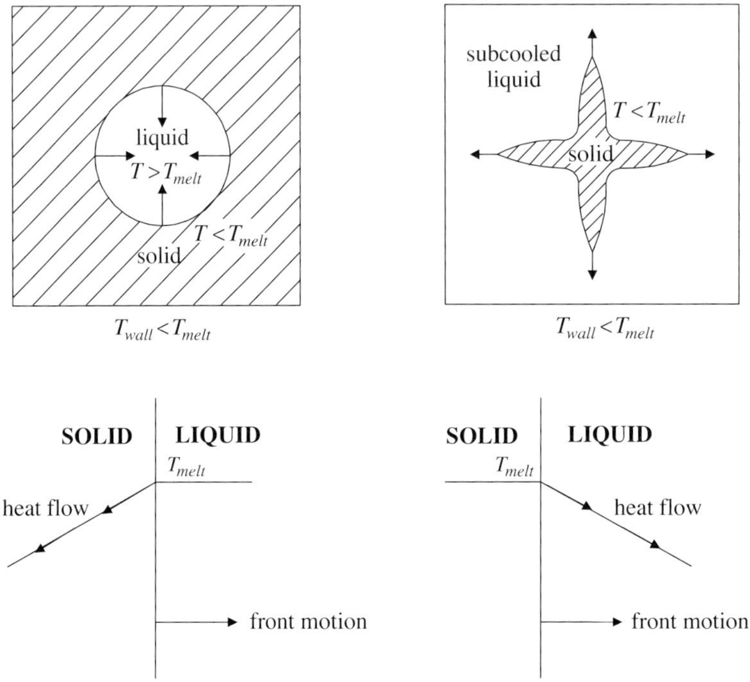

Figure 1. *Two ways of Rejecting Latent Heat are Depicted*

uid. Both of the sketches in the left half of Figure 2 indicate a displacement of the planar front to the right, a local region where the speed of the front runs above the speed of the planar front. To maintain this increased speed, the rate at which latent heat is rejected must also increase. In the upper left sketch, the displacement weakens the temperature gradient available to do this, in the lower left sketch, it strengthens it. A local increase in the speed of the front then acts to oppose itself in the upper sketch and it acts to reinforce itself in the lower sketch. The reader can see that this is also the case if the displacement were to the left (viz., a local region where the speed runs below the speed of the planar front). In this case, the displacement acts to strengthen the temperature gradient in the upper right sketch, reversing the decrease in speed, while it weakens the temperature gradient in the lower right sketch, reinforcing the slowing down.

The two upper sketches predict that a planar front must surely be stable in case the latent heat is rejected to the solid. Then, a displacement of the front to the right produces a new temperature gradient insufficient to

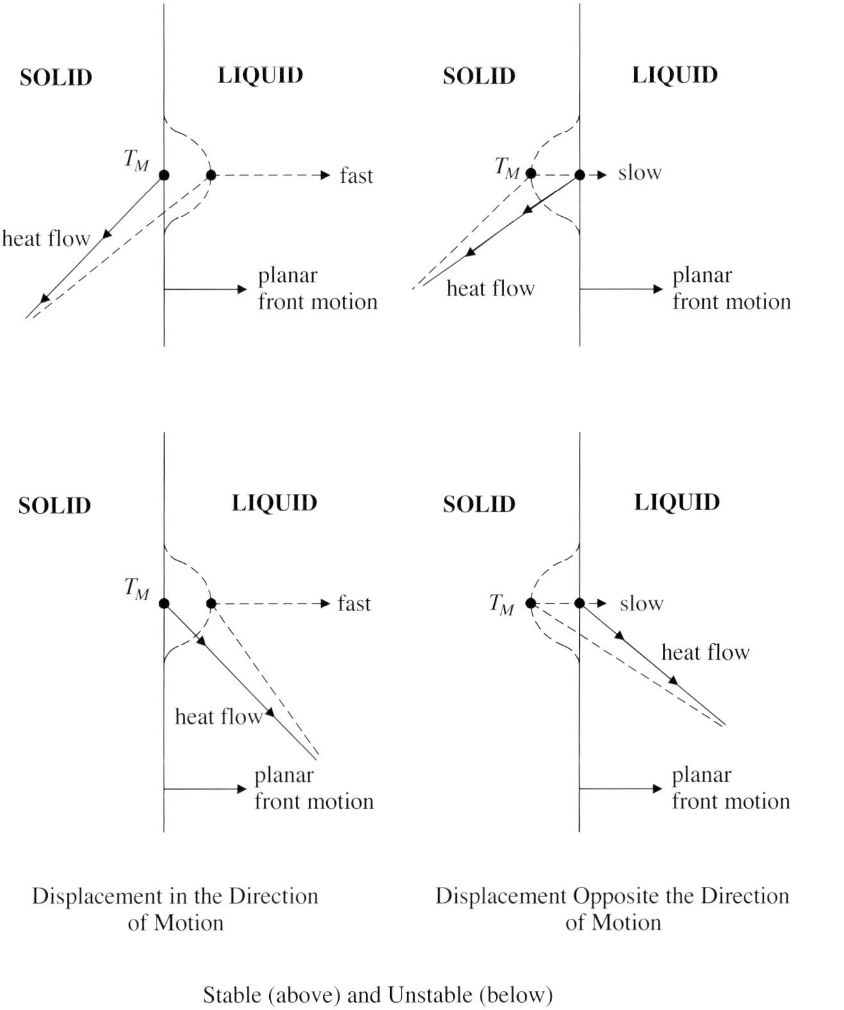

Figure 2. *Displacements of a Solidification Front*

support even its original speed, while a displacement to the left produces a new temperature gradient more than sufficient to support its original speed. The two lower sketches predict only that a front rejecting heat to the liquid may be unstable; the reinforcement may or may not be sufficient to sustain the displacement.

The sketches in Figure 2 do not include the effect of surface tension on the stability of the front. It is stabilizing. It acts in this problem to adjust the melting point of the solid. Although the interface is taken to be at equilibrium, the equilibrium temperature depends on the shape of the interface.

Let T_M denote the melting point of a planar solid surface; then, the melting temperature is reduced at a crest where the solid projects into the liquid, whereas it is increased at a trough where the liquid projects into the solid. The formula for this is

$$T = T_M + \frac{\gamma T_M}{\mathcal{L}} 2H$$

where γ denotes the surface tension, $\mathcal{L}$ denotes the latent heat released on solidification and $2H$ denotes twice the mean curvature of the surface, taken to be negative if the solid projects into the liquid (i.e., at a crest the curvature is negative, at a trough, it is positive, cf. the book by Johns and Narayanan).

The surface tension exercises its influence on the stability of a solidification front in two ways. The first is illustrated in Figure 3, where a wavy displacement of the interface is indicated. At a crest, the temperature falls below T_M; at an adjacent trough it rises above T_M. The faster-moving crest needs to reject heat faster than the slower-moving trough, yet the transverse temperature gradient induced by the displacement carries heat out of the trough into the crest, speeding up the trough while slowing down the crest. This is stabilizing, but how important it is depends on the wavelength of the displacement. For short wavelengths, the effect is strong; for long wavelengths, it is weak. This then is much like the effect of surface tension seen in an earlier lecture. It acts most strongly to stabilize the surface to its shortest wavelength displacements.

The adjustment of the melting point required by surface tension also plays a stabilizing role via its modification of the conclusions drawn from Figure 2. Redrawing that figure, now taking into account the effect of surface tension on the melting point, produces Figure 4, which indicates that a displacement to the right, a crest, weakens the temperature gradient more than before if it is in the solid, whereas it does not strengthen it as much as before, and may even weaken it, if it is in the liquid. Again, the strength of the stabilizing effect depends on the wavelength of the disturbance. By the

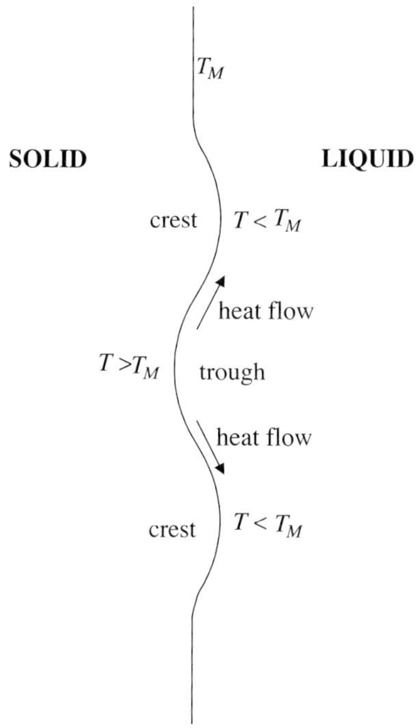

Figure 3. *One Stabilizing Effect of Surface Tension*

same sort of reasoning, surface tension is also stabilizing if the displacement is to the left, a trough, for then the speed of the front is reduced, but the surface temperature is increased.

What we ought to expect, then, on taking surface tension into account, is just what we have found before: A displacement, otherwise unstable, will be stabilized by surface tension if it is of short wavelength, but not if it is of long wavelength.

There is no need to work through the case where the solid is melting and the front is moving into the solid. In this case, latent heat must be supplied and this can be done either by a temperature gradient in the liquid or by a temperature gradient in the solid. The same physical reasoning as above would lead to the conclusion that the first is a picture of a stable melting front, while the second is a picture of a possibly unstable melting front.

The two cases, solidification and melting, indicate that instability corresponds either to heat rejection to a subcooled liquid or to heat supplied by a superheated solid.

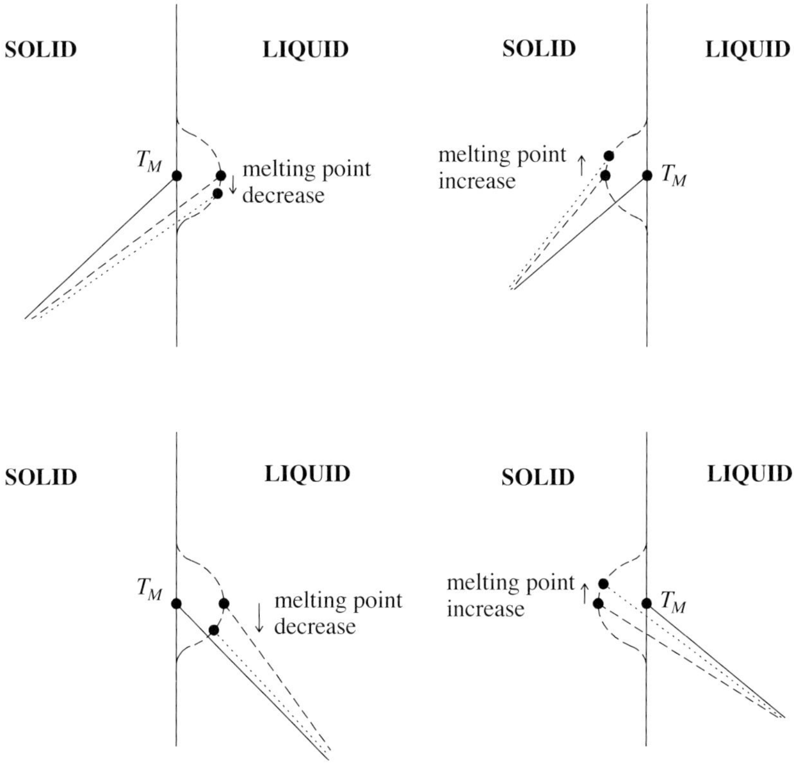

Figure 4. *Figure 2 Redrawn to Take Account of Surface Tension*

1.2 The Model

The Nonlinear Equations

To make this problem simple enough to work out by hand, the densities of the two phases will, in the end, be taken to be equal and hence no flow will be induced by the front motion in either the base or the perturbation problems. Likewise, no account will be taken of the effect of surface curvature on anything beyond the melting point of the solid. The latent heat, for instance, will be that corresponding to a planar interface. All thermophysical properties will be taken to be constant. Taking into account a changing latent heat turns out to be interesting, but we do not do this.

Two cases are of interest. In one, the latent heat is rejected to the solid. In the other it is rejected to the liquid. Lets consider the latter. To make this definite, take the solid to be at rest in the laboratory frame and let the entire solid phase be at its ordinary melting temperature, T_M. Then, introduce a new frame in uniform translation at the velocity $\vec{U} = U\vec{k}$, where U is to be set so that the front lies near the plane $z = 0$. All of the equations we need are form invariant under this change of observer, so every variable we introduce henceforth will be a moving-frame variable.

The latent heat will be removed by supposing that a plane at $z = L_0$, $L_0 > 0$, can be maintained at a fixed temperature T_C, $T_C < T_M$, by some cooling sink.

Let the shape of the front be specified by

$$z = Z(x,\ t)$$

where our problem will be taken to be two dimensional. To determine the function Z, which tells us the position of the interface dividing the solid and the melt, we need to write the equations describing the conduction of heat in the two phases and the equations which tie this to the local motion of the front. These are the nonlinear equations.

They are as follows: First, in the solid phase,

$$\frac{\partial T}{\partial t} = \alpha \nabla^2 T - \vec{v} \cdot \nabla T \tag{1}$$

must hold on $-\infty < z < Z$, while in the liquid phase,

$$\frac{\partial T^*}{\partial t} = \alpha^* \nabla^2 T^* - \vec{v}^* \cdot \nabla T^* \tag{2}$$

must hold on $Z < z < L_0$. Then, at the points along the phase boundary (i.e., at $z = Z$), two requirements must be met. The first is that phase equilibrium be maintained whence

$$T = T_M + \frac{\gamma T_M}{\mathcal{L}}\, 2H = T^* \tag{3}$$

must hold. The second is that the net rate of heat conduction across the surface accounts for the rate at which latent heat is being released due to the motion of the front, namely

$$\lambda \vec{n} \cdot \nabla T - \lambda^* \vec{n} \cdot \nabla T^* = \mathcal{L} \vec{n} \cdot [-\vec{v} + \vec{u}] \tag{4}$$

must hold. This equation is derived in the book by Johns and Narayanan. The latent heat, $\mathcal{L}$ is defined as the difference in the specific enthalpy across the surface, liquid less solid, multiplied by the solid-phase density. The superscript * denotes a liquid-phase variable, λ denotes the thermal conductivity and α denotes the thermal diffusivity. The surface normal, speed and mean curvature can be obtained in terms of the function Z via the formulas

$$\vec{n} = \frac{-Z_x\vec{i} + \vec{k}}{\sqrt{Z_x^2 + 1}} \tag{5}$$

$$u = \frac{Z_t}{\sqrt{Z_x^2 + 1}} \tag{6}$$

and

$$2H = \frac{Z_{xx}}{[Z_x^2 + 1]^{3/2}} \tag{7}$$

To these equations must be added the far-field equations. Away from the interface, then,

$$T(z \to -\infty) = T_M \tag{8}$$

and

$$T^*(z = L_0) = T_C \tag{9}$$

must be satisfied. It is by setting the value of T_C that control over the speed of the solidification front can be achieved.

We introduce no side wall conditions, taking the solidification front to range from $-\infty$ to $+\infty$ in the transverse variable x. Due to this, all non zero transverse wave numbers must be admitted.

Turning to the velocities in the solid and the liquid phases, we can go a long way by declaring only that $\vec{v}$ and $\vec{v}^*$ must satisfy

$$\nabla \cdot \vec{v} = 0 = \nabla \cdot \vec{v}^*$$

in the solid and in the melt and

$$\vec{n} \cdot [\vec{v} - \vec{u}]\rho = \vec{n} \cdot [\vec{v}^* - \vec{u}]\rho^*$$

along the interface dividing the two phases. The solid-phase velocity is set once and for all by requiring the solid to be at rest in the laboratory frame, namely

$$\vec{v} = -\vec{U} = -U\vec{k}$$

Then, at the front, there obtains

$$\vec{n} \cdot \vec{v}^* = -\vec{n} \cdot \vec{k}U\frac{\rho}{\rho^*} - u\frac{[\rho - \rho^*]}{\rho^*} \tag{10}$$

where $u = \vec{n} \cdot \vec{u}$, and this determines the liquid-phase velocity normal to the front in terms of U and u, u being the local speed of the front in the moving frame.

1.3 The Base Solution

The nonlinear equations have a steady one-dimensional solution in which a planar solidification front remains at rest in the moving frame. This solution is called the base solution.

In the laboratory frame, let a planar front be moving normal to itself at a constant speed U in the $\vec{k}$ direction, where U now denotes the base state front speed in the laboratory frame. Then, in the moving frame and at the front, u must be zero and v_z^* must be $-\dfrac{\rho}{\rho^*}\,U$. Hence, $\vec{v}^*$ must be

$$-\frac{\rho}{\rho^*}U\vec{k}$$

throughout the subcooled liquid, as $\nabla \cdot \vec{v}^*$ must be zero there.

To find a steady one-dimensional solution, where ρ and ρ^* are taken to be equal, whence $\vec{v}$ and $\vec{v}^*$ both must be $-\vec{U}$, equations (1) and (2) must be solved and they reduce in this case to

$$0 = \alpha \frac{d^2T}{dz^2} + U\frac{dT}{dz} \tag{11}$$

and

$$0 = \alpha^* \frac{d^2T^*}{dz^2} + U\frac{dT^*}{dz} \tag{12}$$

To fix the coordinate system, let the plane dividing the solid and the liquid in the base state be the $x, y-$ plane. Denoting the base-state variables by the subscript zero, the interface is then the plane

$$z = Z_0 = 0$$

and, at the interface, the normal, the normal speed and the mean curvature are given by

$$\vec{n}_0 = \vec{k},\ u_0 = 0 \ \ and \ \ 2H_0 = 0$$

There is a simple solution to the nonlinear equations. It is

$$T_0 = T_M$$

and

$$T_0^* = A_0^* + B_0^* e^{-\frac{U}{\alpha^*}z}$$

where

$$A_0^* = T_M - B_0^* \ \ and \ \ B_0^* = \frac{T_C - T_M}{e^{-\frac{U}{\alpha^*}L_0} - 1} > 0$$

The solid–liquid interface lies in the plane $z = Z_0 = 0$. It is at rest in the moving frame.

Two derivatives of the base solution carry over to the perturbation problem. They are

$$\frac{dT_0^*}{dz}(z=0) = -\frac{U}{\alpha^*}\ B_0^* < 0 \tag{13}$$

and

$$\frac{d^2T_0^*}{dz^2}(z=0) = \frac{U^2}{\alpha^{*2}}\ B_0^* > 0 \tag{14}$$

The heat balance across the front can be used to determine U in terms of T_C, and there obtains

$$\frac{UL_0}{\alpha^*} = -\ln\left[1 + \frac{T_C - T_M}{\mathcal{L}\alpha^*/\lambda^*}\right]$$

whereupon U increases from zero to infinity as T_C decreases from T_M to $T_M - \mathcal{L}\alpha^*/\lambda^*$. By using this result, a simple formula for B_0^* can be obtained. It is

$$B_0^* = \frac{\mathcal{L}\alpha^*}{\lambda^*}$$

1.4 The Perturbation Equations

To obtain the perturbation equations, the rules explained in the book by Johns and Narayanan can be used. To do this, first turn to the domain. On the solid side of the interface, the term $\vec{v} \cdot \nabla T$ in equation (1) simply leads to $-U\dfrac{\partial T_1}{\partial z}$. But on the liquid side, the term $\vec{v}^* \cdot \nabla T^*$ in equation (2) leads to the sum of two terms, namely, to

$$\vec{v}_1^* \cdot \nabla T_0^* + \vec{v}_0^* \cdot \nabla T_1^*$$

which, in view of the base solution, is just

$$v_{z_1}^* \frac{dT_0^*}{dz} - U\frac{\partial T_1^*}{\partial z}$$

To omit the term $v_{z_1}^* \dfrac{dT_0^*}{dz}$, notice that equation (10), namely

$$\vec{n} \cdot \vec{v}^* = -\vec{n} \cdot \vec{k} U \frac{\rho}{\rho^*} - u \frac{[\rho - \rho^*]}{\rho^*}$$

must hold at the interface, and substitute into this the expansions

$$\vec{n} = \vec{k} - \epsilon Z_{1x} \vec{i}$$

$$\vec{v}^* = \vec{v}_0^* + \epsilon \left[\vec{v}_1^* + Z_1 \frac{\partial \vec{v}_0^*}{\partial z} \right]$$

$$u = \epsilon u_1 = \epsilon Z_{1t}$$

and

$$\vec{v}_0^* = -U \vec{k}$$

to obtain

$$v_{z_1}^* = -Z_{1t} \frac{\rho - \rho^*}{\rho^*}$$

This leads to the important conclusion that $v_{z_1}^*$ must be zero at the interface in the case of interest where ρ and ρ^* are taken to be equal.

Now, the perturbation velocity throughout the subcooled liquid is driven by the motion of the front, as $v_{z_1}^*$ at the front acquires its value $-Z_{1t} \dfrac{\rho - \rho^*}{\rho^*}$. By taking $\rho = \rho^*$, this source of flow is eliminated, and although this, by itself, does not eliminate convection, as $\vec{v}^*$ can still be perturbed, such a perturbation dies out on its own, receiving no input from the rest of the problem. Our work, then, begins once this contribution is small enough to be omitted. As $\vec{v}^*$ is not then perturbed, the perturbation equations on the domain come down to

$$\frac{\partial T_1}{\partial t} = \alpha \nabla^2 T_1 + U \frac{\partial T_1}{\partial z}, \quad -\infty < z < 0 \tag{15}$$

and

$$\frac{\partial T_1^*}{\partial t} = \alpha^* \nabla^2 T_1^* + U \frac{\partial T_1^*}{\partial z}, \quad 0 < z < L_0 \tag{16}$$

At the control surface, $z \to -\infty$,

$$T_1 = 0 \tag{17}$$

must hold, while as $z = L_0$,

$$T_1^* \to 0 \tag{18}$$

must be satisfied.

Now, turn to the interface dividing the two fluids. Two equations must be satisfied there, equations (3) and (4) (i.e., the two phases must be in equilibrium across the front and the latent heat released there must be conducted away). Satisfying these requirements, to first order, requires

$$T_1 + Z_1 \frac{dT_0}{dz} = \frac{\gamma T_M}{\mathcal{L}} \, 2H_1 = T_1^* + Z_1 \frac{dT_0^*}{dz} \tag{19}$$

and

$$\lambda \left[\frac{dT_1}{dz} + Z_1 \frac{d^2 T_0}{dz^2} \right] - \lambda^* \left[\frac{dT_1^*}{dz} + Z_1 \frac{d^2 T_0^*}{dz^2} \right] = \mathcal{L} u_1 = \mathcal{L} \frac{\partial Z_1}{\partial t} \tag{20}$$

to hold where $\frac{dT_0}{dz} = 0 = \frac{d^2 T_0}{dz^2}$, but $\frac{dT_0^*}{dz}$ is negative while $\frac{d^2 T_0^*}{dz^2}$ is positive. The simple form of the two heat conduction terms is due, first, to the fact that $Z_0 = 0$ and then to the fact that T_0 depends only on z, whereupon

$$\vec{n} \cdot \nabla T = [\, \vec{k} - \epsilon Z_{1_x} \vec{i}\,] \cdot \left[\vec{k} \frac{dT_0}{dz} + \epsilon \Big[\nabla T_1 + Z_1 \vec{k} \frac{d^2 T_0}{dz^2} \Big] \right]$$

$$= \frac{dT_0}{dz} + \epsilon \left[\frac{dT_1}{dz} + Z_1 \frac{d^2 T_0}{dz^2} \right]$$

and

$$\vec{n} \cdot \vec{U} = -\epsilon Z_{1_x} \vec{i} \cdot \vec{U} = 0$$

Equations (19) and (20), holding at the reference interface, introduce the base solution into the perturbation problem, and hence they lie at the heart of the stability question.

Equations (15) through (20) present the perturbation problem as an initial-value problem for T_1, T_1^* and Z_1. To turn it into an eigenvalue problem, substitute

$$T_1 = \hat{T}_1(z) e^{ikx} e^{\sigma t}$$

$$T_1^* = \hat{T}_1^*(z) e^{ikx} e^{\sigma t}$$

and

$$Z_1 = \hat{Z}_1 e^{ikx} e^{\sigma t}$$

into it to obtain the equations that $\hat{T}_1$, $\hat{T}_1^*$ and $\hat{Z}_1$ must satisfy. They are

$$\sigma \hat{T}_1 = \alpha \left[\frac{d^2}{dz^2} - k^2 \right] \hat{T}_1 + U \frac{d\hat{T}_1}{dz}, \quad -\infty < z < 0 \tag{21}$$

$$\sigma \hat{T}_1^* = \alpha^* \left[\frac{d^2}{dz^2} - k^2 \right] \hat{T}_1^* + U \frac{d\hat{T}_1^*}{dz}, \quad 0 < z < L_0 \tag{22}$$

$$\hat{T}_1 = 0, \quad z \to -\infty \tag{23}$$

$$\hat{T}_1^* = 0, \quad z = L_0 \tag{24}$$

$$\hat{T}_1 + \hat{Z}_1 \frac{dT_0}{dz} = -\frac{\gamma T_M}{\mathcal{L}} k^2 \hat{Z}_1 = \hat{T}_1{}^* + \hat{Z}_1 \frac{dT_0^*}{dz}, \quad z = 0 \tag{25}$$

and

$$\lambda \left[\frac{d\hat{T}_1}{dz} + \hat{Z}_1 \frac{d^2 T_0}{dz^2} \right] - \lambda^* \left[\frac{d\hat{T}_1^*}{dz} + \hat{Z}_1 \frac{d^2 T_0^*}{dz^2} \right]$$

$$= \mathcal{L} \hat{u}_1 = \mathcal{L} \sigma \hat{Z}_1, \ z = 0 \tag{26}$$

The two equations, denoted equation (25), are the link in this lecture to a common thread of equations running through most interfacial instability problems. In all these equations, which hold at an interface, most of them being phase-equilibrium equations, gradients of the base solution appear, as does the factor γk^2. These are the ingredients out of which pictures explaining the instability can be constructed.

This eigenvalue problem has solutions other than $\hat{T}_1$, $\hat{T}_1^*$ and $\hat{Z}_1$ all zero only for certain values of σ. These values of σ, the eigenvalues, determine whether a disturbance of wave number k grows or dies out.

The simplest thing to do is to determine whether or not there are any values of k^2 for which $\sigma = 0$ is a solution. These values of k^2 are then critical, dividing the disturbances into two classes: those to which the system is stable and those to which it is not. Then setting $\sigma = 0$ to determine whether a critical value of k^2 can be found, we must solve

$$\alpha\left[\frac{d^2}{dz^2}-k^2\right]\hat{T}_1+U\frac{d\hat{T}_1}{dz}=0,\ -\infty<z<0$$

$$\alpha^*\left[\frac{d^2}{dz^2}-k^2\right]\hat{T}_1^*+U\frac{d\hat{T}_1^*}{dz}=0,\ 0<z<L_0$$

$$\hat{T}_1=0,\ z=-\infty$$

$$\hat{T}_1^*=0,\ z=L_0$$

$$\hat{T}_1=-\frac{\gamma T_M}{\mathcal{L}}\ k^2\hat{Z}_1,\ z=0$$

$$\hat{T}_1^*+\hat{Z}_1\frac{dT_0^*}{dz}=-\frac{\gamma T_M}{\mathcal{L}}\ k^2\hat{Z}_1,\ z=0$$

and

$$\lambda\frac{d\hat{T}_1}{dz}-\lambda^*\left[\frac{d\hat{T}_1^*}{dz}+\hat{Z}_1\frac{d^2T_0^*}{dz^2}\right]=0,\ z=0$$

This is an eigenvalue problem. Where it has a solution, other than $\hat{T}_1=0$, $\hat{T}_1^*=0$ and $\hat{Z}_1=0$, determines a value of k^2 such that $\sigma=0$ is a solution to the perturbation eigenvalue problem.

By writing

$$\hat{T}_1=A_1e^{m_+z}$$

and

$$\hat{T}_1^*=A_1^*e^{m_+^*z}+B_1^*e^{m_-^*z}$$

where, again,

$$2m_\pm=-\frac{U}{\alpha}\ \pm\left[\frac{U^2}{\alpha^2}+4k^2\right]^{1/2}$$

and

$$2m^*_\pm = -\frac{U}{\alpha^*} \pm \left[\frac{U^2}{\alpha^{*2}} + 4k^2\right]^{1/2}$$

the first three equations can be satisfied and the remaining four equations reduce to a set of linear, homogeneous, algebraic equations in A_1, A_1^*, B_1^* and $\hat{Z}_1$. The first three of these can be solved for A_1, A_1^* and B_1^* in terms of $\hat{Z}_1$, namely

$$A_1^* = -B_1^* e^{[m^*_- - m^*_+]L_0}$$

$$A_1 = -\frac{\gamma T_M}{\mathcal{L}}\, k^2 \hat{Z}_1$$

and

$$B_1^* = \frac{\dfrac{-\gamma T_M}{\mathcal{L}} k^2 - \dfrac{dT_0^*}{dz}}{1 - e^{[m^*_- - m^*_+]L_0}}\, \hat{Z}_1$$

and the results can be substituted into the fourth equation to obtain

$$\left[-\lambda m_+ \frac{\gamma T_M}{\mathcal{L}} k^2 + \lambda^* M^* \left[\frac{-\gamma T_M}{\mathcal{L}} k^2 - \frac{dT_0^*}{dz}\right] - \lambda^* \frac{d^2 T_0^*}{dz^2}\right] \hat{Z}_1 = 0 \qquad (27)$$

where m_+ is positive and where, for all values of k^2, the factor M^*, where M^* is given by $-\frac{\left[-m^*_+ e^{[m^*_- - m^*_+]L_0} + m^*_-\right]}{1 - e^{[m^*_- - m^*_+]L_0}}$, is of one sign and it is positive.

A solution to equation (27), other than $\hat{Z}_1 = 0$, can be found if and only if the factor multiplying $\hat{Z}_1$ vanishes. This factor is the sum of four terms, and, in view of equations (13) and (14), the first, second and fourth are negative, whereas the third is positive. This, then, presents the possibility that a value of k^2 may be found such that the four terms add to zero, whereupon the factor multiplying $\hat{Z}_1$ vanishes and, by this, $\hat{Z}_1$ itself need not vanish.

The critical value of k^2 is the root of the equation

$$-\lambda m_+ \frac{\gamma T_M}{\mathcal{L}}\, k^2 + \lambda^* M^* \left[-\frac{\gamma T_M}{\mathcal{L}}\, k^2 - \frac{dT_0^*}{dz}\right] - \lambda^* \frac{d^2 T_0^*}{dz^2} = 0 \qquad (28)$$

Now, the base solution comes into the calculation of $k^2_{critical}$ via both the first and the second derivatives of the temperature at the interface. The front moves to the right in the laboratory frame, whereupon the convection is to the left in the moving frame. This makes the second derivative positive when heat is rejected to the liquid. The second derivative is not a source of the instability. In terms of its algebraic sign, it joins the terms multiplied by the surface tension in the equation determining $k^2_{critical}$. It is the first derivative of the temperature at the interface that determines whether or not an advancing solidification front can be unstable. As the source of this derivative can be traced back to the phase-equilibrium requirement across the interface, this not only confirms, but also justifies, the predictions drawn from the pictures at the outset.

This brings the solidification instability problem to a close. The reader can work out the case when latent heat is removed from the solid phase and not from the liquid and show that no instability occurs.

2 Evaporative Instability - Linear Theory

2.1 Introduction

Evaporative convection[3] arises in a variety of natural and industrial processes, such as the drying of lakebeds, heat pipe technology and dry-eye syndrome. The phenomenon of evaporative convection leads to an interfacial instability where an erstwhile flat surface becomes undulated as a control variable, such as temperature drop, exceeds a critical value. In order to study this, the techniques of linear stability analysis are employed. A follow-up section is presented which uses weak nonlinear methods with the objective of obtaining global quantities such as heat transport and changes in evaporation rate as a result of the instability.

2.2 The physical model

The physical model consists of a system where a liquid of depth, d, underlies its own vapor of depth, d*, and is heated from the liquid side (Fig.5).

The base state is where there is a flat interface between the liquid and its vapor, and its stability is in question. The liquid and vapor depths are assumed constant by suitably adjusting the liquid feed and vapor removal rates. The phase-change rate at the interface can be controlled by adjusting the vapor pressure at the top plate wall. The phase-change rate is not constant and it can change upon perturbation. Consequently, the input

[3]The reader is referred to the paper titled "The physics of evaporative instability in bilayer systems:Linear thory," Phys Fluids 16,4644 (2004) and its companion on nonlinear analysis, Phys FLuids 16, 4653, (2004) for more details and other references.

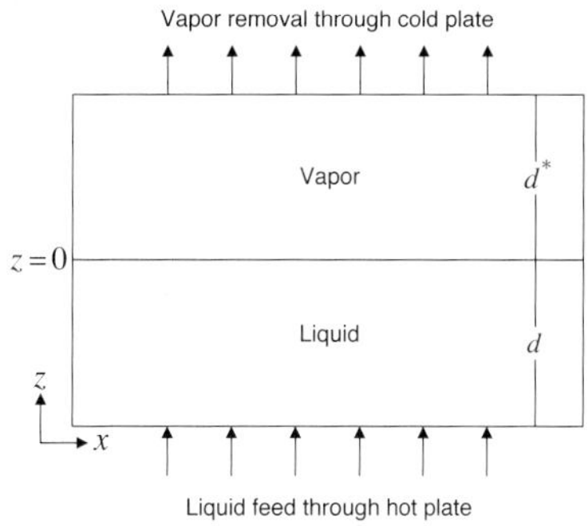

Figure 5. The physical model.

variables are the liquid and vapor depths, the temperatures of the bottom and top plates and the pressure at the top plate. Furthermore, the pressure at the top plate can be adjusted so the phase-change rate at the interface is equal to zero. In the current analysis of the problem, we assume that the liquid and its vapor are in thermodynamic equilibrium because we will only consider either zero or very small phase-change rates at the interface in the base state.

2.3 Physics of the phase-change problem without convection

The physics of evaporation are somewhat complicated. When investigating the problem, it is helpful to look at the different aspects of the problem separately when possible. For this reason, both the liquid and the vapor are momentarily assumed to be fluid dynamically passive and the problem is examined strictly from a heat transfer point of view. As a consequence of the passive fluid dynamics assumption in both domains, pressure perturbations will not arise in either fluid domain. Therefore, the vapor pressure at the interface must be equal to the input pressure at the cold plate above the vapor. As the temperature is in equilibrium with pressure, the temperature perturbations will also vanish along the interface. This, of course, will change later when the fluid motion is taken into account. Interesting conclusions can be drawn simply by looking at the temperature profiles in the base state.

Figure 6 illustrates a phase-change problem with zero evaporation rate where the heat necessary for the evaporation is supplied from the liquid side and both fluids are fluid dynamically passive. The dotted wave represents a perturbed interface and the dotted lines represent the temperature gradients

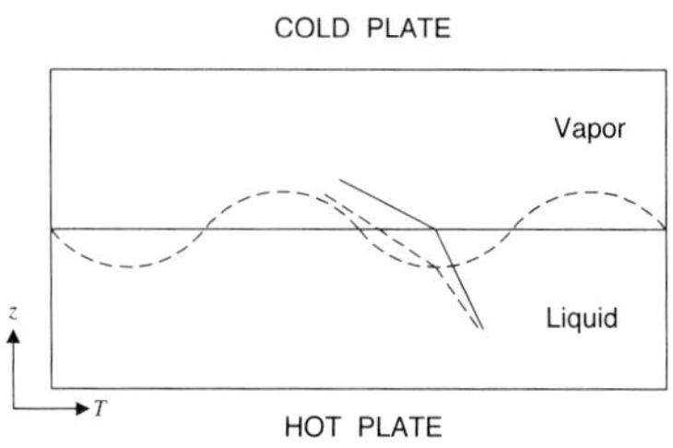

Figure 6. Physics of pure heat transfer.

in both phases at the interface in the perturbed state. The temperature at the perturbed interface is the same as at the flat interface because at low evaporation rates the temperature is assumed to be in equilibrium with the pressure at the interface, and there are no pressure perturbations in the perturbed state as a consequence of the passive fluid dynamics assumption. When the interface is perturbed, a trough is closer to the heat source and further away from the heat sink. The temperature gradient on the liquid side gets sharper whereas the temperature gradient on the vapor side becomes less sharp, i.e., when the interface is perturbed more heat is transferred to a point at a trough and less heat is taken away from that location, encouraging evaporation at the trough and making it grow deeper into the liquid. This makes the interface unstable. The picture argument presented here suggests that when the interface is unstable it is unstable for all wavenumbers. Even the slightest temperature gradient causes instability provided the liquid is the hotter phase. Observe that in this simplified picture argument, forces and force balances are ignored. Of course, physical reality requires that both must be taken into account and that is what is done next.

2.4 Physics of the phase-change problem with convection

When we include the fluid dynamics in both phases, but still leave out gravity and Marangoni convection, things become a little more complicated. A deflected interface does not have the uniform temperature as an erstwhile flat interface. Unlike the case where we have left out the fluid dynamics of both phases, there are now perturbations allowed in the pressure fields as deflecting surfaces induce non-zero curvature which in turn causes pressure variations along the interface. Given a disturbance, phase-change will take place at the interface that will cause perturbations in pressure along the interface. Since the pressure of the vapor and the temperature of the vapor are in thermodynamic equilibrium at the interface, the temperature will also vary along the interface. A trough is at a higher temperature than a crest as it is closer to the heat source. The interface temperature in the base state was equal to the saturation temperature since we began with thermodynamic equilibrium at the interface and no phase-change was taking place. Upon perturbation, a trough is at a temperature higher than the saturation temperature, meaning that evaporation will take place at a trough. Consequently, at a crest where the temperature is colder than the saturation temperature the vapor will condense into its own liquid. There will be upward flow at the troughs and downward flow at the crests as indicated in Figure 7. Again, instability occurs at each wavenumber, but this time a critical temperature difference is necessary to cause the instability for each wavenumber. As in other phase-change problems, such as solidification, the surface tension stabilizes the interface at small wavelengths. Unfortunately, the surface tension itself is a function of the temperature and it requires us to include surface-tension-driven convection in our model. We will account for this Marangoni effect, as well as for gravity and buoyancy-driven convection in the mathematical model.

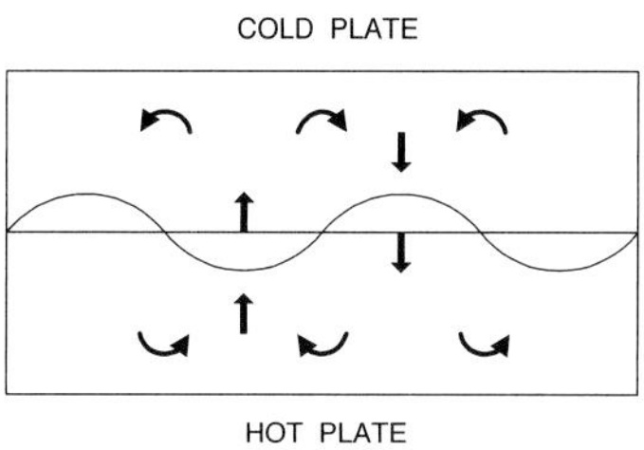

Figure 7. Pure phase-change problem with convection.

2.5 The mathematical model

The equations that model the physics in unscaled form are given by the Navier Stokes equations in each phase, i.e. by

$$\rho\left(\frac{\partial\vec{v}}{\partial t}+\vec{v}\cdot\nabla\vec{v}\right)=-\nabla P+\rho\vec{g}+\mu\nabla^2\vec{v}$$

and

$$\rho^*\left(\frac{\partial\vec{v}^*}{\partial t}+\vec{v}^*\cdot\nabla\vec{v}^*\right)=-\nabla P^*+\rho^*\vec{g}+\mu^*\nabla^2\vec{v}^*$$

by the energy equations in each phase, i.e. by

$$\frac{\partial T}{\partial t} + \vec{v} \cdot \nabla T = \kappa \nabla^2 T$$

and

$$\frac{\partial T^*}{\partial t} + \vec{v}^* \cdot \nabla T^* = \kappa^* \nabla^2 T^* ,$$

and by the continuity equation in each phase assuming that the fluids are incompressible. In the equations above, $\vec{v}$, P and T are velocity, pressure and temperature fields, respectively, and the asterisk denotes the vapor phase. It is assumed that the bottom and top plate temperatures are kept constant and therefore $T(-d) = T_{hot}$ and $T^*(d^*) = T_{cold}$ hold. In the equations, ρ, μ, and κ are the density, viscosity and thermal diffusivity, respectively.

When we assume that the liquid and its vapor are in equilibrium and that no phase-change is taking place across the interface, we need not feed and remove fluid through the top and bottom plates. Therefore, $v_z(-d) = 0$ and $v_z^*(d^*) = 0$ hold. The no-slip condition applies along the top and bottom plates, and gives rise to $v_x(-d) = 0$ and $v_x^*(d^*) = 0$. The interfacial mass balance equation is given by

$$\rho(\vec{v} - \vec{u}) \cdot \vec{n} = \rho^* (\vec{v}^* - \vec{u}) \cdot \vec{n}$$

where

$$\vec{n} = \frac{-\frac{\partial Z}{\partial x}\vec{i} + \vec{k}}{\left(1 + \left(\frac{\partial Z}{\partial x}\right)^2\right)^{1/2}}$$

is the unit outward normal and

$$\vec{u} \cdot \vec{n} = \frac{\frac{\partial Z}{\partial t}}{\left(1 + \left(\frac{\partial Z}{\partial x}\right)^2\right)^{1/2}}$$

is the interface speed.

At the interface, the tangential components of velocities of both fluids are equal to each other, so $\vec{v} \cdot \vec{t} = \vec{v}^* \cdot \vec{t}$ holds where

$$\vec{t} = \frac{\vec{i} + \frac{\partial Z}{\partial x}\vec{k}}{\left(1 + \left(\frac{\partial Z}{\partial x}\right)^2\right)^{1/2}}$$

is the unit tangent vector.

The interfacial tension and its gradient with respect to the temperature at the interface come into the picture through the force balance. By taking the dot product of the stress balance with the unit normal and unit tangent vectors separately, we get the normal and tangential stress balance equations. The force balance is given by

$$\left(\rho\vec{v}(\vec{v}-\vec{u})-\vec{\vec{T}}\right)\cdot\vec{n}+\gamma 2H\vec{n}+\nabla_s\gamma\vec{t}=\left(\rho^*\vec{v}^*(\vec{v}^*-\vec{u})-\vec{\vec{T}}^*\right)\cdot\vec{n}$$

where $\vec{\vec{T}}=-P\vec{\vec{I}}+\vec{\vec{S}}$ is the total stress, γ is the interfacial tension and

$$2H=\frac{\frac{\partial^2 Z}{\partial x^2}}{\left(1+\left(\frac{\partial Z}{\partial x}\right)^2\right)^{3/2}}$$

is the surface mean curvature.

In addition, we need the interfacial energy balance, viz.

$$\left(\rho\left(\hat{H}+\frac{1}{2}(\vec{v}-\vec{u})^2\right)(\vec{v}-\vec{u})+\vec{q}-\vec{\vec{S}}\cdot(\vec{v}-\vec{u})\right)\cdot\vec{n}$$
$$=\left(\rho^*\left(\hat{H}^*+\frac{1}{2}(\vec{v}^*-\vec{u})^2\right)(\vec{v}^*-\vec{u})+\vec{q}^*-\vec{\vec{S}}^*\cdot(\vec{v}^*-\vec{u})\right)\cdot\vec{n}$$

Here $\hat{H}$ is the enthalpy per unit mass, $\vec{\vec{S}}$ is the extra stress, and $\vec{q}$ is the heat flux. At this point, we have only 11 boundary conditions, and we complete the boundary conditions with our choice of equilibrium conditions for the temperature and pressure at the interface. The choice of conditions at the interface between the liquid and the vapor poses a challenge in phase-change problems. Ordinarily, the continuity of temperature is used along with some other thermodynamic equilibrium condition or a relation derived from the kinetic theory is used instead. When the evaporation rate is very small we may assume that there is thermodynamic equilibrium at the phase-change boundary and the temperature of both fluids are equal to each other at the interface, that is $P^*=f(T^*)$ and $T=T^*$.

To study the stability of the interface, arbitrary disturbances of small amplitude are applied and for a given set of input variables, the growth and decay time constants of these disturbances or equivalently the critical temperature difference across both layers that result in marginal stability are determined. The plan, then, is to linearize the above equations about a known base state and to find the onset of the interfacial instability from the perturbed model. Hereafter, the variables of the base state are denoted with the subscript 0, and the variables of the perturbed state with the subscript 1. Thus, the temperature, when perturbed, is described by

$$T = T_0 + \epsilon \left(T_1 + \frac{dT_0}{dz} z_1 \right)$$

where ϵ is the size of the small amplitude of the disturbance on the base state and z_1 is the mapping of the perturbed configuration onto the reference configuration. Its meaning is explained in the book by Johns and Narayanan (cf. reference in the foot note at the beginning of this section) and in their second chapter. At the interface it is simply the perturbation of the surface deflection to first order, Z_1, a variable to be determined during the course of the calculation. We can further expand T_1 and other subscript 'one' variables using a normal mode expansion. For example,

$$T_1 = \hat{T_1}(z) e^{\sigma t} e^{i\omega x}.$$

Here, σ is the inverse time constant while ω is a wavenumber associated with the given perturbation. Observer that we use the notation ω for the wavenumber in the section on evaporation while the earlier section on solidification uses the notation "k" for the wavenumber. A wavenumber arises because the system is taken to be infinite in lateral extent. The same expansion is used for both components of velocity and pressure in both phases. The perturbation equations are then brought into scaled form using the following scales: length, d; velocity, $\bar{v} = \frac{\kappa}{d}$; time, $\frac{d}{\bar{v}}$; temperature, $\bar{T} = \frac{T - T^{sat}}{\Delta T}$; pressures in the liquid and in the vapor are scaled as $\frac{\mu \bar{v}}{d}$. In temperature scale, $\Delta T = T_{hot} - T_{cold}$. After we bring our equations into a dimensionless form, we first solve for the base state temperature profiles. Then we proceed with the perturbation analysis and present the dimensionless perturbation equations.

2.6 The base state solution and the perturbed equations

In the base state there is no flow in either phase if the evaporation rate at the interface is set equal to zero, thus $\vec{v}_0 = \vec{v}_0^* = \vec{0}$. From now on, we will drop the overbar of the scaled temperature. The temperature profiles in the liquid and its vapor in dimensionless form become

$$T = -\frac{k^*}{k\delta + k^*} z \quad and \quad T^* = -\frac{k}{k\delta + k^*} z$$

where $\delta = \frac{d^*}{d}$.

After we perturb the domain and boundary equations in the manner given in the previous section, we arrive at the following equations in the domain in dimensionless form:

$$\left(\frac{d^2}{dz^2} - w^2\right) v_{x1} - iwP_1 = \sigma \frac{1}{Pr} v_{x1}$$

$$\left(\frac{d^2}{dz^2} - w^2\right) v_{z1} + RaT_1 - \frac{\partial P_1}{\partial z} = \sigma \frac{1}{Pr} v_{z1}$$

$$\left(\frac{d^2}{dz^2} - w^2\right) T_1 - \frac{dT_0}{dz} v_{z1} = \sigma T_1$$

and

$$iwv_{x1} + \frac{dv_{z1}}{dz} = 0$$

for the lower phase and

$$\left(\frac{d^2}{dz^2} - w^2\right) v_{x1}^* - iw\frac{\mu}{\mu^*} P_1^* = \sigma \frac{1}{Pr} \frac{\nu}{\nu^*} v_{x1}^*$$

$$\left(\frac{d^2}{dz^2} - w^2\right) v_{z1}^* + \frac{\alpha^*}{\alpha} \frac{\nu}{\nu^*} RaT_1^* - \frac{\mu}{\mu^*} \frac{\partial P_1^*}{\partial z} = \sigma \frac{1}{Pr} \frac{\nu}{\nu^*} v_{z1}^*$$

$$\left(\frac{d^2}{dz^2} - w^2\right) T_1^* - \frac{\kappa}{\kappa^*} \frac{dT_0^*}{dz} v_{z1}^* = \sigma \frac{\kappa}{\kappa^*} T_1^*$$

and

$$iwv_{x1}^* + \frac{dv_{z1}^*}{dz} = 0$$

for the upper phase.

Here, Pr and Ra stand for the dimensionless Prandtl and Rayleigh numbers, respectively. Pr is defined as $\frac{\nu}{\kappa}$, that is, as the ratio of the liquid kinematic viscosity to the liquid thermal diffusivity, and Ra is defined as $\frac{\alpha g \Delta T}{\nu \kappa} d^3$ where α is the negative liquid thermal expansion coefficient, g is the magnitude of gravitational acceleration, and ΔT is the total temperature difference applied across both fluid layers.

At the lower plate, $z = -1$, the liquid is subject to $v_{z1} = 0$, $v_{x1} = 0$, and $T_1 = 0$ while the conditions at the top plate become $v^*_{z1} = 0$, $v^*_{x1} = 0$, and $T^*_1 = 0$.

At the interface the interfacial mass balance turns into

$$v_{z1} - \frac{\rho^*}{\rho} v^*_{z1} = \sigma \left(1 - \frac{\rho^*}{\rho}\right) Z_1$$

where Z_1 is the surface deflection.

The continuity of temperature and the no-slip conditions become

$$T_1 + \frac{dT_0}{dz} Z_1 = T^*_1 + \frac{dT^*_0}{dz} Z_1 \quad and \quad v_{x1} = v^*_{x1}.$$

The normal and tangential stress balances assume the following dimensionless form:

$$Ca(P_1 - P^*_1) - 2Ca\left(\frac{dv_{z1}}{dz} - \frac{\mu^*}{\mu}\frac{dv^*_{z1}}{dz}\right) - (B_0 + w^2)Z_1 = 0$$

and

$$\frac{\mu^*}{\mu}\left(\frac{dv^*_{x1}}{dz} + iwv^*_{z1}\right) - \left(\frac{dv_{x1}}{dz} + iwv_{z1}\right)$$
$$-iwMa\left(T_1 + \frac{dT_0}{dz} Z_1\right) = 0$$

where $B_0 = \dfrac{(\rho - \rho^*)gd^2}{\gamma}$, $Ca = \dfrac{\mu\kappa}{\gamma d}$ and $Ma = \dfrac{\gamma_T \Delta T}{\mu\kappa} d$ are the Bond, Capillary and Marangoni numbers, respectively.

The interfacial energy balance is

$$v_{z1} - E\left(\frac{k^*}{k}\frac{dT^*_1}{dz} - \frac{dT_1}{dz}\right) = \sigma Z_1$$

where E stands for the Evaporation number and is given by $E = \dfrac{C_p \Delta T}{\lambda}$. C_p and λ are the liquid heat capacity and the latent heat of evaporation, respectively. The equilibrium condition at the interface is

$$\Pi_{KE} P^*_1 - \Pi_{PE} Z_1 = \theta\left(T^*_1 + \frac{dT^*_0}{dz} Z_1\right)$$

where $\Pi_{KE} = \dfrac{\rho}{\rho^*}\dfrac{\nu\kappa}{\lambda d^2}$ and $\Pi_{PE} = \dfrac{gd}{\lambda}$ are two dimensionless parameters from the linearized Clapeyron equation, and $\theta = \dfrac{\Delta T}{T^*_0}$ defines a scaled overall temperature drop; T^*_0 being the vapor base state interface temperature in degrees Kelvin.

Results of calculations and explanation

The first goal of the calculations is to determine the conditions at the onset of instability. In order to do this, set σ to zero. The unscaled eigenvalue in the linearized problem is the temperature drop. The Chebyshev spectral tau method is used to solve the resulting eigenvalue problem that determines the critical temperature difference. The calculations were done assuming the physical properties of water and water vapor are at the saturation temperature under 1 atm, i.e., at 100°C. The physical properties used in the calculations are given in Table 1.

For reasons of comparison we present graphs of the ratio of the critical temperature difference for a particular physical case to the critical temperature difference of another comparable case against the wavenumber. The definitions of these ratios are shown in Table 2 where the subscripts pc, Ma, g, and Ra stand for phase-change, Marangoni convection, gravity and buoyancy-driven convection, respectively. For example, ΔT_{pc-Ma} denotes the critical temperature difference for the case where there is phase-change and Marangoni convection, and R^1 therefore is the ratio of ΔT_{pc-Ma} to the critical temperature difference for the case of pure Marangoni convection, ΔT_{Ma}.

First, we present the results for the case where the critical temperature difference for a pure phase-change problem is compared to the critical temperature difference for a case where Marangoni effect is added to the phase-change problem. This comparison is given in Table 3 and the readers can see for themselves that the addition of the Marangoni effect does little to change the threshold of evaporative instability.

Now, Marangoni convection is driven by the temperature variations along the interface. In order to understand the effect of phase-change on temperature gradients at a deflected interface let us first consider such an interface in the absence of gravity and observe that the deflection should make the temperature vary along the interface. As discussed earlier, the temperature at a trough is expected to be higher now than at a crest as it is closer to the heat source. When perturbed, the liquid starts to evaporate into its own vapor at a trough since the temperature at the trough is now higher than the saturation temperature. Consequently, condensation takes place at the crests. When the liquid evaporates at the trough, the otherwise hot trough temperature goes down since part of the energy input now is used for phase-change whereas at the crest the heat is released, and the otherwise low crest temperature goes up. This phase-change action thereby reduces the temperature perturbations along the interface in phase-change problems reducing the surface tension gradient-driven convection, i.e., the

Table 1. Physical properties of water and water vapor at $100°C$.

water	water vapor
$\rho = 960 \frac{kg}{m^3}$	$\rho^* = 0.6 \frac{kg}{m^3}$
$\mu = 2.9 \times 10^{-4} \frac{kg}{m\ sec}$	$\mu^* = 1.3 \times 10^{-5} \frac{kg}{m\ sec}$
$k = 6.8 \times 10^{-1} \frac{J}{m\ sec\ °C}$	$k^* = 2.5 \times 10^{-2} \frac{J}{m\ sec\ °C}$
$K = 1.7 \times 10^{-7} \frac{m^2}{sec}$	$K^* = 2.0 \times 10^{-5} \frac{m^2}{sec}$
$\alpha = 6.0 \times 10^{-4} \frac{1}{°C}$	$\alpha^* = 6.0 \times 10^{-3} \frac{1}{°C}$
$\lambda = 2.3 \times 10^{6} \frac{J}{kg}$	$\gamma = 5.8 \times 10^{-2} \frac{N}{m}$
$\gamma_T = 2 \times 10^{-4} \frac{N}{m°C}$	

Marangoni convection, which therefore ultimately never really plays a significant role on the stability of the phase-change problem. We can also predict the reduction of the temperature perturbations at the interface by considering the linearized thermodynamic equilibrium relation at the interface. The Π terms are very small for most latent heats and this makes the temperature perturbations very small.

Moreover when we look at Figure 8, we observe that for large wavenumbers the pure Marangoni problem is considerably more unstable than the

Table 2. Definitions of R^1, R^2 and R^3.

R^1	$\dfrac{\Delta T_{pc-Ma}}{\Delta T_{Ma}}$
R^2	$\dfrac{\Delta T_{pc-Ma-g}}{\Delta T_{pc-Ma}}$
R^3	$\dfrac{\Delta T_{pc-Ma-Ra}}{\Delta T_{pc-Ma}}$

Table 3. Comparison of critical temperature differences.

Wavenumber, w	$\Delta T_{pc}(^{\circ}C)$	$\Delta T_{pc-Ma}(^{\circ}C)$
0.1	0.03	0.03
0.5	4.35	4.35
1.0	6.05	6.04
1.5	7.39	7.39
2.0	9.43	9.43

phase-change problem with Marangoni convection. At low wave numbers the reverse is true. Before we go on and explain these observations let us make clear the restrictions behind the calculations that generated Figure 8.

In producing Figure 8 we have performed calculations taking two similar systems where a liquid is underlying its own vapor at zero gravity and we

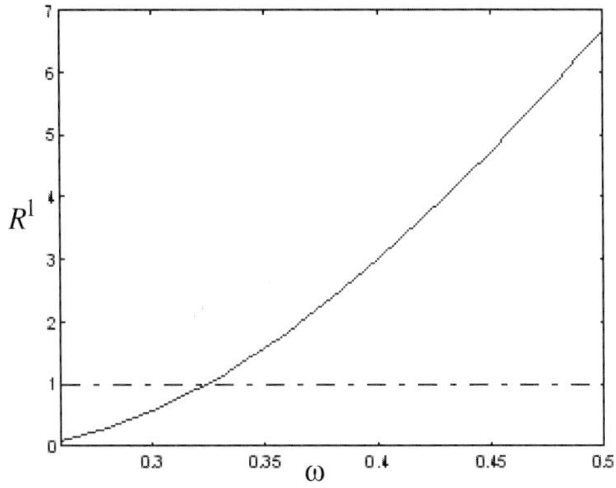

Figure 8. R^1 vs. ω.

apply a temperature gradient across both layers by heating the systems from the liquid side. In one case, we pretend that there is no phase-change across the interface between the liquid and its vapor. In this case, the flow can only be caused by surface-tension gradients along the interface. In the other case, we allow phase-change across the interface. To make a fair comparison, as there is initially no flow in the pure Marangoni problem, we adjust the pressure on the vapor side in the second system in order to set the phase-change rate to zero. As a result, phase-change is only possible through perturbations at the interface. Accordingly, we have two systems with no-flow in their basic states. They are two similar systems with only one difference; one has the possibility of phase-change upon perturbation while the other has no such possibility at all. We then plot the ratio, R^1, versus wavenumber, ω, in Figure 8. For small wavenumbers, the pure Marangoni problem is more stable than the phase-change problem with Marangoni convection for a water and water-vapor system. Small wavenumbers imply that there are reduced transverse thermal gradients thereby stabilizing the Marangoni effect but allowing the phase change instability to continue unabated. For large wavenumbers, we observe the strong action of thermal and momentum diffusion. In many physical situations and laboratory experiments, gravity plays an important role. Now, gravity comes equipped with the burden of buoyancy-driven convection; however, the effects of Rayleigh convection can be avoided in applications that involve very thin liquid and vapor layers. Gravity simply pulls the perturbed interface back to its original position. In other words, gravity plays a stabilizing effect on all wavenumbers. Even so, its effect on the stability of the interface is more obvious for small wavenumbers, small wave numbers imply weaker surface curvature, and that explains why the critical temperature difference is higher for small wavenumbers when the gravity is turned on. This is depicted by the curve labeled R^2 in Figure 9. Working with very thin fluid layers is difficult in a laboratory environment. Thus, eventually we have to account for the effect of buoyancy on the problem. For that reason, Rayleigh convection is introduced into the problem. It is well known from the Rayleigh problem that in the absence of phase-change buoyancy-driven convection can easily be stabilized for very small and very large wavenumbers. Consequently, in Figure 9 we observe that when R^2 and R^3 are plotted against the wavenumber, both plots understandably merge for small wavenumbers. For large enough wavenumbers, in the absence of buoyancy-driven convection, the stability offered by gravity is swept away by the stabilizing effect of the phase-change mechanism and the curve, depicted by R^2, merges with the unity line. This occurs because the surface curvature and in turn phase-change becomes more important compared to gravity at large wavenumbers. In contrast, R^3 diverges to become less stable because of the destabilizing character of buoyancy-driven convection. Yet for even larger wavenumbers, R^3 ultimately merges with the unity line, not shown in Figure 9.

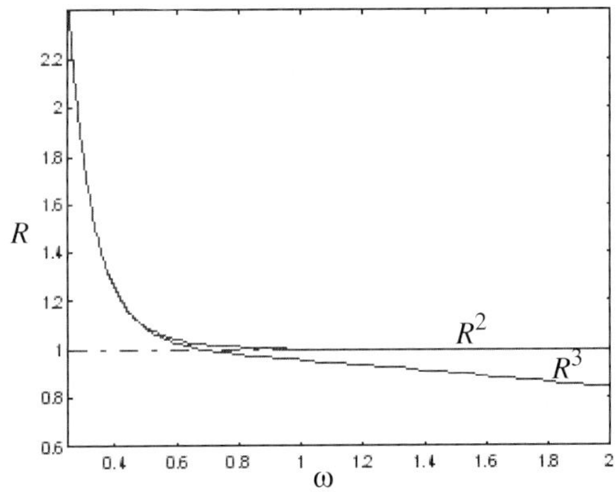

Figure 9. R^2 and R^3 vs. ω.

In order to demonstrate the crucial role of the active vapor layer in the stability of a bilayer phase-change problem, we vary the depth of the layer for a constant liquid depth and calculate the critical temperature difference for a phase-change problem with Marangoni and Rayleigh convection. The

Table 4. Definitions of R^4, R^5, and R^6.

R^4	$\dfrac{\Delta T_{critical}\ of\ a\ system\ with\ \delta = 0.2}{\Delta T_{critical}\ of\ a\ system\ with\ \delta = 0.1}$
R^5	$\dfrac{\Delta T_{critical}\ of\ a\ system\ with\ \delta = 0.3}{\Delta T_{critical}\ of\ a\ system\ with\ \delta = 0.1}$
R^6	$\dfrac{\Delta T_{critical}\ of\ a\ system\ with\ \delta = 0.4}{\Delta T_{critical}\ of\ a\ system\ with\ \delta = 0.1}$

results are plotted in Figure 10 where the ratio of the critical temperature difference for systems with different values of δ is graphed against wavenumber. The ratios are defined in Table 4.

Figure 10 shows us that increasing the vapor depth stabilizes the interface. The deeper the vapor layer is, the more unstable it becomes due to Rayleigh convection which is scaled as the cube of the depth, however, the important stabilizing role of the increased vapor depth is still there; the stabilization comes about because of the increase in the vapor flow and meanwhile the destabilizing effect of heat transfer is weakened.

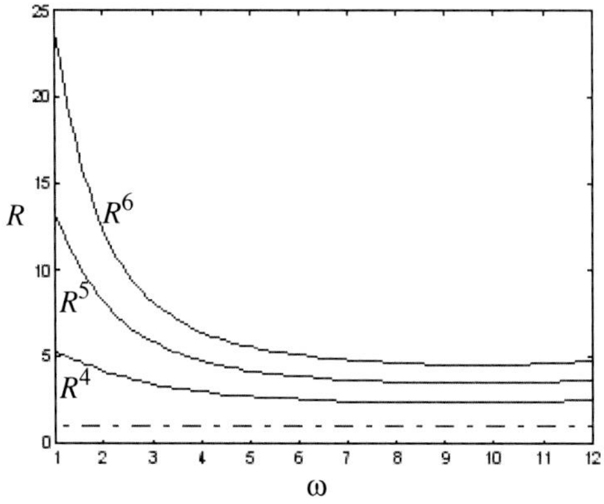

Figure 10. R^4 through R^6 vs. wavenumber.

To understand the stability mechanism in Figure 10, let us first look at the problem from a heat conduction point of view. The heat that is conducted to the interface is used for phase-change and kinetic energy; with the balance being conducted from the interface through the vapor. The interface temperature in the base state is always equal to the saturation temperature. Its value depends upon the pressure at the top plate and the top plate pressure is set to make the evaporation rate in the base state to be zero. A change in depths must influence the pressure at the top plate required to keep the phase-change rate equal to zero and thence the saturation temperature at the interface. The proximity of the top plate to the interface therefore affects the temperature gradients at the interface in both phases. The further away the top plate, or in other words, the greater the ratio of the vapor depth to the liquid depth, the smaller the temperature gradients in the base state. We learned from our preliminary analysis of the phase-change problem in the absence of fluid dynamics that the heat transfer mechanism destabilizes the interface. Thus, from a heat transfer point of view, the deeper the vapor depth the less the destabilizing effect of the heat transfer. From a fluid dynamic point of view, the deeper the vapor layer the less the top plate impedes the stabilizing flow in the vapor. Hence, the interface becomes more stable for two reasons. In other words, the upper wall proximity plays two roles here. The flows in separate phases play different roles on the stability. And it is the vapor flow's stabilizing role which makes the passive vapor assumption questionable.

2.7 Stabilizing effect of the vapor flow

In order to understand how the vapor flow stabilizes the system, it is best to look at the velocity profiles at the onset of instability of a pure phase-change problem.

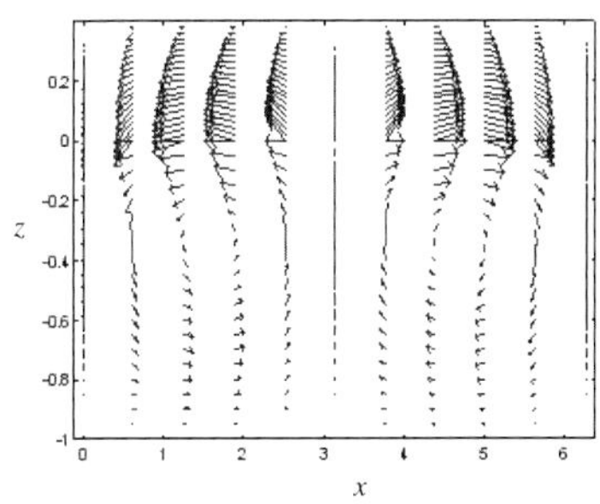

Figure 11. The velocity profiles in both phases at the onset of instability.

In Figure 11 the velocity profiles are shown for a case where the vapor depth to the liquid depth ratio is taken to be equal to 0.4. The position, $z = 0$, represents the liquid-vapor interface. To determine the direction of the flow at a trough (or at a crest) we calculate the value of the vertical component of the velocity in either phase at the interface relative to the computed value of Z_1. This is done from the eigenvector that corresponds to the critical temperature difference. Then, we find the sign of $\frac{v_{z1}}{Z_1}$ (or $\frac{v_{z1}^*}{Z_1}$) where v_{z1} (or v_{z1}^*) is the liquid (or vapor) velocity at the interface and Z_1 is the surface deflection. This ratio turns out to be negative, implying that when the deflection is negative, then the flow is in the positive direction. In other words, there is an up-flow at the troughs, and down-flow at the crests. The fluid flows upset the temperature gradients in both phases at the interface and more so in the vapor because of the stronger flow at the interface in the vapor. Once the phase-change takes place upon perturbation, the flow in the liquid will bring the warmer liquid from below attempting to make the trough hotter whereas the stronger flow in the vapor tries to convect this heat away from the trough. Observe the difference in the magnitudes of the velocities and the nature of the profiles between the two phases. The horizontal component of the velocity in the vapor is very strong compared to the liquid velocity. The vapor brings the warm fluid from the trough to the colder crest, making its temperature go back up again. In other words, the liquid flow convects heat to the interface and destabilizes the interface while the heat conduction from the bottom plate to the interface also encourages instability. The vapor flow offers stability though the heat conduction in the vapor encourages instability. The velocity profiles of the perturbed flow tell us that the liquid flow plays a destabilizing role whereas the vapor flow plays a stabilizing role. To give further credibility to the statement that the vapor flow is stabilizing whereas the liquid flow is destabilizing, we can perform a little test. By deliberately changing the viscosity of the vapor we can strengthen or weaken the vapor flow. Thus, if we increase the vapor viscosity by a small amount, the vapor flow should be impeded and the system will become unstable. Indeed, an increase in the vapor viscosity destabilizes the interface for all wavenumbers. Accordingly, an increase in the liquid viscosity stabilizes the interface because the destabilizing liquid flow is impeded. It is useful to remember that in both phases, the flow is caused by the temperature perturbations at the interface. This is obvious from the energy equation at the interface wherein one sees that perturbed temperature gradients must result in perturbed flows at the interface. This flow at the interface generates pressure perturbations, which in turn feed-back to the temperature through the thermodynamic equilibrium at the interface. To summarize, the stability of a phase-change problem is determined by the competitive nature of heat conduction and heat convection, where the relative values of physical properties of both the liquid and the vapor affect the magnitude of the heat transfer mechanisms.

Another property that plays a pivotal role is the latent heat of evaporation. The higher the latent heat the more stable the interface because then the problem is more like the bilayer problem where there is no phase-change taking place across the interface. Since gravity is left out of our argument for the time being and the temperature perturbations settle down quickly in phase-change problems, it is not difficult to conclude that the problem becomes more stable.

The phase-change problem in the absence of gravity becomes unstable at all wavenumbers given that a certain critical temperature difference is applied across the fluid layers. Thus, if we were to conduct experiments in space, we would have to cut-off the small wavenumbers by limiting the size of the container so that the system is not unstable from the beginning of the experiment. However, when gravity is present, there exists a critical wavenumber at which the interface becomes unstable, as depicted in Figure 12.

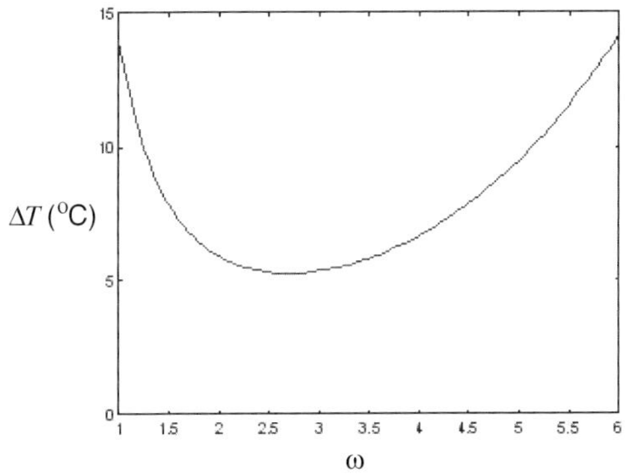

Figure 12. The critical temperature difference versus the wavenumber for water-water vapor system where $d = 3\text{mm}$ and $d^* = 1.5\text{mm}$.

In summary, we have attempted to explain the physics of the phase-change problem at the onset of instability in a comprehensive manner using a straightforward linear stability calculation. We conclude that the consideration of the vapor fluid dynamics is vital for the study of stability of the phase-change problem and leaving it out is an invalid assumption. The linear calculations tell us what the flow and temperature profiles are at the onset of instability. This information is enough for us to understand the physics of the instability mechanisms, however, one needs to support the theory by conducting experiments. In order to know what to expect during the course of an experiment we need to turn to the nonlinear stability analysis. The nonlinear analysis will give us the information on the type of the instability, in other words, it will tell us whether the instability is subcritical or supercritical. Earlier in this section we defined the input variables to be the phase depths, the pressure at the top plate and the overall temperature drop between the hot and cold plates. The critical conditions for the onset of the instability are determined along with the critical wavenumber. Advancing the temperature drop beyond the critical value will not only reveal the magnitudes of the flow and temperature profiles, but will also give the liquid feed rate and the vapor removal rate, as well as the enhancement in heat transport. In order to find the answer to the behavior of the two-phase problem beyond the instability we turn to the nonlinear problem in the next section.

3 Evaporative instability in bilayer systems - Nonlinear theory

3.1 Introduction

In the prior section, 2, linear stability was used to analyze the phase-change between a liquid and its own vapor where the vapor, in addition to the liquid, was assumed to be fluid dynamically active. We concluded there that the active vapor layer plays a significant role in determining the stability of the phase-change problem. In this section we study what happens once the interface becomes unstable. The objective is to construct the steady solution beyond the onset of the instability and to find the magnitude of the state variables at the given conditions as this is of experimental importance.

The change in the heat transfer at the bottom plate is one way to detect the onset of instability in many convective instability problems, provided that the change is measurable. In order to do this, we need to analyze the problem beyond the critical point via nonlinear calculations to learn the magnitude of the change in the heat transfer. Knowing the magnitude of the perturbed domain variables will also tell us, for example, how the flows in both domains change when we change any one of the input variables. The linear calculations only reveal the velocities of both fluids relative to each other for a specific set of input variables, whereas the nonlinear calculations give us the chance to compare different systems with different input variables. This way, we can better understand the effect of each input variable on the instability.

3.2 The physical model

The physical model is as in the earlier section where a liquid of depth, d, underlies its own vapor of depth, d*, and is heated from the liquid side

The base state is one where there is a flat interface between the liquid and its vapor; it is its stability that is in question. For a given phase-change rate, the liquid and vapor depths are assumed constant by suitably adjusting the liquid feed and vapor removal rates. The phase-change rate at the interface can be controlled by adjusting the vapor pressure at the top plate. The phase-change rate is not constant as it can change upon perturbation. Consequently, the input variables in the base state are the liquid and vapor depths, the temperatures of the bottom and top plates, and the pressure at the top plate. Furthermore, the pressure at the top plate can be adjusted so that the phase-change rate at the interface in the base state is equal to zero. In the current analysis of the problem, we assume that the liquid and its vapor are in thermodynamic equilibrium because we will only consider zero phase-change rate at the interface in the base state. There is no great loss of generality by making this assumption, and the algebraic manipulations do become substantially simpler than allowing a finite evaporation rate in the base state. Once the instability occurs and we go beyond the critical point, the bottom liquid and top vapor removal rates must change unless we turn off the feed and removal valves at both plates. In that case, when we go a little beyond the critical point we will see the signature of the instability on the velocity and temperature profiles and, of course, the heat transport as well as the top plate pressure. All other input variables such as the average fluid depths are kept constant and cannot change on account of the fact that the fluid compartment volumes must be constant the fluids being assumed to be incompressible.

3.3 Mathematical model

To get an idea of the nonlinear features, we will investigate the phase-change problem in the absence of gravity and Marangoni convection in this section. While the prior section accounted for these features when the linearized instability was investigated, we shall concern ourselves with the pure evaporation problem when considering the nonlinear analysis as we want to understand the essence of the physics without excessive complication and computation. The problem is studied in two spatial dimensions, x and z.

The nonlinear equations in dimensionless form are given again below for the convenience of the reader. The following scales are used: length, d; velocity, $\bar{v} = \frac{\kappa}{d}$; time, $\frac{d}{\bar{v}}$; pressure, $\bar{P} = \frac{\mu\bar{v}}{d}$ and temperature, $\tilde{T}\frac{T - T_{cold}}{\Delta T}$ where $\Delta = T_{hot} - T_{cold}$. Hereafter, we will drop the hat on the scaled temperature. The momentum equations in each phase become

$$\frac{1}{Pr}\left(\frac{\partial\vec{v}}{\partial t} + \vec{v}\cdot\nabla\vec{v}\right) = -\nabla P + \nabla^2\vec{v} \tag{29}$$

and

$$\frac{\nu}{\nu^*}\frac{1}{Pr}\left(\frac{\partial\vec{v}^*}{\partial t} + \vec{v}^*\cdot\nabla\vec{v}^*\right) = -\frac{\mu}{\mu^*}\nabla P^* + \nabla^2\vec{v}^* \tag{30}$$

$$\nabla^2 T = \left(\frac{\partial T}{\partial t} + \vec{v}\cdot\nabla T\right) \tag{31}$$

and

$$\nabla^2 T^* = \frac{\kappa}{\kappa^*}\left(\frac{\partial T^*}{\partial t} + \vec{v}^*\cdot\nabla T^*\right) \tag{32}$$

The continuity equation in each phase hold assuming that the fluids are incompressible. In the equations above, $\vec{v}$, P and T are scaled velocity, pressure and temperature fields, respectively, and the asterisk denotes the vapor phase. The symbols, ν, κ, μ and Pr stand for the kinematic viscosity, the thermal diffusivity, the viscosity and the Prandtl number, respectively.

At the top and bottom plates, the temperatures are kept constant, and no-slip and no mass transfer conditions are applied. We complete the mathematical model by giving the interfacial conditions at the phase-change boundary. The interfacial mass balance, written at the interface $z = Z(x,t)$, is given by

$$(\vec{v} - \vec{u})\cdot\vec{n} = \frac{\rho^*}{\rho}(\vec{v}^* - \vec{u})\cdot\vec{n} \tag{33}$$

where $\vec{n} = \dfrac{-\frac{\partial Z}{\partial x}\vec{i} + \vec{k}}{\left(1 + \left(\frac{\partial Z}{\partial x}\right)^2\right)^{1/2}}$ is the unit outward normal, ρ is the density,

and $\vec{u}\cdot\vec{n} = \dfrac{\frac{\partial Z}{\partial t}}{\left(1 + \left(\frac{\partial Z}{\partial x}\right)^2\right)^{1/2}}$ is the interface speed.

At the interface, the tangential components of velocities of both fluids are equal to each other, so $\vec{v} \cdot \vec{t} = \vec{v}^* \cdot \vec{t}$ holds where $\vec{t} = \dfrac{\vec{i} + \frac{\partial Z}{\partial x}\vec{k}}{\left(1 + \left(\frac{\partial Z}{\partial x}\right)^2\right)^{1/2}}$ is the unit tangent vector.

The interfacial tension comes into the picture through the force balance. This balance equation has normal and tangential components and is given by

$$\begin{aligned} & Ca\left(\frac{1}{Pr}\vec{v}(\vec{v} - \vec{u}) - \vec{\vec{T}}\right) \cdot \vec{n} + 2H\vec{n} \\ & = Ca\left(\frac{\rho^*}{\rho}\frac{1}{Pr}\vec{v}^*(\vec{v}^* - \vec{u}) - \frac{\mu^*}{\mu}\vec{\vec{T}}\right) \cdot \vec{n} \end{aligned} \tag{34}$$

where $\vec{\vec{T}} = -P\vec{\vec{I}} + \vec{\vec{S}}$ is the total stress, $Ca = \dfrac{\mu\kappa}{\gamma d}$ is the Capillary number, γ is the interfacial tension and $2H = \dfrac{\frac{\partial^2 Z}{\partial x^2}}{\left(1 + \left(\frac{\partial Z}{\partial x}\right)^2\right)^{3/2}}$ is the surface mean curvature.

In addition, we need the interfacial energy balance, viz.

$$\begin{aligned} & \left(-1 + K_{PC}\left(\frac{1}{2}(\vec{v} - \vec{u})^2 - \frac{1}{2}(\vec{v}^* - \vec{u})^2\right)\right)(\vec{v} - \vec{u}) \cdot \vec{n} \\ & \qquad - E\left(\nabla T \cdot \vec{n} - \frac{k^*}{k}\nabla T^* \cdot \vec{n}\right) \\ & - V_{PC}\left(\vec{\vec{S}} \cdot (\vec{v} - \vec{u}) - \frac{\mu^*}{\mu}\vec{\vec{S}}^* \cdot (\vec{v}^* - \vec{u})\right) \cdot \vec{n} = 0 \end{aligned} \tag{35}$$

Here, $K_{PC} = \dfrac{\kappa^2}{Ld^2}$, $E = \dfrac{C_P \Delta T}{\hbar}$, $V_{PC} = \dfrac{\nu\kappa}{Ld^2}$, L is the latent heat per unit mass, and $\vec{\vec{S}}$ is the extra stress tensor.

At the interface, the temperatures of both fluids are equal to each other and there exists thermodynamic equilibrium at the phase-change boundary, that is

$$\Pi_{KE}(P^* - P_{BASE}) = \ln\left(\frac{T^*}{T_{BASE}}\right) \tag{36}$$

where $\Pi_{KE} = \dfrac{\rho}{\rho^*}\dfrac{\nu\kappa}{\hbar d^2}$, and P_{BASE} and T_{BASE} are the scaled base state pressure and temperature of the fluid.

3.4 Nonlinear analysis

In the nonlinear analysis one goes beyond the critical state by advancing a control variable such as the temperature drop across the plates. The scaled temperature drop, however, appears as E, the Evaporation number, therefore it is E that will be advanced from its critical value. Once E is increased all of the state variable will respond and it is their response that interests us. Whether the response is linearly proportional to the change or not is also of interest and will be determined. To this end we introduce perturbation expansions about the base state for each response state variable.

3.5 Perturbation expansions

The expansion of the state variables about the base state introduces a spatial mapping as the domain in the current configuration is an unknown and must be determined. However, this spatial mapping is not known; it is not unique and it cannot be determined save at the interface where it turns out that the normal stress balance offers help in its determination. The reader can see the second chapter in the book by Johns and Narayanan to understand more about this mapping. The domain variables and their derivatives are expanded along the mapping as follows:

$$u = u_0 + (\lambda - \lambda_C)^\alpha \left(u_1 + z_1 \frac{\partial u_0}{\partial z} \right)$$
$$+\frac{1}{2}(\lambda - \lambda_C)^{2\alpha} \left(u_2 + 2z_1 \frac{\partial u_1}{\partial z} + z_1^2 \frac{\partial^2 u_0}{\partial z^2} + z_2 \frac{\partial u_0}{\partial z} \right)$$
$$+\frac{1}{6}(\lambda - \lambda_C)^{3\alpha} \left(u_3 + 3z_1 \frac{\partial u_2}{\partial z} + 3z_2 \frac{\partial u_1}{\partial z} \right.$$
$$\left. +3z_1 z_2 \frac{\partial^2 u_0}{\partial z^2} + 3z_1^2 \frac{\partial^2 u_1}{\partial z^2} + z_1^3 \frac{\partial^3 u_0}{\partial z^3} + z_3 \frac{\partial u_0}{\partial z} \right) + \cdots$$

and

$$\frac{\partial u}{\partial z} = \frac{\partial u_0}{\partial z} + (\lambda - \lambda_C)^\alpha \left(\frac{\partial u_1}{\partial z} + z_1 \frac{\partial^2 u_0}{\partial z^2} \right)$$
$$+\frac{1}{2}(\lambda - \lambda_C)^{2\alpha} \left(\frac{\partial u_2}{\partial z} + 2z_1 \frac{\partial^2 u_1}{\partial z^2} + z_1^2 \frac{\partial^3 u_0}{\partial z^3} + z_2 \frac{\partial^2 u_0}{\partial z^2} \right)$$
$$+\frac{1}{6}(\lambda - \lambda_C)^{3\alpha} \left(\frac{\partial u_3}{\partial z^2} + 3z_1 \frac{\partial^2 u_2}{\partial z^2} + 3z_2 \frac{\partial^2 u_1}{\partial z^2} \right.$$
$$\left. +3z_1 z_2 \frac{\partial^3 u_0}{\partial z^3} + 3z_1^2 \frac{\partial^3 u_1}{\partial z^3} + z_1^3 \frac{\partial^4 u_0}{\partial z^4} + z_3 \frac{\partial^2 u_0}{\partial z^2} \right) + \cdots$$

In the above expansions, z_1, z_2 and z_3 specify the mappings from the current state in the (x, z)-domain to the reference quiescent state in the (x_0, z_0)-domain at first, second and third orders, respectively. The subcripts '0' are dropped from the coordinate variables 'x' and 'z' in the reference state. Higher derivatives with respect to z and derivatives in other directions follow in a like manner. It has been shown that the derivatives of the spatial mappings z_1, z_2, z_3, etc. disappear due to algebraic cancelations and never need to be of any concern. In the above expansions, λ is the dimensionless control variable, and λ_C is its value at which the system becomes unstable, namely the critical value. In our problem λ will be replaced by the parameter E. We go on and set the $\frac{\partial}{\partial t}$ terms equal to zero and investigate the nature of the steady-state solutions to the problem once we cross the threshold value. To do this, we go beyond the critical point by an amount $\in$, i.e. $\lambda = \lambda_C + \in$. Note that here $\in$ represents the deviation from the critical state while in the earlier section it was the arbitrary amplitude of the disturbances to the base state. In both cases it is taken to be small. The control variable is now defined as $E = E_C + \in$. We substitute the expansions of the domain variables into the nonlinear equations and find the values of α, along with the amplitude of the solutions to first order in $\in$. To do this we use the method of dominant balance at every order of the calculation and go to as many orders as needed. The equations are long and the process of balancing the orders is tedious; thus, in order to assist the reader, we give an overview of what we are going to do in the calculation.

3.6 Overview

Suppose the steady nonlinear problem is in the form of $Nu = f$, where N is a nonlinear operator containing the control variable λ, u is the solution vector and b is an inhomogeneity vector that drives the problem. The vector u contains the velocity components, the pressure and the temperature, in addition to the surface position, Z, which is taken to be the last element of the vector u. The base problem must satisfy the nonlinear equation and its solution is trivially known. In our application, the base state is a pure conductive state, so in each fluid the velocity is zero and the temperature field is linear. The first order problem is a homogenous problem and this means that α cannot be determined at this order. The first order problem is written as:

$$Lu_1 = 0. \tag{37}$$

Here, the operator L derived from the $O(\varepsilon^\alpha)$ equations by linearization of N and contains the control variable λ. The solution, the eigenvector

u_1, is of course known only up to an arbitrary multiplicative constant, A, which is the amplitude of the solution to first order. At first order we can determine u_1 and the critical value of λ, but not α and A. It is our aim to determine what A, as well as α, are through the dominant balance method, and therefore we proceed to the next order. The second order problem is in the form

$$Lu_2 = f_1(u_1, u_1). \tag{38}$$

The solvability condition is applied to this system of equations. But, in our problem the operator L is not self-adjoint. Hence, to make the algebra simpler, the elements of the vector u_2 are solved in terms of its last element, i.e. Z_2 and Z_1^2, where Z_2 is the second order correction to the surface position Z. Solvability is then applied to the differential equation for Z_2, which is actually the normal stress balance at the liquid-vapor interface as the differential operator for Z at every order is self-adjoint. The normal stress balances in the first and second orders look as follows:

$$\frac{d^2 Z_1}{dx^2} - aZ_1 = 0 \tag{39}$$

and

$$\frac{d^2 Z_2}{dx^2} - aZ_2 = b_2(Z_1^2) \tag{40}$$

We find in this problem that, because Z_1 is a multiple of $\cos(\omega x)$, the solvability condition is automatically satisfied at the second order; thus we move on to the third order problem. In our problem, at third order, we find that a balance can be struck between the various terms for a specific value of α. We then introduce the specified value of α into the equations and apply the solvability condition on the normal stress balance, which is now in the form of

$$\frac{d^2 Z_3}{dx^2} - aZ_3 = b_3(Z_1^3, Z_1 Z_2). \tag{41}$$

By doing so, we get an equation that determines the value for A^2. Hence, in the dominant balance method, we find the value of α and A at the same order. The value of α tells us what type of bifurcation we should expect to see once the instability sets in. For example, if α is equal to one the bifurcation is trans-critical; and if α is equal to 1/2 it is a pitchfork. To determine whether the pitchfork is forward or backward we focus on the sign of A^2. If it is positive the pitchfork is forward, else it is backward. In the latter case, E must be expanded as $E = E_C - \in$.

3.7 Perturbed equations

Using the method outlined above, we expand the nonlinear domain equations given earlier up to the third order. Thus far we have pretended that our physical system is infinite in lateral extent, but the pure phase-change problem is unlike the standard Benard problem, which produces a nonzero critical wavenumber. For that reason, in this section we will choose an arbitrary wavenumber, and will proceed with the calculations using that wavenumber. As a result, we have tacitly assumed that the fluids are confined within a laterally finite system with slippery and insulating walls. In what follows all of the orders are evaluated on the reference domain and the reader is reminded that the subcript "0" will be dropped from the "x" and "z" coordinate variables.

Returning to the expansions of the governing equations, we first write the momentum equations in both phases as

$$\begin{aligned}(E-E_C)^{\alpha}(-\nabla P_1+\nabla^2\vec{v}_1)\\ \frac{1}{2}(E-E_C)^{2\alpha}\left(-\nabla P_2+\nabla^2\vec{v}_2-\frac{2}{Pr}\vec{v}_1\cdot\nabla\vec{v}_1\right)\\ +\frac{1}{6}(E-E_C)^{3\alpha}\left(-\nabla P_3+\nabla^2\vec{v}_3\right.\\ \left.-\frac{3}{Pr}(\vec{v}_1\cdot\nabla\vec{v}_2+\vec{v}_2\cdot\nabla\vec{v}_1)\right)=0\end{aligned} \tag{42}$$

and

$$\begin{aligned}(E-E_C)^{\alpha}(-\frac{\mu}{\mu^*}\nabla P_1^*+\nabla^2\vec{v}_1^*)\\ \frac{1}{2}(E-E_C)^{2\alpha}\left(-\frac{\mu}{\mu^*}\nabla P_2^*+\nabla^2\vec{v}_2^*-\frac{\nu}{\nu^*}\frac{2}{Pr}\vec{v}_1^*\cdot\nabla\vec{v}_1^*\right)\\ +\frac{1}{6}(E-E_C)^{3\alpha}\left(-\frac{\mu}{\mu^*}\nabla P_3^*+\nabla_0^2\vec{v}_3^*\right.\\ \left.-\frac{\nu}{\nu^*}\frac{3}{Pr}(\vec{v}_1^*\cdot\nabla\vec{v}_2^*+\vec{v}_2^*\cdot\nabla\vec{v}_1^*)\right)=0\end{aligned} \tag{43}$$

$$\begin{aligned}\nabla_0^2T_0+(E-E_C)^{\alpha}\left(\nabla^2T_1-\vec{v}_1\cdot\nabla T_0)\right)\\ +\frac{1}{2}(E-E_C)^{2\alpha}(\nabla^2T_2-\vec{v}_2\cdot\nabla T_0-2\vec{v}_1\cdot\nabla T_1)\\ +\frac{1}{6}(E-E_C)^{3\alpha}\left(\nabla^2T_3-\vec{v}_3\cdot\nabla T_0\right.\\ \left.-3(\vec{v}_2\cdot\nabla T_1+\vec{v}_1\cdot\nabla T_2)\right)=0\end{aligned} \tag{44}$$

$$\nabla^2 T_0^* + (E - E_C)^\alpha \left(\nabla^2 T_1^* - \frac{\kappa}{\kappa^*}\vec{v}_1^* \cdot \nabla_0 T_0^*)\right)$$
$$+\frac{1}{2}(E - E_C)^{2\alpha}(\nabla^2 T_2^* - \frac{\kappa}{\kappa^*}\vec{v}_2^* \cdot \nabla T_0^* - 2\frac{\kappa}{\kappa^*}\vec{v}_1^* \cdot \nabla_0 T_1^*)$$
$$+\frac{1}{6}(E - E_C)^{3\alpha}\left(\nabla^2 T_3^* - \frac{\kappa}{\kappa^*}\vec{v}_3^* \cdot \nabla T_0^*\right.$$
$$\left. -3\frac{\kappa}{\kappa^*}(\vec{v}_2^* \cdot \nabla T_1^* + \vec{v}_1^* \cdot \nabla T_2^*)\right) = 0 \tag{45}$$

The continuity equations become

$$(E - E_C)^\alpha \nabla_0 \cdot \vec{v}_1 + \frac{1}{2}(E - E_C)^{2\alpha} \nabla \cdot \vec{v}_2$$
$$+\frac{1}{6}(E - E_C)^{3\alpha} \nabla \cdot \vec{v}_3 = 0 \tag{46}$$

and

$$(E - E_C)^\alpha \nabla \cdot \vec{v}_1^* + \frac{1}{2}(E - E_C)^{2\alpha} \nabla \cdot \vec{v}_2^*$$
$$+\frac{1}{6}(E - E_C)^{3\alpha} \nabla \cdot \vec{v}_3^* = 0 \tag{47}$$

The unit normal and the unit tangential vectors, upon perturbation up to the third order, become

$$\vec{n} = \vec{k} + (E - E_C)^\alpha \left(-\frac{\partial Z_1}{\partial x}\vec{i}\right)$$
$$+\frac{1}{2}(E - E_C)^{2\alpha}\left(-\frac{\partial Z_2}{\partial x}\vec{i} - \left(\frac{\partial Z_1}{\partial x}\right)^2 \vec{k}\right)$$
$$+\frac{1}{6}(E - E_C)^{3\alpha}\left(\left(3\left(\frac{\partial Z_1}{\partial x}\right)^3 - \frac{\partial Z_3}{\partial x}\right)\vec{i} - 3\frac{\partial Z_1}{\partial x}\frac{\partial Z_2}{\partial x}\vec{k}\right) \tag{48}$$

and

$$\vec{t} = \vec{i} + (E - E_C)^\alpha \frac{\partial Z_1}{\partial x}\vec{k}$$
$$+\frac{1}{2}(E - E_C)^{2\alpha}\left(-\left(\frac{\partial Z_1}{\partial x}\right)^2 \vec{i} + \frac{\partial Z_2}{\partial x}\vec{k}\right)$$
$$+\frac{1}{6}(E - E_C)^{3\alpha}\left(-3\frac{\partial Z_1}{\partial x}\frac{\partial Z_2}{\partial x}\vec{i} + \left(\frac{\partial Z_3}{\partial x} - 3\left(\frac{\partial Z_1}{\partial x}\right)^3\right)\vec{k}\right) \tag{49}$$

At the lower plate, $z = -1$, the boundary conditions $\vec{v} \cdot \vec{n} = 0$, $\vec{v} \cdot \vec{t} = 0$, and $T = 1$ are expanded as

$$(E - E_C)^{\alpha} V_{z1} + \frac{1}{2}(E - E_C)^{2\alpha} V_{z2} + \frac{1}{6}(E - E_C)^{3\alpha} V_{z3} = 0, \tag{50}$$

$$(E - E_C)^{\alpha} V_{x1} + \frac{1}{2}(E - E_C)^{2\alpha} V_{x2} + \frac{1}{6}(E - E_C)^{3\alpha} V_{x3} = 0, \tag{51}$$

and

$$T_0 + (E - E_C)^{\alpha} T_1 + \frac{1}{2}(E - E_C)^{2\alpha} T_2 + \frac{1}{6}(E - E_C)^{3\alpha} T_3 = 1, \tag{52}$$

while the conditions at the top plate $\vec{v}^* \cdot \vec{n} = 0$, $\vec{v}^* \cdot \vec{t} = 0$, and $T^* = 0$ become

$$(E - E_C)^{\alpha} V_{z1}^* + \frac{1}{2}(E - E_C)^{2\alpha} V_{z2}^* + \frac{1}{6}(E - E_C)^{3\alpha} V_{z3}^* = 0, \tag{53}$$

$$(E - E_C)^{\alpha} V_{x1}^* + \frac{1}{2}(E - E_C)^{2\alpha} V_{x2}^* + \frac{1}{6}(E - E_C)^{3\alpha} V_{x3}^* = 0, \tag{54}$$

and

$$T_0^* + (E - E_C)^{\alpha} T_1^* + \frac{1}{2}(E - E_C)^{2\alpha} T_2^* + \frac{1}{6}(E - E_C)^{3\alpha} T_3^* = 0. \tag{55}$$

For reasons of brevity, we do not provide all of the expansions of all interfacial conditions save three that are crucial for the analysis of the problem. One of those three is the interfacial energy balance and, after simplifications, it assumes the form

$$
\begin{aligned}
&-E\left(\frac{\partial T_0}{\partial z}-\frac{k^*}{k}\frac{\partial T_0^*}{\partial z}\right)\\
&+(E-E_C)^{\alpha}\left(-v_{z1}-E\left(\frac{\partial T_1}{\partial z}-\frac{k^*}{k}\frac{\partial T_1^*}{\partial z}\right)\right)\\
&+\frac{1}{2}(E-E_C)^{2\alpha}\Bigg[-\left[v_{z2}+2\frac{\partial v_{z1}}{\partial z}Z_1-2v_{x1}\frac{\partial Z_1}{\partial x}\right]\\
&-E\Bigg[\frac{\partial T_2}{\partial z}+2\frac{\partial^2 T_1}{\partial z^2}Z_1-2\frac{\partial T_1}{\partial x}\frac{\partial Z_1}{\partial x}\\
&-\frac{k^*}{k}\left(\frac{\partial T_2^*}{\partial z}+2\frac{\partial^2 T_1^*}{\partial z^2}Z_1-2\frac{\partial T_1^*}{\partial x}\frac{\partial Z_1}{\partial x}\right)\Bigg]\\
&-4V_{PC}\left(\frac{\partial v_{z1}}{\partial z}-\frac{\mu^*}{\mu}\frac{\partial v_{z1}^*}{\partial z}\right)\Bigg]\\
&+\frac{1}{6}(E-E_C)^{3\alpha}\Bigg[-\Bigg[v_{z3}+3\frac{\partial v_{z2}}{\partial z}Z_1\\
&+3\frac{\partial v_{z1}}{\partial z}Z_2+3\frac{\partial^2 v_{z1}}{\partial z^2}Z_1^2-3v_{x1}\frac{\partial Z_2}{\partial x}-3v_{x2}\frac{\partial Z_1}{\partial x}\\
&-6\frac{\partial v_{x1}}{\partial z}Z_1\frac{\partial Z_1}{\partial x}\Bigg]+3K_{PC}v_{z1}(v_{z1}^2+v_{x1}^2-v_{z1}^{*2}-v_{z1}^{*2})\\
&-E\Bigg[\frac{\partial T_3}{\partial z}+3\frac{\partial^2 T_2}{\partial z^2}Z_1+3\frac{\partial^2 T_1}{\partial z^2}Z_2+3\frac{\partial^3 T_1}{\partial z^3}Z_1^2\\
&-3\frac{\partial T_1}{\partial x}\frac{\partial Z_2}{\partial x}-3\frac{\partial T_2}{\partial x}\frac{\partial Z_1}{\partial x}-6\frac{\partial^2 T_1}{\partial z\partial x}Z_1\frac{\partial Z_1}{\partial x}\\
&-\frac{k^*}{k}\Bigg(\frac{\partial T_3^*}{\partial z}+3\frac{\partial^2 T_2^*}{\partial z^2}Z_1+3\frac{\partial^2 T_1^*}{\partial z^2}Z_2+3\frac{\partial^3 T_1^*}{\partial z^3}Z_1^2\\
&-3\frac{\partial T_1^*}{\partial x}\frac{\partial Z_2}{\partial x}-3\frac{\partial T_2^*}{\partial x}\frac{\partial Z_1}{\partial x}-6\frac{\partial^2 T_1^*}{\partial z\partial x}Z_1\frac{\partial Z_1}{\partial x}\Bigg)\Bigg]\\
&-V_{PC}\Bigg[6v_{z1}\left[\frac{\partial v_{z2}}{\partial z}-2\left(\frac{\partial v_{z1}}{\partial x}+\frac{\partial v_{x1}}{\partial z}\right)\frac{\partial Z_1}{\partial x}+2\frac{\partial^2 v_{z1}}{\partial z^2}Z_1\right]\\
&-12v_{x1}\frac{\partial v_{z1}}{\partial z}\frac{\partial Z_1}{\partial x}+6\frac{\partial v_{z1}}{\partial z}\left(v_{z2}+2\frac{\partial v_{z1}}{\partial z}Z_1\right)\\
&-\frac{\mu^*}{\mu}\Bigg[6v_{z1}^*\left[\frac{\partial v_{z2}^*}{\partial z}-2\left(\frac{\partial v_{z1}^*}{\partial x}+\frac{\partial v_{x1}^*}{\partial z}\right)\frac{\partial Z_1}{\partial x}+2\frac{\partial^2 v_{z1}^*}{\partial z^2}Z_1\right]\\
&-12v_{x1}^*\frac{\partial v_{z1}^*}{\partial z}\frac{\partial Z_1}{\partial x}+6\frac{\partial v_{z1}^*}{\partial z}\left(v_{z2}^*+2\frac{\partial v_{z1}^*}{\partial z}Z_1\right)\Bigg]\Bigg]\Bigg]=0 \qquad (56)
\end{aligned}
$$

Another one is the thermodynamic equilibrium relation between the vapor pressure and the vapor temperature;

$$\begin{aligned}
&\Pi_{KE}(P_0^* - P_{sat}) - ln\left(\frac{T_0^*}{T_{sat}}\right) \\
&+(E-E_C)^{\alpha}\left(\Pi_{KE}P_1^* - \frac{1}{T_{sat}}\left(T_1^* + \frac{dT_0^*}{dz}Z_1\right)\right) \\
&+\frac{1}{2}(E-E_C)^{2\alpha}\left[\Pi_{KE}\left(P_2^* + 2\frac{\partial P_1^*}{\partial z}Z_1\right)\right. \\
&\left.-\frac{1}{T_{sat}}\left(T_2^* + 2\frac{\partial T_1^*}{\partial z}Z_1 + \frac{dT_0^*}{dz}Z_2\right) + \left(\frac{1}{T_{sat}}\right)^2\left(T_1^* + \frac{dT_0^*}{dz}Z_1\right)^2\right] \\
&+\frac{1}{6}(E-E_C)^{3\alpha}\left[\Pi_{KE}\left(P_3^* + 3\frac{\partial P_2^*}{\partial z}Z_1 + 3\frac{\partial P_1^*}{\partial z}Z_2 + 3\frac{\partial^2 P_1^*}{\partial z^2}Z_1^2\right)\right. \\
&-\frac{1}{T_{sat}}\left(T_3^* + 3\frac{\partial T_2^*}{\partial z}Z_1 + 3\frac{\partial T_1^*}{\partial z}Z_2 + 3\frac{\partial^2 T_1^*}{\partial z^2}Z_1^2 + \frac{dT_0^*}{dz}Z_3\right) \\
&+3\left(\frac{1}{T_{sat}}\right)^2\left(T_2^* + 2\frac{\partial T_1^*}{\partial z}Z_1 + \frac{dT_0^*}{dz}Z_2\right)\left(T_1^* + \frac{dT_0^*}{dz}Z_1\right) \\
&\left.-2\left(\frac{1}{T_{sat}}\right)^3\left(T_1^* + \frac{dT_0^*}{dz}Z_1\right)^3\right] = 0 \qquad (57)
\end{aligned}$$

We also give the third order normal stress balance equation because we will use it to find the values of α and A. The third order normal stress balance is given by

$$
\begin{aligned}
& Ca(P_0 - P_0^*) \\
&+(E-E_C)^{\alpha}\left(Ca\left(P_1 - P_1^* - 2\frac{\partial v_{z1}}{\partial z} + 2\frac{\mu^*}{\mu}\frac{\partial v_{z1}^*}{\partial z}\right) + \frac{\partial^2 Z_1}{\partial x^2}\right) \\
&+\frac{1}{2}(E-E_C)^{2\alpha}\left[Ca\left[P_2 + 2\frac{\partial P_1}{\partial z}Z_1 - P_2^* - 2\frac{\partial P_1^*}{\partial z}Z_1\right.\right. \\
&\left.-2\left(\frac{\partial v_{z2}}{\partial z} + 2\frac{\partial^2 v_{z1}}{\partial z^2}Z_1\right) + 2\frac{\mu^*}{\mu}\left(\frac{\partial v_{z2}^*}{\partial z} + 2\frac{\partial^2 v_{z1}^*}{\partial z^2}Z_1\right)\right] \\
&\left.+2\frac{Ca}{Pr}v_{z1}(v_{z1} - v_{z1}^*) + \frac{\partial^2 Z_2}{\partial x^2}\right] \\
&+\frac{1}{6}(E-E_C)^{3\alpha}\left[Ca\left[P_3 + 3\frac{\partial P_2}{\partial z}Z_1 + 3\frac{\partial P_1}{\partial z}Z_2\right.\right. \\
&+3\frac{\partial^2 P_1}{\partial z^2}Z_1^2 - P_3^* - 3\frac{\partial P_2^*}{\partial z}Z_1 - 3\frac{\partial P_1^*}{\partial z}Z_2 - 3\frac{\partial^2 P_1^*}{\partial z^2}Z_1^2 \\
&-2\left[\frac{\partial v_{z3}}{\partial z} + 3\frac{\partial^2 v_{z2}}{\partial z^2}Z_1 + 3\frac{\partial^2 v_{z1}}{\partial z^2}Z_2 + 3\frac{\partial^3 v_{z1}}{\partial z^3}Z_1^2\right. \\
&\left.+12\frac{\partial v_{z1}}{\partial z}\left(\frac{\partial Z_1}{\partial x}\right)^2\right] + 2\frac{\mu^*}{\mu}\left[\frac{\partial v_{z3}^*}{\partial z} + 3\frac{\partial^2 v_{z2}^*}{\partial z^2}Z_1\right. \\
&\left.\left.+3\frac{\partial^2 v_{z1}^*}{\partial z^2}Z_2 + 3\frac{\partial^3 v_{z1}^*}{\partial z^3}Z_1^2 + 12\frac{\partial v_{z1}^*}{\partial z}\left(\frac{\partial Z_1}{\partial x}\right)^2\right]\right] \\
&+3\frac{Ca}{Pr}\left(v_{z2} + 2\frac{\partial v_{z1}}{\partial z}Z_1 - 2v_{x1}\frac{\partial Z_1}{\partial x}\right)(v_{z1} - v_{z1}^*) \\
&+3\frac{Ca}{Pr}\left(v_{z2} + 2\frac{\partial v_{z1}}{\partial z}Z_1 - 2v_{x1}\frac{\partial Z_1}{\partial x} - v_{z2}^* - 2\frac{\partial v_{z1}^*}{\partial z}Z_1\right. \\
&\left.\left.+2v_{x1}^*\frac{\partial Z_1}{\partial x}\right)v_{z1} + \frac{\partial^2 Z_3}{\partial x^2} - 9\frac{\partial^2 Z_1}{\partial x^2}\left(\frac{\partial Z_1}{\partial x}\right)^2\right] = 0 \qquad (58)
\end{aligned}
$$

Recall that in all the expanded equations, E will be replaced by $E_C + \in$. The first order problem is an eigenvalue problem and it was solved in the earlier section. The solution to the first order problem is used at every subsequent order along with the solvability condition. We learn from the first order equations that the eigenvector u_1 in terms of Z_1 is

$$u_1 = A\hat{u}_1(\hat{Z}_1)\cos\omega x.$$

Recall that ω is an input and it is determined by the width of the compartment in the lateral direction. We cannot determine the value of α either from the first or the second order equations. As noted in the earlier section, the solvability condition is automatically satisfied at the second order. The solution to the second order, u_2, will go as

$$u_2 = Bu_{2a}(\hat{Z}_2)\cos\omega x + A^2 u_{2b}(\hat{Z}_1^2) + A^2 u_{2c}(\hat{Z}_1^2)\cos 2\omega x. \tag{59}$$

It is at the third order that we learn a is equal to $\frac{1}{2}$. At this order, there exists a value for A other than zero. A appears as A^2, and the value of A^2 is obtained using the solvability condition on the third order normal stress balance. The third order solution, u_3, depends upon the form of u_1, u_{2P} as well as Z_3, i.e.

$$u_3 = Cu_{3a}(Z_3)\cos\omega x + A^3 u_{3b}(Z_1^3, Z_1 Z_2)\cos\omega x$$

$$+A^3 u_{3c}(Z_1^3, Z_1 Z_2)\cos 3\omega x + Au_{3d}(Z_1)\cos\omega x$$

3.8 Discussion

The Chebyshev spectral tau method is used to solve the eigenvalue problem that determines the critical temperature difference and the value of the unknown multiplicative constant A . Once we know what A is, we can calculate the actual change in the heat transfer rate at the bottom plate when we advance ΔT a little beyond the critical point. Our system is finite in the lateral extent with periodic conditions. The heat transfer up to the second order is calculated as

$$\int_0^\lambda \vec{q}\cdot\vec{n} = -k\int_0^\lambda\left(\frac{dT_0}{dz} + \in^{\frac{1}{2}}\frac{d\hat{T}}{dz}\cos\omega x + \frac{1}{2}\in\left(\frac{d\hat{T}_2}{dz} + \frac{d\hat{\hat{T}}_2}{dz}\cos 2\omega x\right)\right)dx$$

Clearly, $\cos\omega x$ and $\cos 2\omega x$, when integrated over the width, give zero. We are more interested in the percentage of change of the heat transfer, so one can plot $\frac{\vec{q}\cdot\vec{n}}{\vec{q_0}\cdot\vec{n_0}}$ against $\frac{E}{E_C}$ in a system for a given liquid and vapor depths and wavenumber ω.

We learn that once the interface is unstable, more heat is transferred through the bottom plate into the system. Thus, if we wanted to know when the system becomes unstable, we only have to measure the amount of heat that is needed to keep the bottom plate at a constant temperature. And, the heat needed changes when the instability sets in.

In summary, we have performed a weak nonlinear calculation for the pure phase-change problem using the dominant balance method. It shows that the instability occurs as a forward pitchfork bifurcation for a pure phase-change problem. The nonlinear analysis of the problem gives us the magnitude of the change in the domain variables and global variables such as the heat transport. This information can be used to detect the onset of instability in stability problems. Knowing the magnitude of the domain variables helps one to better determine the experimental method to be used and what to expect in the course of the experiment. Calculations show that the heat transport enhancement is of the order of a percent and therefore measurable.